METHODS IN MOLECULAR BIOLOGY™

Series Editor
John M. Walker
School of Life Sciences
University of Hertfordshire
Hatfield, Hertfordshire, AL10 9AB, UK

For further volumes:
http://www.springer.com/series/7651

Therapeutic Applications of Ribozymes and Riboswitches

Methods and Protocols

Edited by

Daniel Lafontaine and Audrey Dubé

Department of Biology, Université de Sherbrooke, Sherbrooke, QC, Canada

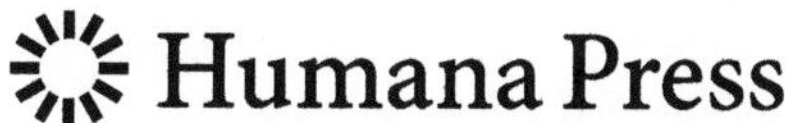

Editors
Daniel Lafontaine
Department of Biology
Université de Sherbrooke
Sherbrooke, QC, Canada

Audrey Dubé
Department of Biology
Université de Sherbrooke
Sherbrooke, QC, Canada

ISSN 1064-3745 ISSN 1940-6029 (electronic)
ISBN 978-1-62703-729-7 ISBN 978-1-62703-730-3 (eBook)
DOI 10.1007/978-1-62703-730-3
Springer New York Heidelberg Dordrecht London

Library of Congress Control Number: 2013951224

Printed on acid-free paper

Humana Press is a brand of Springer
Springer is part of Springer Science+Business Media (www.springer.com)

Preface

The RNA field is currently undergoing a revolution. In the past years, several unexpected functions have been uncovered for RNA molecules including phosphodiester backbone cleavage and gene expression regulation, both in a cis- and trans-acting fashion. Upon those discoveries, the idea that RNA could be used as a molecular tool was rapidly considered. For instance, considerable efforts have been put to use RNA molecules either as drugs or as drug targets. Today, therapeutic applications employing RNA molecules include the use of ribozymes as silencing tools, while riboswitches are involved in the development of novel antimicrobial agents. This volume presents essential protocols for the development of RNA-based therapeutic strategies using ribozymes and riboswitches as a way to treat various diseases.

Ribozymes are autocatalytic RNAs mainly found in viroids and viral genomes. Small ribozymes such as the one found in the hepatitis delta virus are important for the replication of the host. The main characteristic of ribozymes which is to specifically recognize and cleave RNA molecules was rapidly applied to target undesirable messenger RNAs. Thus, over the years, various ribozyme candidates have been engineered, optimized, and characterized to cleave target mRNAs with high specificity and efficiency. The degradation of targeted mRNAs and the associated shutdown expression are the direct consequences of the ribozyme cleavage activity.

Several chapters of this book describe recent and promising techniques designed to identify accessible regions in RNA molecules. This step is the premise for the development of good ribozyme candidates. Technical approaches such as SHAPE and nuclease probing, analyzed jointly with bioinformatic tools, are still relevant methods. However, the challenge is now to apply those methods simultaneously on a genomic scale and to determine the accessibility of each nucleotide directly in vivo. Moreover the catalytic activity of ribozyme candidates has to be evaluated both in vitro and in vivo. This is important because it is known that ribozymes possessing a great in vitro cleavage activity do not necessarily show the best activities in vivo. Upon the design and evaluation of ribozyme candidates, the choice of the expression system and the cellular model will greatly depend on the target RNA.

In order to offer a general idea of protocols used, specific strategies associated to various biological targets will be presented. For example, researchers have now been able to construct ribozymes against mRNAs important for prions, viruses, and cancer progression. Also, recent experiments have shown the efficiency of delivery systems, which are nevertheless still limited by the inherent cellular instability of RNA. Relevant modifications to increase the half-life of ribozymes in specific conditions are also important issues that researchers will have to improve in the next few years.

Riboswitches are genetic regulatory elements mostly retrieved in the 5′ untranslated regions of bacteria. The capacity of riboswitches to specifically recognize small metabolites and theirs derivatives makes them an attractive target to develop novel antimicrobial agents.

Antibiotics against riboswitches are promising for therapeutic use since none of those RNA switches have been identified so far in humans. The first step in the antibiotic development consists in the identification of riboswitches controlling essentials genes. It is also important to characterize the regulation mechanism used by each targeted riboswitch. Bioinformatic analysis used in combination with biochemical and activity assays is the principal method needed to validate a riboswitch candidate. Upon riboswitch confirmation, the determination of the most promising therapeutic molecule can then be performed. Various methods to characterize affinity properties and intrinsic riboswitch characteristics were developed in the last few years. Efficient small molecules targeting resistant bacterial riboswitches constitute a new class of antibiotic, which is exciting considering that only one new antibiotic class has been found since 1985. New challenges, including rapid, efficient, and high-throughput methods to easily identify riboswitches and screen a diversity of ligands, have now emerged from those recent discoveries.

This book offers a complete overview of protocols used in the development of RNA molecule as drugs and drug target. All chapters describe a recent and precise RNA molecule approaches or studies in the development of an RNA therapeutic tool. We are convinced that these methods will help researchers from various domains of life sciences, including clinicians, biochemists, and virologists.

We want to thank all authors who participated in the production of this book. Their contributions will hopefully inspire many researches in the development of new therapeutic applications implicating RNA molecules or directly targeting harmful RNA.

Sherbrooke, QC, Canada

Daniel Lafontaine
Audrey Dubé

Contents

Contributors

NATHAN J. BAIRD • *Laboratory of RNA Biophysics and Cellular Physiology, National Heart, Lung and Blood Institute, Bethesda, MD, USA*

PREETI BAJAJ • *Department of Insect Resistance, International Center for Genetic Engineering and BiotechnologyNew DelhiIndia*

VELMURUGAN BALARAMAN • *Department of Biology, Eck Institute for Global Health, University of Notre Dame, Notre Dame, IN, USA*

RUTH BRENK • *Johannes Gutenberg-Universität Mainz, Institut für Pharmazie und Biochemie, Mainz, Germany*

DONALD H. BURKE • *Department of Molecular Microbiology & Immunology, University of Missouri, Columbia, MO, USA; Department of Biochemistry, University of Missouri, Columbia, MO, USA*

GIOVANNI BUSSI • *Scuola Internazionale Superiore di Studi Avanzati, Trieste, Italy*

JAMES R. CARTER • *Department of Biology, Eck Institute for Global Health, University of Notre Dame, Notre Dame, IN, USA*

FRANCESCO COLIZZI • *Scuola Internazionale Superiore di Studi Avanzati, Trieste, Italy*

FRÉDÉRIC COUTURE • *Département de chirurgie et service de urologie, Institut de pharmacologie de Sherbrooke, Université de Sherbrooke, Sherbrooke, QC, Canada*

PETER DALDROP • *Division of Biological Chemistry and Drug Discovery, College of Life Sciences, University of Dundee, Dundee, UK*

FRANÇOIS D'ANJOU • *Département de chirurgie et service de urologie, Institut de pharmacologie de Sherbrooke, Université de Sherbrooke, Sherbrooke, QC, Canada*

ROBERT DAY • *Département de chirurgie et service de urologie, Institut de pharmacologie de Sherbrooke, Université de Sherbrooke, Sherbrooke, QC, Canada*

ROXANE DESJARDINS • *Département de chirurgie et service de urologie, Institut de pharmacologie de Sherbrooke, Université de Sherbrooke, Sherbrooke, QC, Canada*

AMELL EL KORBI • *INRS-Institut Armand-Frappier, Laval, QC, Canada; Institute for Research in Immunology and Cancer (IRIC), Université de Montréal, Montréal, QC, Canada*

ADRIAN R. FERRÉ-D'AMARÉ • *Laboratory of RNA Biophysics and Cellular Physiology, National Heart, Lung and Blood Institute, Bethesda, MD, USA*

CASEY C. FOWLER • *Department of Biochemistry and Biomedical Sciences, McMaster University, Hamilton, ON, Canada; Department of Chemistry and Chemical Biology, McMaster University, Hamilton, ON, Canada; Michael D. DeGroote Infectious Disease Research Institute, McMaster University, Hamilton, ON, Canada*

MALCOLM J. FRASER JR • *Department of Biology, Eck Institute for Global Health, University of Notre Dame, Notre Dame, IN, USA*

ANNE GATIGNOL • *Virus-Cell Interactions Laboratory, Lady Davis Institute for Medical Research, McGill University, Montréal, QC, Canada; Department of Microbiology & Immunology, McGill University, Montréal, QC, Canada; Department of Experimental Medicine, McGill University, Montréal, QC, Canada*

Christian Hammann • *Ribogenetics @ Biochemistry Lab, Molecular Life Sciences Research Center, School of Engineering and Science, Jacobs University, Bremen, Germany*
Jin-Sook Jeong • *Department of Pathology and Medical Research Center for Cancer Molecular Therapy, Dong-A University College of Medicine, Busan, South Korea*
Xiaohong Jiang • *School of Public Health, University of California, Berkeley, CA, USA*
Daniel A. Lafontaine • *RNA Group, Department of Biology, Faculty of Science, Université de Sherbrooke, Sherbrooke, QC, Canada*
Anne-Marie Lamontagne • *RNA Group, Department of Biology, Faculty of Science, Université de Sherbrooke, Sherbrooke, QC, Canada*
Margaret J. Lange • *Department of Molecular Microbiology & Immunology, University of Missouri, Columbia, MO, USA*
Seong-Wook Lee • *Department of Molecular Biology, Institute of Nanosensor and Biotechnology, Dankook University, Yongin, South Korea*
Michel V. Lévesque • *RNA Group/Groupe ARN, Département de Biochimie, Université de Sherbrooke, Sherbrooke, QC, Canada*
Matthew Levy • *Department of Biochemistry, Albert Einstein College of Medicine, Bronx, NY, USA*
Yingfu Li • *Department of Biochemistry and Biomedical Sciences, McMaster University, Hamilton, ON, Canada; Department of Chemistry and Chemical Biology, McMaster University, Hamilton, ON, Canada; Michael D. DeGroote Infectious Disease Research Institute, McMaster University, Hamilton, ON, Canada*
Fenyong Liu • *School of Public Health, University of California, Berkeley, CA, USA*
Sangwei Lu • *School of Public Health, University of California, Berkeley, CA, USA*
Christina E. Lünse • *Life and Medical Sciences Institute, University of Bonn, Bonn, Germany*
Matthias Mack • *Institute for Technical Microbiology, Mannheim University of Applied Sciences, Mannheim, Germany*
Günter Mayer • *Life and Medical Sciences Institute, University of Bonn, Bonn, Germany*
Ronald Micura • *Institute of Organic Chemistry, Center for Chemistry and Biomedicine, Leopold Franzens University, Innsbruck, Austria*
Mohammad Reza Naghdi • *INRS-Institut Armand-Frappier, Laval, QC, Canada*
Pruksa Nawtaisong • *Department of Biology, Eck Institute for Global Health, University of Notre Dame, Notre Dame, IN, USA*
Jonathan Ouellet • *INRS-Institut Armand-Frappier, Laval, QC, Canada*
Danielle Biscaro Pedrolli • *Institute for Technical Microbiology, Mannheim University of Applied Sciences, Mannheim, Germany*
Jean-Pierre Perreault • *RNA Group/Groupe ARN, Département de Biochimie, Université de Sherbrooke, Sherbrooke, QC, Canada*
Jonathan Perreault • *INRS-Institut Armand-Frappier, Laval, QC, Canada*
Robert J. Scarborough • *Virus-Cell Interactions Laboratory, Lady Davis Institute for Medical Research, McGill University, Montréal, QC, Canada; Department of Microbiology & Immunology, McGill University, Montréal, QC, Canada; Department of Experimental Medicine, McGill University, Montréal, QC, Canada*
Dipankar Sen • *Department of Chemistry, Simon Fraser University, Burnaby, BC, Canada; Department of Molecular Biology and Biochemistry, Simon Fraser University, Burnaby, BC, Canada*
Navjot Singh • *Wadsworth Center, New York State Department of Health, Albany, NY, USA*

MARIE F. SOULIÈRE • *Institute of Organic Chemistry, Center for Chemistry and Biomedicine, Leopold Franzens University, Innsbruck, Austria*
NARESH SUNKARA • *School of Public Health, University of California, Berkeley, CA, USA*
JASON M. THOMAS • *Department of Chemistry, Simon Fraser University, Burnaby, BC, Canada; Department of Molecular Biology and Biochemistry, Simon Fraser University, Burnaby, BC, Canada*
VALERIE T. TRIPP • *Department of Chemistry, College of William & Mary, Williamsburg, VA, USA*
JOSEPH T. WADE • *Wadsworth Center, New York State Department of Health, Albany, NY, USA; Department of Biomedical Sciences, University at Albany, Albany, NY, USA*
AMY YAN • *Department of Biochemistry, Albert Einstein College of Medicine, Bronx, NY, USA*
DOUGLAS D. YOUNG • *Department of ChemistryCollege of William & Mary, Williamsburg, VA, USA*
HUA-ZHONG YU • *Department of Chemistry, Simon Fraser University, Burnaby, BC, Canada; Department of Molecular Biology and Biochemistry, Simon Fraser University, Burnaby, BC, Canada*

Chapter 1

Identification of Regulatory RNA in Bacterial Genomes by Genome-Scale Mapping of Transcription Start Sites

Navjot Singh and Joseph T. Wade

Abstract

The ability to map transcription start sites is critical for studies of gene regulation and for identification of novel RNAs. Conventional RNA-seq is often insufficient for identification of transcription start sites due to low coverage and/or RNA processing events. We have developed a highly sensitive, genome-scale method for detection of transcription start sites in bacteria. This method uses deep sequencing of cDNA libraries to identify transcription start sites with strand specificity at nucleotide resolution. Here, we describe the application of this method for transcription start site identification in *Escherichia coli*.

Key words Transcription start site, RppH, 5′ RNA-seq, 5′ RACE

1 Introduction

Noncoding RNAs regulate a wide variety of processes in bacteria by base-pairing with mRNA targets and controlling mRNA stability, transcription, and/or translation. Such regulatory RNAs have been implicated in regulating virulence gene expression in many pathogenic species and hence represent promising targets for antibacterial therapy. Deep sequencing approaches have revolutionized our ability to identify novel RNAs in bacteria, including regulatory RNAs [1, 2]. Standard RNA-seq methods identify sequence from any portion of an RNA, but precise identification of RNA 5′ ends is often impossible using these methods, especially for RNAs of low abundance. Furthermore, standard RNA-seq does not distinguish between transcription start sites and sites of RNA fragmentation, even for abundant transcripts. 5′ RACE is a targeted approach that is widely used for identification of individual RNA 5′ ends [3]. We have developed a genome-scale 5′ RACE method (5′ RNA-seq) that identifies all transcription start sites for an RNA sample of interest. Unlike 5′ RACE, 5′ RNA-seq distinguishes between 5′ ends of fragmented RNAs and transcription start sites. The 5′ RNA-seq method relies on the fact that nascent RNAs are

Daniel Lafontaine and Audrey Dubé (eds.), *Therapeutic Applications of Ribozymes and Riboswitches: Methods and Protocols*, vol. 1103, DOI 10.1007/978-1-62703-730-3_1,

triphosphorylated at their 5′ end whereas fragmented RNAs are not. By sequencing two libraries, one generated from RNAs with either 5′ tri- or monophosphates and the other generated from only RNAs with monophosphates, it is possible to infer sites of transcription initiation in a strand-specific manner at nucleotide resolution.

Nascent transcripts in many bacterial species are rapidly dephosphorylated at their 5′ ends by cellular pyrophosphohydrolases, such as *Escherichia coli* RppH [4]. Hence, 5′ monophosphorylated RNAs include both degraded and full-length transcripts. We have used 5′ RNA-seq to map transcription start sites in *E. coli*, using Δ*rppH* mutants that lack the pyrophosphohydrolase. Working in a pyrophosphohydrolase mutant background greatly reduces the number of false negatives (i.e., transcription start sites that are not identified). We have used 5′ RNA-seq to map novel transcription start sites within intergenic regions and within genes, in both sense and antisense orientations.

An equivalent method, dRNA-seq, has been described previously [5]. dRNA-seq relies on specific degradation of 5′ monophosphorylated but not 5′ triphosphorylated RNAs by an exonuclease. dRNA-seq has not been tested in a pyrophosphohydrolase mutant background. Hence, existing dRNA-seq datasets are likely to have many false negatives.

2 Materials

2.1 RNA Purification and Processing

1. LB medium (1 % NaCl, 1 % tryptone, 0.5 % yeast extract). Sterilize by autoclaving.
2. *E. coli* K-12 Δ*rppH* strain, e.g., BW25113 Δ*rppH* [6].
3. Kanamycin (Sigma-Aldrich) stock at a concentration of 50 mg/mL (1,000×) in water. Filter sterilize with a 0.22 μm filter. Store at 4 °C.
4. RNA freeze buffer (Ethanol + 5 % Acid Phenol (Ambion)).
5. 1× TBS (make 10× stock: 500 mM Tris–HCl, 1,500 mM NaCl, pH 7.4).
6. RNA lysis buffer (2 % SDS, 4 mM EDTA).
7. Acid Phenol/Chloroform (Ambion).
8. Chloroform:isoamyl alcohol (24:1).
9. 75 % EtOH made with diethylpyrocarbonate-treated water (DEPC-dH_2O).
10. TE (100 mM Tris–HCl pH 8.0, 10 mM EDTA).
11. DNase I (RNase Free; NEB).
12. 3 M Sodium acetate (NaOAc), pH 5.5.

13. NanoDrop spectrophotometer or equivalent for measuring nucleic acid concentrations.
14. Ribo-Zero ribosomal RNA (rRNA) removal kit for gram negative bacteria (Epicentre).
15. Calf intestinal alkaline phosphatase (CIP; NEB).
16. RNA 5′ polyphosphatase (Epicentre).

2.2 Adaptor Ligation to 5′ Ends

1. 5′ RNA Illumina adaptors (Integrated DNA Technologies) with optional barcodes for multiplexing samples (*see* oligonucleotide sequences 1, 2, 3, and 4 in Table 1).
2. 4 % denaturing acrylamide gel: Add 1.6 mL Urea Gel concentrate, 1 mL Urea Gel buffer, 7.4 mL Urea Gel diluent, 100 μL 10 % Ammonium persulfate, 10 μL TEMED (Urea Gel, National Diagnostics). Pour the gel and let polymerize for 30 min.
3. 1× TBE (make 10× TBE stock: 1 M Tris, 0.9 M boric acid, 0.01 M EDTA).
4. Ethidium bromide (MP bio), 10 mg/mL stock. Use 10 μL/100 mL of 1× TBE gel running buffer.
5. RNA extraction buffer (300 mM NaOAC, pH 5.5, 1 mM EDTA).

2.3 First-Round cDNA Synthesis

1. Reverse Transcription primer (Integrated DNA Technologies; *see* oligonucleotide sequence 5, Table 1).
2. 10 mM dNTP mix.
3. SuperScript III Reverse Transcriptase (Invitrogen).

2.4 PCR Amplification of the Sequencing Library

1. Illumina PE PCR Primer 1.01 (Integrated DNA Technologies; *see* oligonucleotide sequence 6, Table 1).
2. Illumina PE PCR Primer 2.01 (Integrated DNA Technologies; *see* oligonucleotide sequence 7, Table 1).
3. 8 % Non-denaturing polyacrylamide gel (for a 10 mL gel, mix 2 mL 40 % acrylamide, 1 mL 10× TBE, 6.89 mL H_2O, 100 μL 10 % ammonium persulfate, 10 μL TEMED).
4. DNA extraction buffer (300 mM NaCl, 10 mM Tris–HCl pH 8.0, 1 mM EDTA).

2.5 Library Validation

1. pGEM-T Vector System (Promega).
2. SP6 sequencing primer (Integrated DNA Technologies; *see* oligonucleotide sequence 8, Table 1).
3. T7 sequencing primer (Integrated DNA Technologies; *see* oligonucleotide sequence 9, Table 1).

Table 1
List of oligonucleotides

	Sequence	
1	ACACUCUUUCCCUACACGACGCUCUUCCGAUCUAUU	Barcoded RNA adaptor for 5′ RACE Illumina library (barcode underlined)
2	ACACUCUUUCCCUACACGACGCUCUUCCGAUCUCAU	Barcoded RNA adaptor for 5′ RACE Illumina library (barcode underlined)
3	ACACUCUUUCCCUACACGACGCUCUUCCGAUCUGCU	Barcoded RNA adaptor for 5′ RACE Illumina library (barcode underlined)
4	ACACUCUUUCCCUACACGACGCUCUUCCGAUCUUGU	Barcoded RNA adaptor for 5′ RACE Illumina library (barcode underlined)
5	$CCTGCTGAACCGCTCTTCCGATCTN_9$	Reverse Transcription primer with part of PE Primer 2.01 at 5′ end
6	AATGATACGGCGACCACCGAGATCTACACTCTTTCCCTACACGACGCTCTTCCGATCT	Illumina PE PCR Primer 1.01
7	CAAGCAGAAGACGGCATACGAGATCGGTCTCGGCATTCCTGCTGAACCGCTCTTCCGATC	Illumina PE PCR Primer 2.01
8	TACGATTTAGGTGACACTATAG	SP6 Sequencing Primer
9	TAATACGACTCACTATAGGG	T7 Sequencing Primer

3 Methods

The overall method is summarized in Fig. 1.

3.1 RNA Purification and Processing

Grow the *E. coli* culture to desired cell density (e.g., OD_{600} 0.5–0.7). Prior to harvesting cells, add 2 mL RNA freeze buffer to 5 mL of bacterial culture. Immediately pellet cells by centrifugation at 4 °C. Discard the supernatant, and wash the pellet once with 1 mL ice-cold

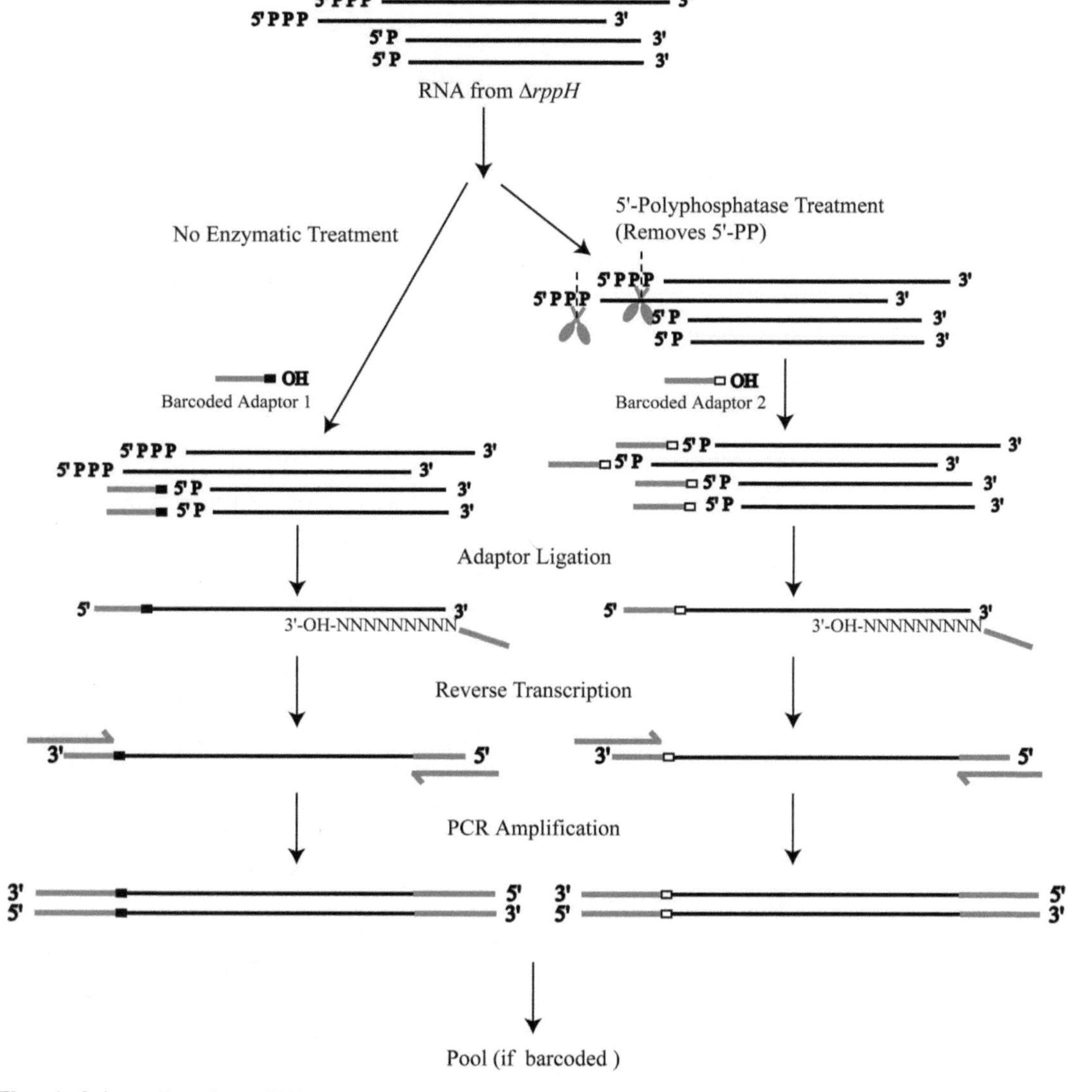

Fig. 1 Schematic of 5′ RNA-seq method. The *right-hand side* illustrates the method for generating pyrophosphatase-treated libraries, and the *left-hand side* illustrates the method for generating pyrophosphatase-untreated libraries. In this example, the two libraries are differentially barcoded and can be pooled and sequenced in a single, multiplexed lane

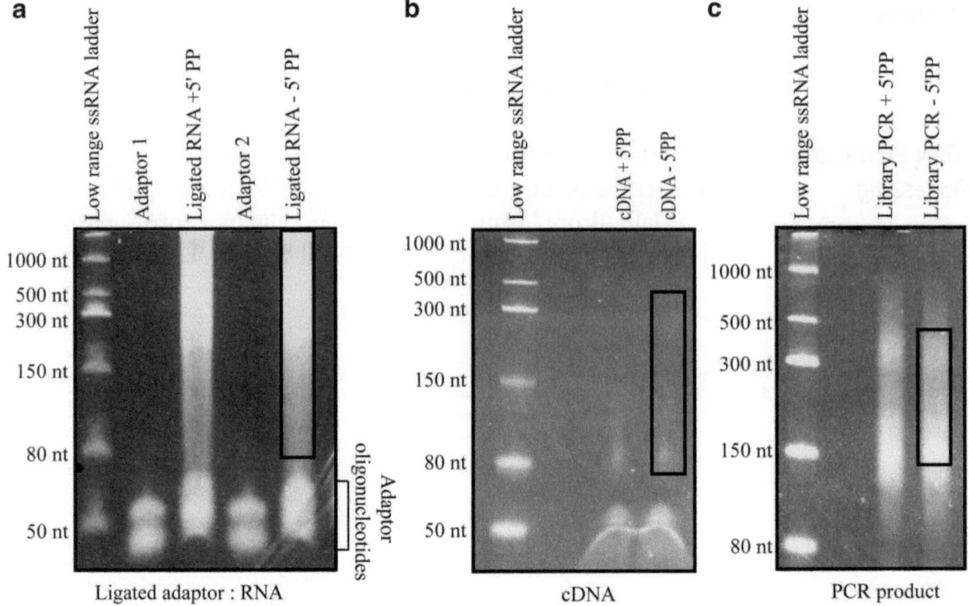

Fig. 2 Gel images for 5′ RNA-seq method. Denaturing acrylamide gels were used for gel purification of (**a**) RNA ligated to adaptors, (**b**) cDNA, and (**c**) final PCR product. In each case, both pyrophosphatase-treated and -untreated libraries are shown. *Black rectangles* indicate the region that was excised from the gel for purification

1× TBS. Resuspend the cell pellet in 500 μL RNA lysis buffer. Transfer the sample to a 1.5 mL microcentrifuge tube. Boil the sample for 3 min. Add 500 μL acid phenol/chloroform and incubate at 65 °C for 6 min. Chill the sample on ice for 5 min. Spin at full speed in a microcentrifuge for 5 min. Transfer the aqueous layer to a new tube. Extract with one volume (~500 μL) of acid phenol/chloroform. Transfer supernatant to a new tube and add one volume of chloroform: isoamyl alcohol (24:1). Transfer supernatant to a new tube and add 1/10 volume of 3 M NaOAc and 2 volumes of 100 % ethanol. Leave at least for 1 h at −20 °C or at least for 15 min at −80 °C. Spin at full speed in a microcentrifuge for 15 min. Wash pellet once with 75 % ethanol prepared with DEPC-treated water. Resuspend the pellet in 100 μL 1× TE. RNA can be frozen at this stage (*see* **Note 1**).

Treat the RNA sample with DNase I by adding 30 μL 10× DNase I buffer, 265 μL DEPC water, and 5 μL (10 units) of DNase I to 100 μL RNA. Incubate at 37 °C for 1 h. Extract the RNA with phenol/chloroform/isoamyl alcohol and ethanol precipitate as described above. Resuspend the RNA in 50–80 μL DEPC-H_2O. Measure the RNA concentration using a Nanodrop spectrophotometer. Yields of 50–80 μg of RNA are typical (*see* **Note 1**). Remove the 23S, 16S, and 5S rRNA from 10 μg total RNA using the Ribo-Zero rRNA removal kit (*see* **Note 2**).

CIP treat the barcoded Illumina RNA adaptor oligonucleotides in a total volume of 50 μL by adding 41 μL of adaptor oligonucleotide (100 μM stock), 5 μL of 10× NEB buffer #3, 2 μL CIP enzyme, and 2 μL dH_2O. Incubate at 37 °C for 30 min. Bring volume to 300 μL with DEPC-H_2O. Phenol/chloroform extract and ethanol precipitate as described above. Resuspend the CIP-treated adaptor oligonucleotides in 20 μL DEPC-H_2O (*see* **Note 3**).

Divide the rRNA-depleted RNA sample equally between two micocentrifuge tubes (~2 μg each). To first tube, add 1 μL of 10× polyphosphatase buffer, 2 μL of 5′ polyphosphatase enzyme, and RNase-free H_2O to a total volume of 10 μL. In the second tube, repeat as above but without the addition of 5′ polyphosphatase enzyme. Incubate both reactions at 37 °C for 1 h (*see* **Note 4**).

3.2 Adaptor Ligation to 5′ Ends

Add 2 μL CIP-treated adaptor, 2 μL DMSO, 2 μL 10× RNA ligase buffer, and 2 μL T4 RNA ligase 1 to the polyphosphatase-treated and -untreated RNA samples. Bring the reaction to 20 μL with DEPC-H_2O. Incubate the reaction at 37 °C for 90 min. Purify the RNA from a 4 % denaturing acrylamide gel. The gel should be sliced from 70 nt to the top of gel (Fig. 2a). Puncture a hole in a 0.5 mL microcentrifuge tube using a needle, and place the gel slice in this tube. Place this tube in a 1.5 mL microcentrifuge tube. Centrifuge the stacked tubes in a microcentrifuge at full speed in a microcentrifuge. This will allow the gel slice to move through the hole into the 1.5 mL tube. Add 300 μL RNA extraction buffer. Elute the RNA by incubation for 3 h at room temperature or overnight at 4 °C. Transfer the sample (both the buffer and gel slices) into a Spin-X filter. Centrifuge the filter in a collection tube at full speed in a microcentrifuge to remove the polyacrylamide. Ethanol precipitate as described above but with the addition of 1 μL of glycogen. Resuspend the pellet in 10 μL DEPC-H_2O (*see* **Note 5**).

3.3 First-Round cDNA Synthesis

To 10 μL RNA (from above), add 1 μL RT adaptor (100 μM stock, oligonucleotide sequence 5, Table 1) and 1 μL dH_2O. Heat samples at 70 °C for 5 min. Chill on ice for 2 min. Add 4 μL of 5× reverse transcription buffer, 2 μL of 10 mM dNTP mix, 1 μL 0.1 M DTT, and 1 μL of Super Script III reverse transcriptase. Incubate at 50 °C for 1 h. Purify the first-strand cDNA from a 4 % denaturing acrylamide gel as described above. Gel purify from 80 to 500 nt (Fig. 2b). Resuspend precipitated RNA in 20 μL dH_2O.

3.4 PCR Amplification of the Sequencing Library

PCR amplify the cDNA with Illumina PE library primers (oligonucleotide sequences 6 and 7, Table 1). Template for PCR can range from 5 to 20 μL, depending on the quantity of cDNA. Template concentration and the number of PCR cycles need to be

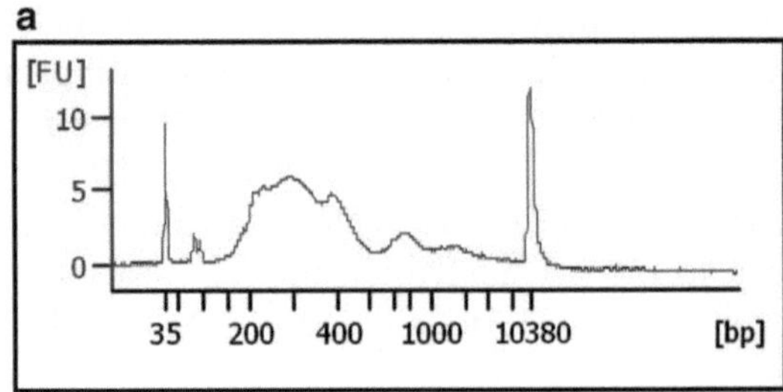

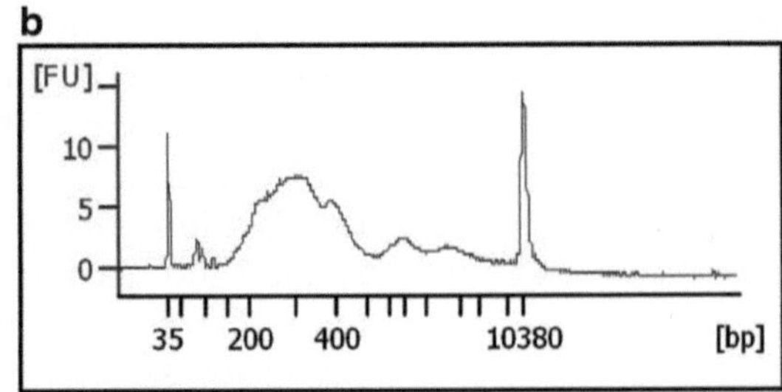

Fig. 3 Bioanalyzer traces for final libraries. Final Bioanalyzer traces for a (**a**) pyrophosphatase-treated and (**b**) -untreated library to illustrate the distribution of fragment sizes. The *spikes* at either end of each trace represent control DNA fragments included in every Bioanalyzer lane

determined empirically (*see* **Note 6**). A typical PCR reaction contains the following reagents: 10 μL 10× Taq polymerase buffer, 10 μL 10 mM dNTP mix, 2.5 μL primer 6 (5 μM stock, Table 1), 2.5 μL primer 7 (10 μM stock, Table 1), 3 μL Taq DNA polymerase, and H_2O to 100 μL. A typical PCR reaction uses the following cycling parameters: 95 °C, 10 min (if using hot-start Taq), 15 cycles (95 °C, 30 s; 62 °C, 30 s; 72 °C, 30 s), 72 °C, 7 min. Purify the PCR product using a MinElute column (Qiagen). Purify the library from a 4 % denaturing acrylamide gel. Purify the products between 150 and 500 nt (Fig. 2c; *see* **Note 7**).

3.5 Library Validation

To check library quality, clone a small aliquot of PCR in pGEM-T vector and save the rest for deep sequencing. For pGEM-T cloning, set up 10 μL reaction by adding up to 3.5 μL library, 5 μL 10× ligation buffer, 0.5 μL pGEM-T vector (25 ng), 1 μL T4 DNA Ligase, and dH_2O to 10 μL. Incubate at 4 °C overnight and transform into competent *E. coli* cells, plating onto ampicillin selective media. Screen the pGEM-T clones using colony PCR with T7 and SP6 sequencing primers. Miniprep and sequence insert for 10–15 clones with T7 or SP6 sequencing primers. Most of the clones should have sequence that corresponds to RNA 5′ ends, with the expected Illumina adaptor sequences at either end. If the library looks correct after quality control, check the size and concentration of a small aliquot using a Bioanalyzer (Agilent). Figure 3 shows the typical size distribution of a library. For multiplexing multiple barcoded libraries (recommended), mix equimolar amounts of each sample with different barcodes. Typically, 10 ng library is sufficient for Illumina sequencing.

3.6 Deep Sequencing Data Analysis

Deep sequencing of libraries using the Illumina HiSeq instrument typically generates 100–200 million sequence reads, the majority of which correspond to library fragments. Each sequence read

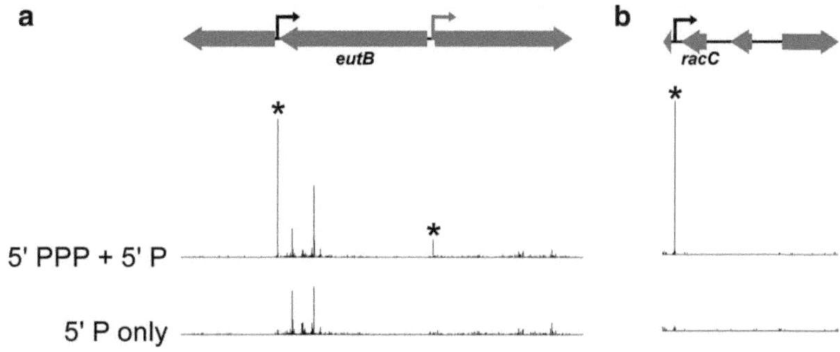

Fig. 4 Representative 5′ RNA-seq data for *Escherichia coli*. Mapped sequence reads from pyrophosphatase-treated (*top trace*) and -untreated (*bottom trace*) libraries are shown for representative genomic regions. The height of the bars in each trace represents the number of sequence reads that mapped to that position. Genes are shown above the traces as *straight arrows*. The *gray*, *bent arrow* represents a transcription start site for a protein-coding gene. The *black*, *bent arrows* represent novel transcription start sites that are most likely for noncoding RNAs. *Asterisks* above the sequence read traces indicate the sites of transcription initiation that are characterized by large number of sequence reads in the pyrophosphatase-treated sample and low number of sequence reads in the untreated sample and thus likely indicate triphosphorylated RNA 5′ ends

from a library fragment will begin with the nucleotide that corresponds to an RNA 5′ end. The genomic coordinates for these RNA 5′ ends can be determined by mapping sequence reads to a reference genome using standard alignment methods such as Bowtie [7]. It is often beneficial to exclude reads that map to multiple locations, i.e., repetitive sequence. For multiplexed, barcoded libraries, reads will need to be separated by barcode and the barcodes removed before alignment to the reference genome. Once reads have been mapped, transcription start sites are identified by comparison of the phosphatase-treated and -untreated libraries. Total mapped read counts should be normalized to account for differences in read numbers for different libraries. We typically identify transcription start sites by requiring a minimum number of sequence reads for a given genomic position (an example threshold would be the total number of uniquely mapped reads/100,000) and based on the ratio of reads in the pyrophosphatase-treated and -untreated libraries (typically we require a ≥10-fold difference). Figure 4 shows example data for *E. coli*, highlighting novel RNAs identified using this approach.

4 Notes

1. RNA integrity can be checked by running a 1 % agarose TBE gel. 23S, 16S, and 5S rRNA bands should be visible.
2. rRNAs are highly abundant. It is important to remove them; otherwise, rRNA will dominate the deep sequencing library. Typically, 2–4 μg RNA remains per sample after rRNA removal. For lower yields, multiple rRNA-depleted samples can be pooled.
3. It is important to CIP treat the adaptors to remove 5′ phosphates; otherwise, adaptor–adaptor ligation contaminates the sequencing library in the subsequent steps. Intact adaptors are not substrates for adaptor–adaptor ligation because they are not phosphorylated at their 5′ end. However, adaptor oligonucleotide degradation can occur, thus generating 5′ monophosphorylated oligonucleotides. CIP removes phosphates from all such degraded adaptor oligonucleotides.
4. Tobacco acid pyrophosphatase (TAP) can also be used instead of 5′ polyphosphatase to generate 5′ monophosphate ends on RNA.
5. It is important to remove the unligated adaptor from the ligated RNA by gel purification because the adaptor can cause contamination in cDNA synthesis reaction.
6. The template concentration and number of PCR cycles will ideally result in near-saturation of the PCR. It is important to limit the PCR cycles to avoid complete saturation which can lead to biased amplification.
7. The PCR product should be cut above 120 nt so as to remove adaptor dimers (~120 bp). An 8 % non-denaturing acrylamide gel can also be used and is more appropriate if the library is to be validated.

References

1. Waters LS, Storz G (2009) Regulatory RNAs in bacteria. Cell 136:615–628
2. Ozsolak F, Milos PM (2011) RNA sequencing: advances, challenges and opportunities. Nat Rev Genet 12:87–98
3. Scotto-Lavino E, Du G, Frohman MA (2006) 5′ End cDNA amplification using classic RACE. Nat Protoc 1:2555–2562
4. Deana A, Celesnik H, Belasco JG (2008) The bacterial enzyme RppH triggers messenger RNA degradation by 5′ pyrophosphate removal. Nature 451:355–358
5. Sharma CM, Hoffmann S, Darfeuille F, Reignier J, Findeiss S, Sittka A, Chabas S, Reiche K, Hackermüller J, Reinhardt R, Stadler PF, Vogel J (2010) The primary transcriptome of the major human pathogen *Helicobacter pylori*. Nature 464:250–255
6. Baba T, Ara T, Hasegawa M, Takai Y, Okumura Y, Baba M, Datsenko KA, Tomita M, Wanner BL, Mori H (2006) Construction of *Escherichia coli* K-12 in-frame, single-gene knockout mutants: the Keio collection. Mol Syst Biol 2:0008
7. Langmead B, Trapnell C, Pop M, Salzberg SL (2009) Ultrafast and memory-efficient alignment of short DNA sequences to the human genome. Genome Biol 10:R25

Chapter 2

Screening Inhibitory Potential of Anti-HIV RT RNA Aptamers

Margaret J. Lange and Donald H. Burke

Abstract

Aptamers targeted to HIV reverse transcriptase (RT) have been demonstrated to inhibit RT in biochemical assays and as in cell culture. However, methods employed to date to evaluate viral suppression utilize time-consuming serial passage of infectious HIV in aptamer-expressing stable cell lines. We have established a rapid, transfection-based assay system to effectively examine the inhibitory potential of anti-HIV RT aptamers expressed between two catalytically inactive hammerhead ribozymes. Our system can be altered and optimized for a variety of cloning schemes, and addition of sequences of interest to the cassette is simple and straightforward. When paired with methods to analyze aptamer RNA accumulation and localization in cells and as packaging into pseudotyped virions, the method has a very high level of success in predicting good inhibitors.

Key words Aptamer, Human immunodeficiency virus (HIV), Reverse transcriptase, Pseudotyping, RNA accumulation

1 Introduction

The current therapeutic regimen for slowing or preventing replication of human immunodeficiency virus 1 (HIV-1) includes a cocktail of compounds that target several viral proteins such as reverse transcriptase (RT) [1, 2]. Despite the success of these therapies, toxicity, noncompliance, and appearance of drug-resistant viral strains continue to present significant problems. Aptamers are structured nucleic acids generated to bind specific molecular targets through an iterative process termed Systematic Evolution of Ligands by EXponential enrichment, or SELEX [3, 4]. These small nucleic acids bind their targets with high-affinity and -specificity, rivaling antibodies. In addition, aptamers can be produced chemically in a scalable process, making them economical, non-immunogenic, and small enough in size to allow efficient entry into biological compartments [5]. Aptamers can also be chemically modified to increase stability and resistance to nuclease-mediated degradation and subjected to conjugation

Daniel Lafontaine and Audrey Dubé (eds.), *Therapeutic Applications of Ribozymes and Riboswitches: Methods and Protocols*, vol. 1103, DOI 10.1007/978-1-62703-730-3_2, © Springer Science+Business Media New York 2014

chemistries to produce more potent, diverse molecular combinations [5]. Along with these benefits, aptamers may be evolved to bind extracellular, intracellular, and cell-surface targets, giving them broad therapeutic potential [2, 5–7]. Notably, aptamers have been selected to bind critical proteins from several viruses and have been shown to impair the replication of the associated viruses, including HIV [3, 4, 8–15], hepatitis B virus (HBV) [16, 17], human cytomegalovirus (CMV) [18], hepatitis C virus (HCV) [19–23], and influenza virus [24, 25]. Thus, the versatility of potential protein targets for aptamers is quite broad.

Anti-HIV RT aptamers have previously been shown to inhibit HIV replication in cell culture [8, 12, 14, 15, 26]. Most of these studies employed a time-consuming system requiring the development of clonal, stable, aptamer-expressing cell lines and long-term serial passage of HIV in the presence of aptamer to demonstrate inhibition. Recently, we developed a fast, streamlined assay to enable screening of multiple aptamers in a variety of expression contexts using a transient transfection-based single-cycle infectivity assay [26]. Our assay utilizes aptamer-expressing plasmids in concert with an EGFP-expressing HIV proviral plasmid and a VSV-G envelope plasmid. Co-transfection of these plasmids directs the production of VSV-G-pseudotyped HIV from cells that also express the aptamer. The assay system allows for determination of both transfection efficiency and infectivity documented by the presence of EGFP-positive cells. The aptamer accumulates within the cells and is packaged within the EGFP-encoding virus, potentially via interaction with the RT portion of the polyprotein. Through aptamer packaging within the virus, the aptamer is carried to a new cell upon infection, where it remains tightly bound the HIV RT, preventing replication and the appearance of EGFP-positive cells. The system allows rapid evaluation of new aptamers and new designs for their expression context. For example, by assaying several different aptamer expression contexts in which highly active or inactivated ribozymes flank the aptamer, we found that ribozyme catalytic activity was not required for inhibition [26]. Intracellular ribozyme cleavage has been reported to be highly inefficient due to the low magnesium environment of the cell [27]. Thus, our data suggested that the uncleaved or partially cleaved transcripts may be the active forms in all tested expression contexts. The ribozyme structures flanking the aptamer, despite their inactivity, likely provide adequate tertiary structural support to favor proper aptamer folding for binding to the RT. The following method details the use of the single-cycle infectivity assay coupled with the catalytically inactive aptamer expression cassette. The method reliably correlates aptamer biological activity to biochemical studies and provides a fast, efficient screening system. Additionally, the system is designed to be highly versatile in terms of adding or subtracting cassette components.

2 Materials

2.1 Generation of Aptamer-Expressing and Control Plasmids

2.1.1 Plasmids and Oligonucleotides

Parental CMV-driven RNA expression vector (*see* **Note 1**), oligonucleotides for preparation of the parental aptamer expression cassette (Integrated DNA Technologies, Coralville, IA) (Table 1).

2.1.2 Cloning Reagents (See Note 1)

Tris–EDTA (TE) buffer (10 mM Tris–HCl pH 8.0, 1 mM EDTA), restriction enzymes with appropriate restriction enzyme buffers (New England Biolabs, Ipswitch, MA), T4 DNA ligase with buffer (New England Biolabs, Ipswitch, MA), Antarctic Phosphatase with buffer (New England Biolabs, Ipswitch, MA), LB-Ampicillin broth (Sigma, St. Louis, MO), LB-Ampicillin agar plates, *Escherichia coli* HB101 competent cells, super optimal broth (SOC) medium (Invitrogen, Carlsbad, CA), thermocycler, 37 °C bacterial culture incubator, 37 °C shaker incubator.

2.2 Determination of Aptamer Bioactivity by Single-Cycle Infectivity Assay

2.2.1 Cell Culture

293FT cells (Invitrogen, Carlsbad, CA) (*see* **Note 2**) maintained in Dulbecco's Modified Eagle's Medium (DMEM) (Sigma, St. Louis, MO) supplemented with 10 % fetal bovine serum (FBS) (Sigma, St. Louis, MO), 500 μg/mL Geneticin (Sigma, St. Louis, MO), 6 mM L-glutamine (Gibco, Life Technologies, Grand Island, NY), 1 mM sodium pyruvate (Gibco, Life Technologies, Grand Island, NY), 0.1 mM non-essential amino acids (Gibco, Life Technologies, Grand Island, NY), and 1× vitamins (Gibco, Life Technologies, Grand Island, NY). Cells should be maintained at 37 °C in 5 % CO_2. For cell splitting and plating for bioassays: TrypLE Express (Gibco, Life Technologies, Grand Island, NY), sterile 1× Dulbecco's Phosphate Buffered Saline (PBS) (Gibco, Life Technologies, Grand Island, NY), 100 mm tissue culture dishes (ISC Bioexpress, Kaysville, UT), 6-well tissue culture plates (ISC Bioexpress, Kaysville, UT), and 96-well ELISA plates (ISC Bioexpress, Kaysville, UT).

2.2.2 Plasmids and Transfection Reagents

Aptamer-expressing plasmids, control plasmids, proviral plasmid pNL4-3-CMV-EGFP (*see* Subheading 3.2) (*see* **Note 3**), vesicular stomatitis virus glycoprotein (VSV-G) plasmid pMD-G (Invitrogen, Carlsbad, CA), polyethylenimine (PEI) transfection reagent [28] (Sigma, St. Louis, MO) (*see* **Note 4**), serum-free DMEM medium, standard DMEM cell culture medium (*see* Subheading 2.2.1), microcentrifuge tubes.

2.2.3 Flow Cytometry Reagents

20 % Paraformaldehyde (*see* **Note 5**), 1× PBS, flow cytometer, microcentrifuge tubes, flow cytometer tubes (if applicable).

2.2.4 Virus Harvesting

0.45 μm syringe filters (Millipore, Billerica, MA), 5 mL syringes, microcentrifuge tubes, −80 °C freezer.

Table 1
Oligonucleotide sequences for assembly of the aptamer expression cassette and primer sequences for amplification of reference genes used in endpoint or qPCR experiments

Name	Sequence
RzModA[a,b]	GCATAATACGACTCACTATAGGGCTAGCGGATTAAACAAG
RzModB[b]	GCGTTTCGTCGCATCCCAGCGACTCATTTCCTTGTTTAATCCGCTAGCCC
RzModC[b]	GCTGGGATGCGACGAAACGCCTTCGGGCGTCCTTGTATCTGAATTCCCTACGAACCC
RzModD[b]	GACAGGGCCCGTTTTCCAGTGTTTTCCCCTTTATCTCCTGGGTTCGTAGGGAATTCAG
RzModE[b]	GGAAAACACTGGAAAACGGGCCCTGTCACGGATTGTGCTTATCCGT
RzModF[b]	GGAGGCGGCCGCTGTTTCGTCCTCACGGACTCATTTCACGGATAAGCACAATCCGTGAC
RzModG[a,b]	TTTTATTAGGAAAGGACAGTGGGAGGCGGCCGCTGTTTCGTCCTC
GAPDH-F	AGAAGGCTGGGGCTCATTTG
GAPDH-R	AGGGGCCATCCACAGTCTTC
GAPDHspliced-F	GGTGAAGGTCGGAGTCAACG
GAPDHspliced-R	GTTGAGGTCAATGAAGGGGTC
U6-F	CGCTTCGGCAGCACATATAC
U6-R	TTCACGAATTTGCGTGTCAT
18S-F	CAGCCACCCGAGATTGAGCA
18S-R	TAGTAGCGACGGGCGGTGTG
HIV Gag-F	TGCTATGTCAGTTCCCCTTGGTTCTCT
HIV Gag-R	AGTTGGAGGACATCAAGCAGCCATGCAAAT

[a]Primers used for amplification of inserts for cloning into aptamer expression cassettes
[b]Restriction enzyme sites are denoted in underlining

2.3 Analysis of Intracellular Aptamer RNA Accumulation and Localization

293FT cells, 6-well tissue culture plates, aptamer-expressing plasmids, control plasmids, PEI transfection reagent (*see* **Note 4**), serum-free DMEM, standard DMEM culture medium (*see* Subheading 2.2.1), 1× PBS, cell scraper, microcentrifuge tubes.

2.3.1 Cells, Plasmids, and Transfection Reagents

2.3.2 RNA Fractionation and Isolation Reagents

TRIzol and TRIzol LS reagent (Invitrogen, Carlsbad, CA) or the Paris RNA Isolation Kit (Ambion, Life Technologies, Grand Island, NY), chloroform (Sigma, St. Louis, MO), TKM buffer (10 mM Tris–HCl pH 7.5, 10 mM KCl, 1 mM $MgCl_2$), Triton X-100, 70 % ethanol, Turbo DNase (Ambion, Life Technologies, Grand Island, NY), nuclease-free water, NanoDrop Spectrophotometer (Thermo-Fisher Scientific, Waltham, MA).

2.3.3 cDNA Synthesis and Real-Time PCR

ImProm II Reverse Transcription System with random hexamer primers (Promega, Madison, WI), primers for amplification of aptamer and aptamer cassette sequences (Integrated DNA Technologies, Coralville, IA), primers for amplification of reference genes for normalization (Table 1) (Integrated DNA Technologies, Coralville, IA), PCR tubes, microcentrifuge tubes, Real Time PCR Plates and Optical Covers (Applied Biosystems, Foster City, CA), Power SybrGreen PCR Master Mix (Applied Biosystems, Foster City, CA), ABI 7500 Real Time RT-PCR Thermocycler (Applied Biosystems, Foster City, CA).

2.4 Analysis of Aptamer Incorporation into Viral Particles

1. Virus-containing supernatant generated from transfection (*see* Subheading 2.2.4), ultracentrifuge, ultracentrifuge-appropriate tubes, TRIzol reagent, chloroform, Turbo DNase, nuclease-free water, 3 M sodium acetate, 100 % ethanol, 70 % ethanol, −80 °C freezer, NanoDrop.

3 Methods

3.1 Generation of Aptamer Expression Cassettes and Controls

The method described in Subheading 3.1 will prepare the constructs to be utilized in the single-cycle infectivity assay (Fig. 1). The cloning process is straightforward and can be easily modified to accommodate a wide range of restriction sites (*see* **Note 1**). The parental aptamer expression cassette generated in this method employs a minimal core (~40 nt) aptamer, flanked by two catalytically inactive ribozymes derived from RzB [29, 30] and sTRSV [31]. The ribozymes on either side of the aptamer allow the aptamer to fold into its active conformation and may decrease the probability of exonuclease degradation [27]. The CMV promoter in these constructs results in high-level expression of aptamer RNA in both the nuclear and cytoplasmic compartments, and this accumulation and localization can be verified using other methods described in this chapter.

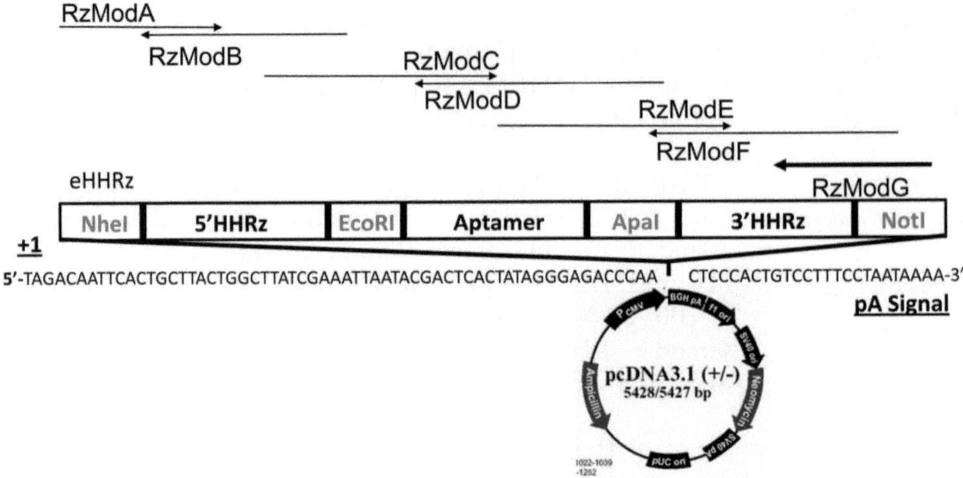

Fig. 1 Cloning schematic for insertion of the catalytically inactive ribozyme cassette into the parental CMV-driven vector

1. Reconstitute the oligonucleotides found in Table 1 at a concentration of 100 μM using TE buffer prepared in nuclease-free water (*see* **Note 1**). Dilute each of the oligonucleotides to 10 μM using nuclease-free water, except for the two outermost oligonucleotides (RzModA and RzModG, Table 1), which should be diluted to 25 μM. In a PCR tube, combine 1 μL of each oligonucleotide with 5 μL 10× Taq Buffer, 1.25 μL 25 mM $MgCl_2$, 1 μL 10 mM dNTPs, 1 U Taq Polymerase, and enough water to bring the final volume to 50 μL. Amplify the reaction using a thermocycler (*see* **Note 6**).

   Initial denaturation (5 min at 95 °C).

   30 cycles of denaturation (30 s at 95 °C), annealing (30 s at 60 °C), and extension (45 s at 72 °C).

   Final extension (10 min at 72 °C).

   Final hold (4 °C).

2. Ethanol precipitate the PCR product from Subheading 3.1, **step 2**, by adding 1/10 volume of 3 M sodium acetate and 3 volumes of 100 % ethanol in a microcentrifuge tube. Precipitate the PCR product for 1 h at −80 °C or overnight at −20 °C. Centrifuge the solution at full speed (16,100 × *g* for Eppendorf 5415D) using a microcentrifuge for 30 min at 4 °C. Remove the supernatant, and gently wash the pellet once using cold 75 % ethanol. Repeat the centrifugation. Remove the supernatant, and air-dry the pellet for approximately 10 min (*see* **Note 7**). Resuspend the pellet in 25–50 μL nuclease-free water, depending on band intensity of the PCR product by ethidium bromide staining in an agarose gel. Further dilution of the product prior to ligation into the cassette may be required to achieve a 3:1 molar ratio of insert to vector.

3. Prepare the parental CMV-driven vector (*see* **Note 1**) and parental aptamer expression cassette for cloning by digesting each component with restriction enzymes NheI-HF and NotI-HF (½ μL each enzyme (10 U), 3 μg DNA, 1× NEB buffer 4, 1× BSA, water). After 1 h at 37 °C, inactivate the enzymes by incubation for 20 min at 65 °C. Subsequent phosphatase treatment of the vector is optional but may help reduce vector religation.
4. Purify the digested vector using a PCR Clean-Up or Gel Extraction Kit. The parental aptamer expression cassette may be purified or ethanol precipitated as in Subheading 3.1, **step 3**. Quantify the purified vector and insert using a NanoDrop spectrophotometer.
5. Ligate the cassette and vector using a molar ratio of 3:1 (insert:vector). Perform ligation in a 20 μL reaction volume using T4 DNA ligase (1× NEB ligase buffer with ATP, ½ μL T4 DNA ligase (200 U), 3:1 insert:vector, water).
6. Transform 10 μL of a 20 μL ligation reaction to 50 μL HB101 competent cells. Incubate on ice for 30 min. Heat shock the cells for 1 min at 42 °C and place back on ice for 5 min. Add 200 μL SOC medium and incubate at 37 °C for 1 h with shaking. Plate 100 μL on LB plates containing ampicillin. Incubate overnight at 37 °C. Pick colonies and screen for the presence of the correct insert (*see* **Note 8**). Confirm by DNA sequencing.
7. After verification by sequencing, proceed to clone in any aptamer or control RNA of choice by amplifying the aptamer or control sequence with primers to append EcoRI and ApaI restriction sites to the 5′ and 3′ ends, respectively (*see* **Note 1**). Proceed as for the cassette cloning (*see* **Note 9**).

3.2 Determination of Aptamer Bioactivity by Single-Cycle Infectivity Assay

The method described in Subheading 3.2 allows for the determination of aptamer bioactivity in the context of the aptamer expression construct made in Subheading 3.1. The method utilizes a single-cycle infectivity assay where aptamer-expressing plasmid is first transfected into the cell, followed by the proviral plasmid pNL4-3-CMV-EGFP and VSV-G envelope plasmid, pMD-G [26]. The HIV-$1_{\text{NL3-4-derived}}$ CMV-EGFP plasmid (pNL4-3-CMV-EGFP) used in single-cycle infectivity assays was kindly provided by Vineet KewalRamani (National Cancer Institute [NCI]—Frederick). This proviral vector lacks the genes encoding *vif*, *vpr*, *vpu*, *nef*, and *env* and has a CMV immediate early promoter-driven EGFP in place of *nef*. As such, the virus can undergo only one round of replication. Thus, the virus is produced in the presence of aptamer, allowing the aptamer to be effectively packaged within the pseudotyped virus, presumably via interaction with the RT [14, 26]. Upon infection of fresh cells, aptamer interaction with the RT prevents DNA synthesis and, thus, the appearance of EGFP-positive cells.

To validate the results obtained from the bioactivity assay, it is strongly suggested to proceed with Subheadings 3.3 and 3.4, which verify that the aptamer accumulates to sufficient levels in total, cytoplasmic, and nuclear compartments of the cell, as well as to verify that the aptamer packages efficiently within the pseudotyped viral particles.

1. Generate high-quality plasmid DNA preparations using standard methods (*see* **Note 10**). Dilute each of the plasmids to a concentration of 100 ng/μL using nuclease-free water.
2. The day prior to transfection, plate 300,000–500,000 293FT cells in 6-well plates, aiming for ~60–70 % confluence the next day (*see* **Note 11**). Media volume should be 2 mL per well. Plate enough wells to accommodate all of the aptamer-expressing plasmids and control plasmids, in addition to "no-plasmid" and "mock" wells.
3. Prepare transfection mixtures for the aptamer-expressing plasmids. Combine 97 μL serum-free medium (DMEM) with 3 μL PEI transfection reagent (3 μL PEI/μg DNA) (*see* **Note 4**). Incubate the mixture at room temperature for 5 min. Add 10 μL (for a total of 1 μg DNA using 100 ng/μL stock) of aptamer-expressing plasmid to the transfection mixture. Filler DNA (1 μg) should be used for "no-plasmid" and "mock" wells (*see* **Note 12**). Flick the tube to mix, and briefly spin the liquid out of the cap. Allow the mixture to incubate at room temperature for 30 min (*see* **Note 13**). Add the mixture dropwise to the 293FT cells in 6-well plates so as to not disrupt transfection complexes or the cells. Gently swirl the plates to mix and place in the cell culture incubator.
4. After 4 h, remove the medium from the cells and gently replace it with 2 mL of fresh medium (*see* **Notes 2** and **4**).
5. Prepare a second transfection mixture for each well, containing the proviral plasmid and envelope plasmid for production of VSV-G-pseudotyped HIV-1 in the presence of aptamer. Amounts may be multiplied depending on sample number to allow for a single mastermix. Combine 48.5 μL serum-free medium with 3 μL PEI transfection reagent for each well. Incubate the mixture at room temperature for 5 min. In a separate tube, combine 48.5 μL serum-free medium with 2.5 μL pNL4-3-CMV-EGFP (250 ng) and 1.5 μL pMD-G (150 ng) for each well (*see* **Notes 14** and **15**). Incubate the mixture at room temperature for 5 min. After the incubation, add the DNA-containing mixture to the PEI-containing mixture. Flick the tube to mix, and briefly spin the liquid out of the cap. Allow the mixture to incubate at room temperature for 30 min (*see* **Note 13**). Add the mixture dropwise to the 293FT cells in 6-well plates, taking care not to disrupt the

cells. Gently swirl the plates to mix and place in the cell culture incubator.

6. After 4 h, remove the medium from the cells and gently replace it with 2 mL of fresh medium (*see* **Note 4**).
7. 48 h after the second media change, harvest the VSV-G-pseudotyped virus by syringe filtering the supernatant using a 0.45 μm syringe filter. To minimize freeze–thaw cycles, aliquot the virus into sterile 1.7 mL tubes, keeping the following applications in mind: infectivity assays (50–150 μL), p24 ELISA (100 μL), and harvesting of viral RNA (1 mL) (*see* **Note 16**). Store the cell-free virus at −80 °C until future use.
8. After harvesting the virus, collect the cells by scraping or trypsinization and transfer them to a fresh 1.7 mL microcentrifuge tube. These cells will be used in Subheading 3.4 for analysis of aptamer accumulation and localization (use scraping method). Perform Subheading 3.4 immediately upon cell harvest. Alternatively, transfection efficiency may be determined by flow cytometry using the cells after fixation for 30 min in 4 % paraformaldehyde (use trypsinization method; *see* Subheading 3.2, **step 12**) (*see* **Note 5**).
9. To determine infectivity of the harvested virus, plate 293FT cells in 6-well plates (as for the transfection in Subheading 3.1, **step 2**) the day prior to infection (*see* **Note 17**).
10. Thaw one aliquot of harvested virus for measuring infectivity and mix well by gently inverting the tube. Use 50–150 μL of the harvested virus (*see* **Note 18**) to infect the 293FT cells plated for infection (Fig. 2). For an individual assay, each sample should be infected with the same volume of viral supernatant.
11. Thaw another aliquot of virus to perform a p24 ELISA for determination of the amount of p24 present in the viral aliquots. Dilutions of 1:10 and 1:100 are recommended for the p24 ELISA to ensure that obtained values are within the range of the standard curve. A protocol for generating a homemade p24 ELISA is available [32]. Commercial kits may also be utilized. These p24 values will be used to normalize the infectivity data to be collected in Subheading 3.2, **step 12**.
12. After 24–48 h, remove the cell culture medium from each well (*see* **Note 19**). Harvest the cells by trypsinization with 200 μL TrypLE Express, wash the well once with 800 μL 1× PBS, and transfer the 1 mL cell suspension to a 1.7 mL microcentrifuge tube. Centrifuge the cells at 300 ×*g* for 10 min at 4 °C. Remove the supernatant, and resuspend the cells in 500 μL 4 % paraformaldehyde (*see* **Note 5**). Incubate the cells for 30 min at room temperature or for 1 h at 4 °C. Wash the cells once with 1× PBS, centrifuge at 300 ×*g* for 10 min, remove the

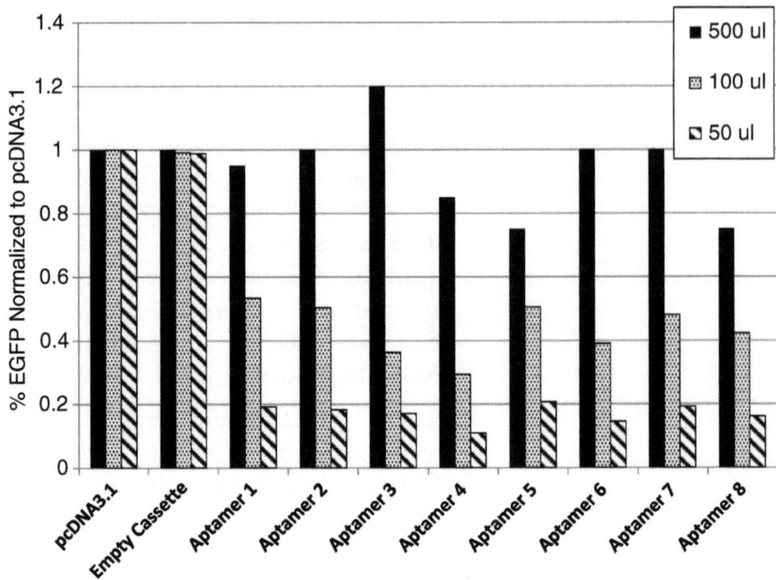

Fig. 2 Single-cycle infectivity assay performed with several different amounts of viral supernatant demonstrating the informative range of the infectivity assay. 293FT cells were transfected with 1 mg aptamer-expressing plasmid or control plasmid (pcDNA3.1 and empty cassette). After 4 h, medium was changed and cells were transfected with proviral and envelope plasmids to produce VSV-G-pseudotyped HIV-1 virions in the presence of aptamer. After 4 h, the medium was changed. Viral supernatant was collected after 48 h by syringe filtration, and fresh 293FT cells were infected with 50, 100, or 500 mL viral supernatant. Infected cells were collected after 24 h, fixed with 4 % paraformaldehyde, and the percentage of EGFP-positive cells was determined by flow cytometry. Data were normalized to pcDNA3.1

supernatant, and resuspend the cells in 500 μL 1× PBS. The cells can now be analyzed using a flow cytometer for the percentage of GFP-positive (infected) cells. Following analysis, infectivity can be normalized to the amount of p24 present per sample.

3.3 Analysis of Intracellular Aptamer RNA Accumulation and Localization

The method described in Subheading 3.3 assesses required components (accumulation and localization) of the single-cycle aptamer bioactivity assay performed in Subheading 3.2. For inhibition to occur, the aptamer must accumulate to significant levels within the cell [26]. If the levels of aptamer are not sufficient to bind the available virus-associated RTs, inhibition is not expected to be observed. This method validates the bioactivity assay by providing additional confirmation of both positive and negative outcomes. If the accumulated RNA levels of aptamer are high, yet inhibition does not occur, it is likely that the specific aptamer utilized may not efficiently bind to and/or inhibit the RT or that it may localize to a nonproductive biological compartment. Binding may occur at many surfaces on the RT [33] and as such may bind in a manner that does not yield inhibition of RT activity. If the levels of aptamer expression are high and inhibition does occur, the assay confirms the positive result. Finally, if the levels of aptamer expression are low and a negative result is obtained, the reasons for low-level aptamer expression should be explored and the expression system should be optimized (*see* **Note 15**). This validation is best used in concert with the determination of aptamer packaging with the virus (Subheading 3.4), which is an additional important component in terms of aptamer-mediated inhibition of viral replication.

1. This protocol will utilize the transfected cells collected by scraping into 1× PBS from Subheading 3.2 (*see* Subheading 3.2, **step 8**). Alternatively, a separate transfection may be performed (*see* **Note 20**).
2. Separate the cells collected in Subheading 3.2, **step 8**, into two 1.7 mL microcentrifuge tubes. One tube will be used for total RNA isolation, while the other will be used for cytoplasmic and nuclear fractionation prior to RNA isolation. Pellet the cells by centrifugation for 10 min at 4 °C and 300 ×*g*. Remove the supernatant.
3. For total RNA isolation, resuspend the cell pellet in TRIzol reagent according to the manufacturer's instructions. Alternatively, the column-based Paris RNA Isolation kit may be used. After completing the TRIzol or Paris RNA Isolation kit protocol, DNase-treat the sample using Turbo DNase (5 μL 10× buffer, 2 μL or 4 U Turbo DNase in an ~50 μL sample) for 1 h at 37 °C (*see* **Note 21**). Inactivate the reaction by adding 11 μL of the provided inactivation resin, mixing well, incubating for 2 min at room temperature, centrifuging at 16,100 ×*g* for 3 min, and transferring the supernatant to a fresh tube. The DNase-free RNA may now be quantified using a Nanodrop spectrophotometer. Make sure that the quality of the isolated RNA is high according to the spectrophotometer (*see* **Note 22**). RNA may be subjected to subsequent precipitations to enhance quality using Subheading 3.1, **step 3**.
4. The Paris RNA Isolation kit may also be used for fractionation and RNA isolation. Alternatively, the following method may be used: For cytoplasmic and nuclear fractionation, resuspend the cell pellet in 500 μL ice-cold TKM buffer and incubate on ice for 5 min. Add 15 μL of 10 % Triton X-100 to lyse the cells while keeping the nuclei intact. Incubate the lysis mixture for 10 min on ice. Centrifuge the samples at 500 ×*g* for 5 min at 4 °C. Transfer the supernatant containing the cytoplasmic material to a fresh tube with care not to disrupt the nuclear pellet. Repeat the lysis process an additional time for the nuclear pellet to ensure that cytoplasmic contaminants are eliminated. After the second lysis process, wash the nuclear pellet once in 500 μL TKM buffer prior to RNA extraction. Isolate cytoplasmic and nuclear RNA

using TRIzol reagent as per the manufacturer's instructions. Due to the liquid component of the cytoplasmic fraction, TRIzol LS reagent should be used (*see* **Note 23**). Complete the remainder of the protocol as instructed in Subheading 3.3, **step 3**.

5. Isolated total, cytoplasmic, and nuclear RNA may now be used for synthesis of cDNA using the ImProm II Reverse Transcription System as per the manufacturer's instructions. Briefly, at least 200 ng of RNA is suggested for synthesis of cDNA using this protocol. Reaction conditions are as follows: 4 mL 5× reaction buffer, 2 μL 25 mM $MgCl_2$, 1 μL 10 mM dNTPs, 1 μL random hexamer primers (0.5 μg), 0.5 μL RNasin (20 U), 1 μL ImProm II reverse transcriptase, and water to 20 μL. Random hexamer primers, oligoDT primers, or gene-specific primers may be used. Always include "no-RT" controls to ensure that any amplification is not due to genomic or plasmid DNA contamination (*see* **Note 21**). Reactions should be performed at 42 °C for 1 h, with inactivation at 99 °C for 5 min. Samples should be kept at −20 °C.
6. The cDNA generated in Subheading 3.3, **step 5**, may be used for subsequent endpoint PCR or qPCR analysis. This protocol describes the quantitative determination of aptamer expression levels in each RNA fraction using the relative quantity qPCR method. Primers for several reference genes to enable normalization can be found in Table 1 (*see* **Note 24**). Additionally, obtain primers specific to the aptamer constant regions or the entire aptamer expression cassette to enable amplification of aptamer sequences within the cDNA (*see* **Note 25**). Determine amplification efficiency for each set of primers using the Pfaffl method prior to performing qPCR [34]. Perform real-time RT-PCR (qPCR) using PowerSybr Green PCR Master Mix with the appropriate primers according to the manufacturer's instructions. The formula for calculating the relative quantity is as follows: CT values for the reference genes (e.g., GAPDH) are subtracted from the CT values of the sample (e.g., $\Delta CT = CT_{aptamer} - CT_{GAPDH}$) to determine the ΔCT. Then, the ΔCT values for the negative control (e.g., pcDNA3.1 vector) are subtracted from the ΔCT values for the samples ($\Delta\Delta CT = \Delta CT_{sample} - \Delta CT_{control}$). The ΔΔCT values are then used to calculate relative quantity (RQ), using the equation $RQ = 2^{-\Delta\Delta CT}$. Thus, graphs are represented as quantity relative to the specified reference gene, with subsequent normalization to sample controls (set to 1).

3.4 Analysis of Aptamer Incorporation into Viral Particles

The method described in Subheading 3.4 provides an additional experimental layer of validation to the single-cycle infectivity assay, verifying positive and negative results. When paired with the method in Subheading 3.3 to determine aptamer accumulation levels and the bioassay in Subheading 3.2, the method has been highly successful in predicting aptamers that will be successful viral inhibitors.

1. Thaw 1 mL aliquots of virus-containing supernatant harvested in Subheading 3.2, **step 7**. Transfer the supernatant to an appropriate high-speed centrifuge tube, and adjust the sample volume with PBS to ensure that each sample contains an equal volume of supernatant prior to centrifugation. Centrifuge the samples at 125,000×*g* for 1 h at 4 °C using a Beckman Coulter TL-100 ultracentrifuge with the TLA 45 rotor. Remove the supernatant, and resuspend the viral pellet in TRIzol Reagent (Invitrogen) as per the manufacturer's instructions. TRIzol LS reagent may be used if a portion of liquid remains over the viral pellet as per the manufacturer's instructions. Addition of 5–10 μg glycogen as a carrier to the aqueous phase during RNA isolation is recommended to improve RNA yields. Alternatively, a column-based viral RNA isolation kit (e.g., Qiagen QIAmp Viral RNA Mini kit) may be used. After completing the TRIzol protocol, proceed as in Subheading 3.3, **step 3**.
2. Isolated viral RNA may now be used for synthesis of cDNA using the ImProm II reverse transcription system as described in Subheading 3.3, **step 5**. For this protocol, ≥200 ng of RNA is suggested for synthesis of cDNA. Always include "no-RT" controls to ensure that any amplification is not due to genomic or plasmid DNA contamination (*see* **Note 21**).
3. The cDNA generated in Subheading 3.4, **step 2**, may be used for subsequent endpoint PCR or qPCR analysis as in Subheading 3.3, **step 6**. Primers for HIV GAG as a reference gene to enable normalization to the viral genome are located in Table 1. Primers specific to the aptamer constant regions or aptamer expression cassette obtained in Subheading 3.3, **step 6**, may be used for amplification of aptamers in viral cDNA.

4 Notes

1. Use of a parental CMV-driven vector: The parental vector utilized in our original study was obtained from the laboratory of Vinayaka R. Prasad [14]. This vector was a modified pcDNA3.1 vector, and the final sequence differs from the original pcDNA3.1 in that several cloning sites have been removed and/or moved. We modified this vector by changing the ribozymes used in the original cassette. While the cloning in this method can utilize pcDNA3.1, slight modifications to the restriction enzymes utilized would be required. For example, cloning of the NheI–NotI aptamer expression cassette into pcDNA3.1 would result in the presence of two ApaI restriction sites (one within the aptamer expression cassette and the

other downstream of the NotI restriction site). Other CMV-driven vectors would be equally acceptable, provided that the restriction sites were modified appropriately. Although we do not anticipate issues with use of the aptamer expression cassette in other CMV-driven vectors, our results are based solely on use of the modified pcDNA3.1 vector.

2. The transfection and infection efficiencies of the 293FT cells decrease with age or passage. We recommend using low-passage 293FT cells (<30 passages) for all transfections and infections. Additionally, 293FT cells are very sensitive to forces generated during removing medium, adding medium, adding transfection mixtures, etc. Extreme care should be taken due to the sensitivity, extreme care should be taken not to disrupt the cells from the plate during such procedures.

3. The pNL4-3-CMV-EGFP proviral plasmid was kindly provided by Vineet KewalRamani (NCI—Frederick). This proviral vector lacks the genes encoding *vif*, *vpr*, *vpu*, *nef*, and *env* and has a CMV immediate early promoter-driven EGFP in place of *nef*.

4. Use of the PEI transfection reagent. Other transfection reagents may also be used. PEI provides highly efficient and consistent transfection of 293FT cells. PEI utilized in this study was made according to the literature [28]. The PEI-DNA transfection mixture should be added dropwise to the cells. Refrain from vortexing or vigorously pipeting the PEI-DNA transfection complex, as these may produce sheer forces that disrupt the PEI-DNA complexes and reduce transfection efficiency. Vortexing after formation of the complex is not recommended. Additionally, dropwise addition of the mixture is crucial to maintaining the complex formation upon addition to the cells. As with many transfection reagents, PEI can be toxic to cells. We have found that using over 6 μL of PEI in a single transfection in a 6-well plate can result in cellular toxicity. Notably, different preparations of PEI will have different toxicity levels, and toxicity should be evaluated prior to the use of the reagent. Due to the associated toxicity, we recommend two separate transfections: one for the aptamer-expressing plasmids and the other for the proviral plasmids. Changing the cell culture medium in between the transfections will prevent accumulation of PEI toxicity within the well.

5. Making of paraformaldehyde solutions for cell fixation: The 4 % paraformaldehyde solution should be diluted from a 20 % paraformaldehyde solution. For 20 % paraformaldehyde stock solution preparation, use the following recipe:

   200 g paraformaldehyde in 800 mL MilliQ water.

   1 mL 10 N NaOH, added dropwise until solubilization occurs.

   Bring up to 1 L with MilliQ water, and heat to 65 °C with stirring to dissolve.

Aliquot in desired volume, and store at −20 °C.

To make 4 % paraformaldehyde working stock.

Mix 100 mL 20 % paraformaldehyde with 50 mL 10× PBS and bring up to 500 mL with MilliQ water.

20 % Paraformaldehyde may be stored at 4 °C for up to 2 weeks or at −20 °C or 20 °C for long-term storage.

The 4 % paraformaldehyde solution should be diluted fresh from the 20 % solution for use.

6. Additional parent aptamer expression cassette for cloning into the parental vector (*see* **Note 1**) may be generated using the outer oligonucleotides (RzModA and RzModG, Table 1) as primers. Dilute the outermost oligonucleotides to 10 μM prior to PCR. Amplify the reaction as in Subheading 3.1, **step 1**.
7. Do not allow the pellet to dry completely. If the pellet is allowed to dry completely, there will be difficulty solubilizing the pellet.
8. To expedite the screening process, PCR may be performed after resuspending the bacterial colony in 10 μL LB-AMP broth and incubating the resuspension for 2–3 h at 37 °C. PCR may be performed with 2 μL of the 10 μL culture. Alternatively, one-half of the bacterial colony may be directly utilized in the PCR reaction.
9. Full-length aptamers (>40 nt) simplify screening of the PCR amplicon by size. Additionally, use of aptamer sequences larger than 100 nt enables easier quantification of aptamer accumulation by real-time RT-PCR. Including segments that are present within all aptamer transcripts allows one set of RT/PCR primers to be used in all RNA quantification assays.
10. Use of maxi preparations of DNA is highly recommended. These preparations typically yield high-quality DNA. Assay variability using lower quality mini preparations of DNA has been encountered. Optimally, DNA should have a 260/280 ratio of ~1.6–1.8 and a 260/230 ratio ≥2.
11. Age or passage number of the 293FT cells affects cellular proliferation speed. Older cells may take longer to reach sufficient levels of confluence than newer cells. This should be taken into consideration when plating prior to experimental procedures.
12. Filler DNA should not affect the experimental outcome. Generally, we utilize a plasmid that expresses mCherry via the CMV promoter as filler (pCMV-mCherry). Salmon sperm DNA may also be used. Experiments should be performed to evaluate the filler DNA prior to introduction into single-cycle infectivity assays using aptamer-expressing plasmids.
13. The mixture may be incubated for 20 min to 2 h without any detrimental effect on transfection efficiency.

14. A mastermix may be prepared for proviral and envelope plasmid transfections. However, prior to addition of DNA to the PEI mixture, the two separate plasmids should be mixed in their appropriate amounts to ensure that each plasmid has equivalent access to the PEI once introduced to the PEI mixture.
15. It is important that aptamer RNA expressed in the cell does not get overwhelmed by the amount of virus produced. If too much virus is produced, the results will be skewed toward a lack of inhibition. We have optimized the amount of proviral and envelope plasmids in our system. We recommend performing proviral titrations in the presence of aptamer to determine the appropriate amount of proviral and envelope plasmid to include based on the amount of aptamer RNA generated from the aptamer-expressing plasmids.
16. Freeze–thaw cycles should be minimized. Care should be taken to ensure that all sample measurements are taken using samples with equal freeze–thaw cycles. Variation in sample concentrations and experimental measurements may be observed with multiple freeze–thaw cycles.
17. For infectivity determination, cells may be plated the day prior to or on the day of infection. Care should be taken to ensure that the confluency will be between 60 and 70 % regardless of the day of plating (*see* **Note 11**).
18. Use of too much or too little virus-containing supernatant will skew the results (Fig. 2). The amount of medium used for infection depends on the amount of virus produced and the confluency of the cells to be infected. Too much virus will saturate the EGFP signal, while too little virus will not provide sufficient positive cells for proper analysis and normalization. The optimal final percentage of EGFP-positive cells for the controls should be 5–10 %.
19. Cells may be harvested for infectivity determination upon the appearance of EGFP-positive cells by fluorescence microscopy. Fluorescence may appear between 24 and 48 h. We recommend at least 24-h infection time.
20. Differences in aptamer accumulation and localization may be observed depending on the presence or the absence of the proviral transfection within the experimental sample. Additionally, if a separate transfection is performed, it should be done side by side with the transfection containing proviral and envelope plasmids to ensure that the cells are in the same state (*see* **Notes 2** and **11**).
21. DNase treatment of the RNA samples is essential to ensure that the observed signal is not because amplification of contaminating genomic or plasmid DNA did not occur.

Amplification of contaminating DNA will give false-positive results and will also result in amplification in the "no-RT" cDNA samples.

22. Care should be taken to ensure the highest quality of RNA. The optimal 260/280 ratio for pure RNA should be ~2.0. Additionally, the 260/230 ratio should be ≥2. Certain RNA isolation protocols may yield low-quality RNA and benefit substantially from subsequent RNA precipitations.
23. TRIzol LS reagent is a more highly concentrated form of TRIzol reagent that accounts for additional "liquid sample" remaining in the tube, assuring the final concentration to be appropriate for nucleic acid or protein isolation.
24. Fractionated RNA samples require specific housekeeping genes. For cytoplasmic samples, spliced GAPDH is recommended (Table 1). For nuclear samples, U6 RNA is recommended (Table 1). Unspliced GAPDH or 18S RNA may be used for total RNA preparations (Table 1). Endpoint PCR to determine cytoplasmic contamination in nuclear samples (and vice versa) is recommended prior to performing qPCR.
25. Aptamer-specific primers may vary with your choice of aptamer due to differences in the content of the constant regions.

References

1. Joshi PJ, Fisher TS, Prasad VR (2003) Anti-HIV inhibitors based on nucleic acids: emergence of aptamers as potent antivirals. Curr Drug Targets Infect Disord 3(4):383–400
2. Gopinath SC (2007) Antiviral aptamers. Arch Virol 152(12):2137–2157. doi:10.1007/s00705-007-1014-1
3. Burke DH, Scates L, Andrews K, Gold L (1996) Bent pseudoknots and novel RNA inhibitors of type 1 human immunodeficiency virus (HIV-1) reverse transcriptase. J Mol Biol 264(4):650–666. doi:10.1006/jmbi.1996.0667
4. Tuerk C, MacDougal S, Gold L (1992) RNA pseudoknots that inhibit human immunodeficiency virus type 1 reverse transcriptase. Proc Natl Acad Sci U S A 89(15):6988–6992
5. Burnett John C, Rossi John J (2012) RNA-based therapeutics: current progress and future prospects. Chem Biol 19(1):60–71. doi:10.1016/j.chembiol.2011.12.008
6. Zhou J, Rossi JJ (2011) Aptamer-targeted RNAi for HIV-1 therapy. Methods Mol Biol 721:355–371. doi:10.1007/978-1-61779-037-9_22
7. Syed MA, Pervaiz S (2010) Advances in aptamers. Oligonucleotides 20(5):215–224. doi:10.1089/oli.2010.0234
8. Cohen C, Forzan M, Sproat B, Pantophlet R, McGowan I, Burton D, James W (2008) An aptamer that neutralizes R5 strains of HIV-1 binds to core residues of gp120 in the CCR5 binding site. Virology 381(1):46–54. doi:10.1016/j.virol.2008.08.025
9. Li N, Wang Y, Pothukuchy A, Syrett A, Husain N, Gopalakrisha S, Kosaraju P, Ellington AD (2008) Aptamers that recognize drug-resistant HIV-1 reverse transcriptase. Nucleic Acids Res 36(21):6739–6751. doi:10.1093/nar/gkn775
10. Neff CP, Zhou J, Remling L, Kuruvilla J, Zhang J, Li H, Smith DD, Swiderski P, Rossi JJ, Akkina R (2011) An aptamer-siRNA chimera suppresses HIV-1 viral loads and protects from helper CD4(+) T cell decline in humanized mice. Sci Transl Med 3(66):66ra66. doi:10.1126/scitranslmed.3001581
11. Held DM, Kissel JD, Thacker SJ, Michalowski D, Saran D, Ji J, Hardy RW, Rossi JJ, Burke DH (2007) Cross-clade inhibition of recombinant human immunodeficiency virus type 1 (HIV-1), HIV-2, and simian immunodeficiency virus SIVcpz reverse transcriptases by RNA pseudoknot aptamers. J Virol 81(10):5375–5384. doi:10.1128/JVI.01923-06
12. Chaloin L, Lehmann MJ, Sczakiel G, Restle T (2002) Endogenous expression of a high-affinity pseudoknot RNA aptamer suppresses

replication of HIV-1. Nucleic Acids Res 30(18):4001–4008
13. Kolb G, Reigadas S, Castanotto D, Faure A, Ventura M, Rossi JJ, Toulme JJ (2006) Endogenous expression of an anti-TAR aptamer reduces HIV-1 replication. RNA Biol 3(4):150–156
14. Joshi P, Prasad VR (2002) Potent inhibition of human immunodeficiency virus type 1 replication by template analog reverse transcriptase inhibitors derived by SELEX (systematic evolution of ligands by exponential enrichment). J Virol 76(13):6545–6557
15. Ramalingam D, Duclair S, Datta SA, Ellington A, Rein A, Prasad VR (2011) RNA aptamers directed to human immunodeficiency virus type 1 Gag polyprotein bind to the matrix and nucleocapsid domains and inhibit virus production. J Virol 85(1):305–314. doi:10.1128/JVI.02626-09
16. Zhang W, Ke W, Wu SS, Gan L, Zhou R, Sun CY, Long QS, Jiang W, Xin HB (2009) An adenovirus-delivered peptide aptamer C1-1 targeting the core protein of hepatitis B virus inhibits viral DNA replication and production in vitro and in vivo. Peptides 30(10):1816–1821. doi:10.1016/j.peptides.2009.07.006
17. Feng H, Beck J, Nassal M, Hu KH (2011) A SELEX-screened aptamer of human hepatitis B virus RNA encapsidation signal suppresses viral replication. PLoS ONE 6(11):e27862. doi:10.1371/journal.pone.0027862
18. Kaiser N, Lischka P, Wagenknecht N, Stamminger T (2009) Inhibition of human cytomegalovirus replication via peptide aptamers directed against the nonconventional nuclear localization signal of the essential viral replication factor pUL84. J Virol 83(22):11902–11913. doi:10.1128/JVI.01378-09
19. Umehara T, Fukuda K, Nishikawa F, Sekiya S, Kohara M, Hasegawa T, Nishikawa S (2004) Designing and analysis of a potent bi-functional aptamers that inhibit protease and helicase activities of HCV NS3. Nucleic Acids Symp Ser 48:195–196. doi:10.1093/nass/48.1.195
20. Kikuchi K, Umehara T, Fukuda K, Kuno A, Hasegawa T, Nishikawa S (2005) A hepatitis C virus (HCV) internal ribosome entry site (IRES) domain III-IV-targeted aptamer inhibits translation by binding to an apical loop of domain IIId. Nucleic Acids Res 33(2):683–692. doi:10.1093/nar/gki215
21. Jones LA, Clancy LE, Rawlinson WD, White PA (2006) High-affinity aptamers to subtype 3a hepatitis C virus polymerase display genotypic specificity. Antimicrob Agents Chemother 50(9):3019–3027. doi:10.1128/AAC. 01603-05
22. Kikuchi K, Umehara T, Nishikawa F, Fukuda K, Hasegawa T, Nishikawa S (2009) Increased inhibitory ability of conjugated RNA aptamers against the HCV IRES. Biochem Biophys Res Commun 386(1):118–123. doi:10.1016/j.bbrc.2009.05.135
23. Romero-Lopez C, Diaz-Gonzalez R, Barroso-delJesus A, Berzal-Herranz A (2009) Inhibition of hepatitis C virus replication and internal ribosome entry site-dependent translation by an RNA molecule. J Gen Virol 90(Pt 7):1659–1669. doi:10.1099/vir.0.008821-0
24. Cheng C, Dong J, Yao L, Chen A, Jia R, Huan L, Guo J, Shu Y, Zhang Z (2008) Potent inhibition of human influenza H5N1 virus by oligonucleotides derived by SELEX. Biochem Biophys Res Commun 366(3):670–674. doi:10.1016/j.bbrc.2007.11.183
25. Park SY, Kim S, Yoon H, Kim KB, Kalme SS, Oh S, Song CS, Kim DE (2011) Selection of an antiviral RNA aptamer against hemagglutinin of the subtype H5 avian influenza virus. Nucleic Acid Ther 21(6):395–402. doi:10.1089/nat.2011.0321
26. Lange MJ, Sharma TK, Whatley AS, Landon LA, Tempesta MA, Burke DH, Johnson MC (2012) Robust suppression of HIV replication by intracellularly expressed reverse transcriptase aptamers is independent of ribozyme processing. Mol Ther 20:2304–2314. doi:10.1038/mt.2012.158
27. Khvorova A, Lescoute A, Westhof E, Jayasena SD (2003) Sequence elements outside the hammerhead ribozyme catalytic core enable intracellular activity. Nat Struct Biol 10(9):708–712. doi:10.1038/nsb959
28. Reed SE, Staley EM, Mayginnes JP, Pintel DJ, Tullis GE (2006) Transfection of mammalian cells using linear polyethylenimine is a simple and effective means of producing recombinant adeno-associated virus vectors. J Virol Methods 138(1–2):85–98. doi:10.1016/j.jviromet. 2006.07.024
29. Saksmerprome V, Roychowdhury-Saha M, Jayasena S, Khvorova A, Burke DH (2004) Artificial tertiary motifs stabilize trans-cleaving hammerhead ribozymes under conditions of submillimolar divalent ions and high temperatures. RNA 10(12):1916–1924. doi:10.1261/rna.7159504
30. Roychowdhury-Saha M, Burke DH (2007) Distinct reaction pathway promoted by non-divalent-metal cations in a tertiary stabilized hammerhead ribozyme. RNA 13(6):841–848. doi:10.1261/rna.339207
31. Burke DH, Greathouse ST (2005) Low-magnesium, trans-cleavage activity by type III, tertiary stabilized hammerhead ribozymes with stem 1 discontinuities. BMC Biochem 6:14. doi:10.1186/1471-2091-6-14

32. Aiken C (2009) Cell-Free Assays for HIV-1 Uncoating. Methods Mol Biol 485:41–53. doi:10.1007/978-1-59745-170-3_4
33. Ditzler MA, Bose D, Shkriabai N, Marchand B, Sarafianos SG, Kvaratskhelia M, Burke DH (2011) Broad-spectrum aptamer inhibitors of HIV reverse transcriptase closely mimic natural substrates. Nucleic Acids Res 39(18):8237–8247. doi:10.1093/nar/gkr381
34. Pfaffl MW (2001) A new mathematical model for relative quantification in real-time RT-PCR. Nucleic Acids Res 29(9):e45

Chapter 3

Design and Evaluation of Clinically Relevant SOFA-HDV Ribozymes Targeting HIV RNA

Robert J. Scarborough, Michel V. Lévesque, Jean-Pierre Perreault, and Anne Gatignol

Abstract

Nucleic acid therapies targeting HIV replication have the potential to be used in conjunction with or in place of the standard small-molecule therapies. Among the different classes of nucleic acid therapies, several ribozymes (Rzs, RNA enzymes) have been developed to target HIV RNA. The design of Rzs targeting HIV RNA is complicated by the sequence diversity of viral strains and the structural diversity of their target sites. Using the SOFA-HDV Rz as an example, this chapter describes methods that can be used to design Rzs for controlling HIV replication. We describe how to (1) identify highly conserved Rz target sites in HIV RNA; (2) generate a set of Rzs with the potential to be used as therapeutics; and (3) screen these Rzs for activity against HIV production.

Key words HIV, Ribozymes, Sequence conservation, Antisense, RNA therapeutics

1 Introduction

The treatment of human immunodeficiency virus (HIV) infection with combination small-molecule therapy is effective in preventing the development of acquired immune deficiency syndrome (AIDS) [1, 2]. Although the emergence of resistant virus can often be managed through the proper administration and monitoring of combination therapy, the cumulative toxicological effects of chronic, changing, and lifelong small-molecule therapy will always present a major health problem for HIV-infected individuals [3]. Several small-RNA therapeutics specifically targeting HIV RNA have been developed. These molecules have the potential to be used either in a combination gene therapy approach [4] or as a complement to current small-molecule therapies using appropriate delivery vehicles [5].

Ribozymes (Rzs) represent a small group of catalytic RNA molecules that are widely distributed throughout nature. They can

Daniel Lafontaine and Audrey Dubé (eds.), *Therapeutic Applications of Ribozymes and Riboswitches: Methods and Protocols*, vol. 1103, DOI 10.1007/978-1-62703-730-3_3, © Springer Science+Business Media New York 2014

recognize and cleave an RNA target in *trans* through specific base pairing and intrinsic catalytic activity. The ability of engineered Rzs to cleave their targets without the assistance of cellular proteins makes them excellent candidates for therapeutic applications [6]. Rz motifs that have been modified to target HIV RNA include hammerhead, hairpin, and hepatitis delta virus (HDV) Rzs [7–9]. The HDV Rz is found in the RNA genome of HDV, a satellite RNA of the hepatitis B virus [10]. Modifications around the catalytic core of this motif have been made to generate a *trans*-acting HDV Rz with a specific on/off adaptor (SOFA), providing the necessary target specificity required for their development as therapeutic agents (Fig. 1a) [11–13]. Several SOFA-HDV Rzs have been identified with activity against viral [8, 14, 15] or cellular RNAs [16–18], and general methods for the identification and screening of SOFA-HDV Rzs have been described [19]. In this chapter we describe specific methods to identify conserved SOFA-HDV Rz target sites in HIV RNA (Subheading 3.1), design SOFA-HDV Rz vectors for transient transfection under the control of an RNA polymerase III promoter (Subheading 3.2), and screen a large number of SOFA-HDV Rz vectors for activity against HIV production in human cells (Subheading 3.3).

Estimates of HIV sequence conservation at the nucleotide (nt) level have been used to identify conserved hairpin Rz target sites [20] and to characterize small interfering (si)RNA target sites [21]. In Subheading 3.1, we describe an alternative approach to estimate HIV sequence conservation at the nt level. We then explain how to identify SOFA-HDV Rz target sites in HIV RNA, using open-access and currently available software. Specifically, the methods describe how to obtain sequence alignments from the Los Alamos HIV database, analyze them using Jalview alignment editor [22], and display them in Microsoft Excel spreadsheets in a manner that is convenient for identifying conserved SOFA-HDV Rz target sites. These methods allow higher cutoff values to be set for essential nts surrounding the SOFA-HDV Rz cleavage site, compared to less essential nts in other regions of the target site.

Several different promoters have been used to express anti-HIV Rzs in cells so that their effects on HIV production can be evaluated. In Subheading 3.2 we describe the methods we have used to construct expression vectors with SOFA-HDV Rzs under the control of the human H1 RNA polymerase III promoter using a commercially available vector (psiRNA-hH1GFPzeo, InvivoGen). This vector has previously been used to express SOFA-HDV Rzs targeting cellular RNAs in human cells [16] and contains the GFP::zeo^{R} fusion gene, which can be used to both evaluate the transfection efficiency (GFP) and select for transfected cells in the presence of Zeocin (zeo^{R}). Like the U6 and 7SK promoters, the H1 promoter can be used to express small RNAs with the addition of only 2–4 Us at the 3′ end [23]. Rzs

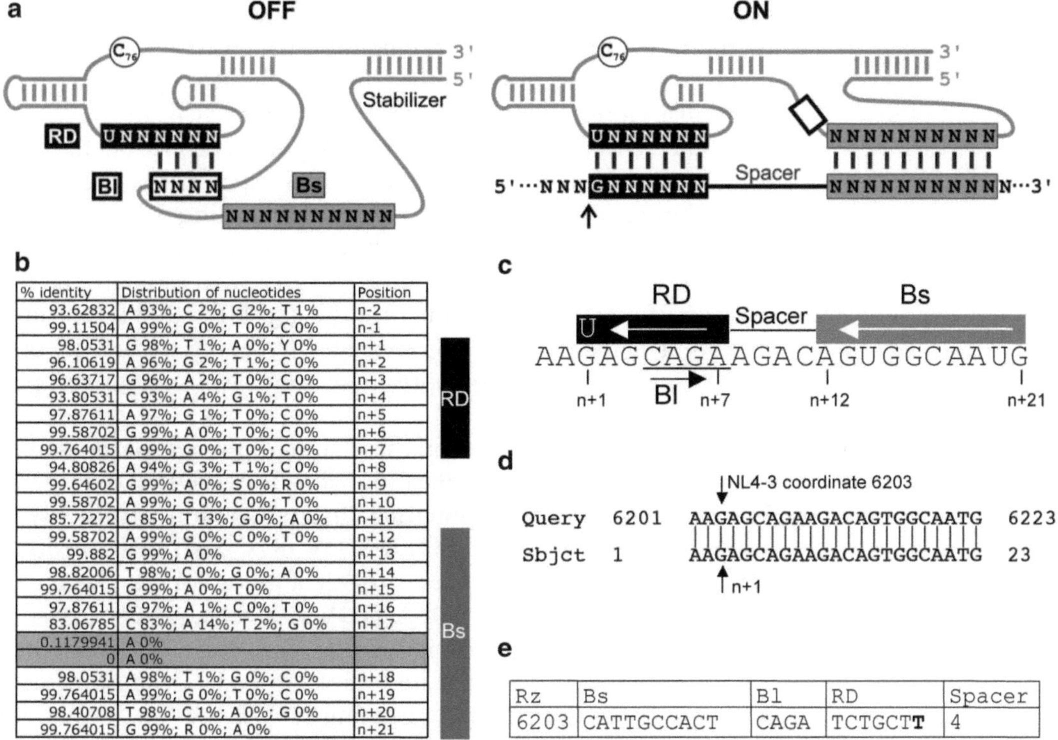

% identity	Distribution of nucleotides	Position
93.62832	A 93%; C 2%; G 2%; T 1%	n-2
99.11504	A 99%; G 0%; T 0%; C 0%	n-1
98.0531	G 98%; T 1%; A 0%; Y 0%	n+1
96.10619	A 96%; G 2%; T 1%; C 0%	n+2
96.63717	G 96%; A 2%; T 0%; C 0%	n+3
93.80531	C 93%; A 4%; G 1%; T 0%	n+4
97.87611	A 97%; G 1%; T 0%; C 0%	n+5
99.58702	G 99%; A 0%; T 0%; C 0%	n+6
99.764015	A 99%; G 0%; T 0%; C 0%	n+7
94.80826	A 94%; G 3%; T 1%; C 0%	n+8
99.64602	G 99%; A 0%; S 0%; R 0%	n+9
99.58702	A 99%; G 0%; C 0%; T 0%	n+10
85.72272	C 85%; T 13%; G 0%; A 0%	n+11
99.58702	A 99%; G 0%; C 0%; T 0%	n+12
99.882	G 99%; A 0%	n+13
98.82006	T 98%; C 0%; G 0%; A 0%	n+14
99.764015	G 99%; A 0%; T 0%	n+15
97.87611	G 97%; A 1%; C 0%; T 0%	n+16
83.06785	C 83%; A 14%; T 2%; G 0%	n+17
0.1179941	A 0%	
0	A 0%	
98.0531	A 98%; T 1%; G 0%; C 0%	n+18
99.764015	A 99%; G 0%; T 0%; C 0%	n+19
98.40708	T 98%; C 1%; A 0%; G 0%	n+20
99.764015	G 99%; R 0%; A 0%	n+21

Rz	Bs	Bl	RD	Spacer
6203	CATTGCCACT	CAGA	TCTGCT**T**	4

Fig. 1 SOFA-HDV Rz and design process. (**a**) Representation of the SOFA-HDV Rz in both OFF (without the target RNA) and ON (with the target RNA) conformation. The key features of the SOFA-HDV Rz are indicated. The recognition domain (RD) is in a *black box*, the blocker (Bl) is in a *white box*, and the biosensor (Bs) is in a *grey box*. The catalytic cytosine C76, which can be mutated to generate an inactive ribozyme, is represented by a *circle*. The stabilizer stem joins the 5′ and 3′ ends, and the spacer is the region between the RD- and the Bs-binding sites in the target RNA. The *arrow* points to the cleavage site. (**b**) The percentage of sequences in an HIV alignment with identity to the consensus nt (*column 1*) and the distribution and identities of all nts (*column 2*) at each position of a potential SOFA-HDV Rz target site (*column 3*) are shown. *Letters* other than A, T, C, and G in *column 2* represent ambiguities (example Y = T or C). (**c**) The target site as RNA, including the $n-2$ and $n-1$ positions and excluding the low-frequency insertions, is shown below the corresponding SOFA-HDV Rz. The Rz RD and Bs are illustrated as *black* and *grey rectangles*, respectively. The Bl sequence is *underlined*. (**d**) An example of the nBLAST output of this target site aligned to HIV strain NL4-3 (M19921) is shown. (**e**) The DNA sequences for the corresponding Rz are illustrated in a table

expressed from these promoters localize predominantly in the nucleus [24] and have been shown to be more active against HIV production compared to Rzs expressed from tRNA and RNA polymerase II promoters [24, 25]. The procedures described include the sequences of the DNA used for cloning SOFA-HDV Rzs into the psiRNA-hH1 vector and methods to quickly generate a large number of constructs for cellular expression.

Several methods have been used to evaluate the activity of small-RNA therapeutics targeting HIV RNA. Although HIV infection models are the most clinically relevant, they are not amenable to transient transfection and require either stable transfection or transduction of the therapeutic RNAs under evaluation. A general method, that has been widely employed to screen anti-HIV RNAs, involves cotransfection of the therapeutic RNA with an HIV molecular clone, followed by an evaluation of virus production in adherent cell culture supernatants. A good correlation has been found between activity in cotransfection experiments and activity in HIV infection models [26, 27]. In Subheading 3.3 we describe a simple and cost-effective method to screen SOFA-HDV Rzs using an HIV reverse transcriptase (RT) assay to estimate the production of infectious virus in cotransfected HEK 293T cell supernatants. The methods are optimized for screening the activity of up to 30 potential Rzs, with appropriate controls, using a microplate scintillation counter.

The methods presented in this chapter outline a specific screening protocol for the identification of SOFA-HDV Rzs targeting HIV RNA with clinical potential. In conjunction with considerations for the design of SOFA-HDV Rzs described previously [19], the procedures outlined provide a guide for the design and evaluation of therapeutic SOFA-HDV Rzs, which may be particularly useful for their development as antivirals. These methods could also be useful for the identification and screening of other antisense-based therapeutics targeting HIV RNA.

2 Materials

2.1 Computer Software

1. HIV LosAlamos database, available for online use at http://www.hiv.lanl.gov.
2. Jalview Alignment Editor Version 2, available for download at http://www.jalview.org/.
3. NCBI BLAST, available for online use at http://blast.ncbi.nlm.nih.gov/Blast.cgi.
4. Ribosubstrate software, available for online use at http://www.riboclub.org/ribosubstrates.

2.2 psiRNA-hH1 SOFA-HDV Rz Expression Vectors

1. psiRNA-hH1GFPzeo (InvivoGen).
2. Zeocin (100 mg/mL, InvivoGen).
3. *BbsI/BpiI*, 10× Green buffer (Fermentas).
4. Transfection grade mini-prep kit (Purelink HiPure miniprep kit, Invitrogen).

2.3 Determination of HIV RT Activity (Modified from Refs. 28, 29)

1. TransIT-LT1 (Mirus).
2. Nonradioactive cocktail: 60 mM Tris–HCl (from 1 M Tris–HCl pH 7.8), 75 mM KCl, 5 mM $MgCl_2$, 1.04 mM EDTA, 1 % Nonidet P-40 (NP-40).
3. Radioactive cocktail: 60 mM Tris–HCl (from 1 M Tris–HCl pH 7.8), 75 mM KCl, 5 mM $MgCl_2$, 1.04 mM EDTA, 10 μg/

mL Polyadenylic acid (Roche), 0.33 μg/mL oligo dT (Invitrogen). Added immediately before use: 8 mM DTT ($C_4H_{10}O_2S_2$, EMD Millipore), 5 μL [^{32}P] dTTP (3,000 Ci/mmol, Perkin Elmer) for each 500 μL reaction volume.

4. Diethylaminoethyl (DEAE) filter mat, printed 96 well grid, 90 × 120 mm (1450-522, Perkin Elmer).
5. Sample bags 90 × 120 mm (1450-432, Perkin Elmer).
6. Microplate scintillation counter (Microbeta TriLux, Perkin Elmer).

3 Methods

3.1 Identification of Conserved SOFA-HDV Rz Target Sites in HIV RNA

1. In your Internet browser, go to the HIV Los Alamos database and select the sequence database.
2. Under the tools drop-down box, select QuickAlign (formerly Epilign and Primalign).
3. Scroll down to "Retrieve alignment(s) based on coordinates," enter the coordinates of a target region in HIV strain HXB2 (Genbank accession number K03455), and select "complete" for Gene/region/protein.
4. Under options select "HIV1" as Organism, "nucleotide" as Sequence type, and "Web alignment (all complete sequences)" as Alignment type. Click submit.
5. The position of the HIV sequence that you entered will be highlighted in an illustration of the HIV genome, and a summary table will be provided. Click on the download button with fasta selected as the format. Save this file to your computer.
6. Open Jalview alignment editor. Close all pop-up windows. Under File, select "input alignment" → "from file" Select the fasta file that contains the alignment.
7. A window will open with the alignment; at the bottom of the window a histogram will be shown above the consensus sequence. This histogram represents the percent (%) of conservation for each nt in the alignment, in reference to the consensus nt.
8. Under view, scroll down to "Autocalculated Annotation," and make sure that "show consensus histogram" and "show consensus logo" are selected (*see* **Note 1**).
9. In the File menu of the window that contains the alignment, select "Export Annotations," choose CSV (Spreadsheet) as format, and export to file. Save this file to your computer with a .txt extension (*see* **Note 2**).

10. Open the .txt file with Microsoft Excel. The Text Import Wizard will ask you to select "the original data type." Select "delimited → characters such as commas or tabs separate each field." Click next.
11. Select "comma" as delimiters (deselect all other delimiters), click next, and click finish.
12. The data from the alignment will open in two rows. Copy these rows, and use paste special (transpose) to convert them into columns. The first column represents the percentage of sequences in the alignment with identity to the consensus nt at each position. The second column represents the distribution of nt identities among the sequences in the alignment at each position (Fig. 1b).
13. Delete all rows for which the value in column 1 is less than 10 % (Fig. 1b, grey rows). This will remove insertions that occur in <10 % of the sequences relative to the consensus sequence (*see* **Note 3**).
14. Set cutoff values for different nts in the Rz target site. The conservation requirements we use for SOFA-HDV Rzs targeting HIV RNA include the following: (1) the nts upstream from the cleavage site ($n-2$ and $n-1$) cannot be G ($n-1$) or CC at >5 %; (2) at the level of the recognition domain (RD)-binding site ($n+1$ to $n+7$), the first nt ($n+1$) must be G at >95 %, and $n+2$ to $n+7$ can be any nt at >85 %; and (3) the biosensor (Bs)-binding site [$n+(9-15)$ to $n+(19-25)$] can be any nt at >75 % (*see* **Note 4**). The RD (black rectangle)- and Bs (grey rectangle)-binding sites are illustrated next to a potential SOFA-HDV Rz target site (Fig. 1b).
15. For each conserved target site, record the $n-2$ to $n+(19-25)$ sequence, excluding low-frequency (<10 %) insertions (Fig. 1c, target site shown as RNA).
16. In your Internet browser, go to the NCBI BLAST homepage. Choose nucleotide blast from the Basic BLAST programs, and enter the accession number of the HIV viral strain that will be used for screening under "Enter Query Sequence" (example: NL4-3, Gene accession number "M19921").
17. Select "Align two or more sequences" under the Query Sequence. Copy and paste the potential target site from **step 15** (Fig. 1c) as DNA into "Enter Subject Sequence." Select "Somewhat similar sequences (blastn)" for Program Selection [30]. Click on "BLAST."
18. Under "Alignments," the potential target site will be aligned to the selected viral strain with its coordinates in that strain (Fig. 1d). Make sure that the displayed "sbjct" length matches the input length (23 nt for this example) (*see* **Note 5**).

19. Record the corresponding DNA sequences of the SOFA-HDV Rz in a table (Fig. 1e), where the Bs and RD sequences are the reverse complement of their corresponding target sites (Fig. 1c), with C replaced by T at the end of the RD (Fig. 1e, bold). The blocker (Bl) is the 4 nt at the end of the RD-binding site (Fig. 1c, underlined), and the spacer is the number of nts between the RD- and Bs-binding sites (Fig. 1c).
20. Go to the Ribosubstrate homepage, and click on "Search for SOFA-HDV Rz substrates." Select "Human NCBI build 36.2 mRNAs" as cDNA bank. Copy and paste the first 6 nt of the RD and all of the nts of the Bs for a specific Rz (Fig. 1e) into the respective fields. Select "Wait for results here," and click submit. Eliminate Rzs that have low scores in "With wobble" or "With wobble and mismatch" tables. We eliminate any Rz that has a score of less than 10 for any human RNA (*see* **Note 6**).

3.2 Design of SOFA-HDV Rz Vectors for Transient Transfection

1. For each SOFA-HDV Rz target site identified, design a specific oligonucleotide with the sequence 5′-TAATACGACTCACTATAGGGCCAGCTAGTTT(Bs)(Bl)CAGGGTCCACCTCCTCGCGGT(RD)GGGCATCCGTTCGCG-3′, using the sequences documented from Subheading 3.1 for Bs, Bl, and RD (Fig. 1e).
2. To make double-stranded (ds) DNA coding for each specific SOFA-HDV Rz, design the common reverse oligonucleotide: 5′-CCAGCTAGAAAGGGTCCCTTAGCCATCCGCGAACGGATGCCC-3′, the overlapping regions of the oligonucleotides (**steps 1** and **2**) are underlined (*see* **Note 7**).
3. To make dsDNA coding for each specific SOFA-HDV Rz with an inactivated Rz backbone, design the common inactive reverse oligonucleotide: 5′-CCAGCTAGAAAGGGTCCCTTA**T**CCATCCGCGAACGGATGCCC-3′, the G-T change, which produces a C-to-A change in the Rz coding sequence, is in bold, and the overlapping regions of the oligonucleotides (**steps 1** and **3**) are underlined (*see* **Note 8**).
4. An irrelevant Rz should be designed as a control. For this we use HBV303 [12], which targets a site in HBV RNA and does not have an effect on HIV production compared to the empty transfection plasmid in experiments described in Subheading 3.3 (Fig. 2). The specific oligonucleotide for this Rz (**step 1**) contains the following sequences from 5′ to 3′: (Bs): GAGACAAGAA, (Bl): GTTT, and (RD): AAACCAT (*see* **Note 9**).

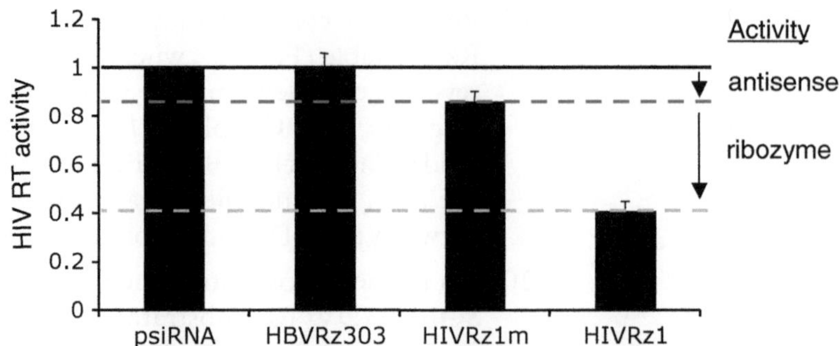

Fig. 2 Inhibition of HIV production by a SOFA-HDV Rz. HEK 293T cells were cotransfected with HIV molecular clone pNL4-3 and different Rz constructs. HIV-1 RT activity was measured in the cell supernatant. Data are normalized to cotransfection with the empty Rz expression plasmid (psiRNA) and include the activity of an irrelevant Rz (HBVRz303), a Rz targeting HIV RNA with an inactivated backbone (HIVRz1m), and the same Rz with an active Rz backbone (HIVRz1). The antisense and ribozyme activity of HIVRz1 are indicated

5. Design cloning primers:

   (A): 5′-TATAAGTTCTGTATGAGTTCACG**GAAGAC**CGACCT↓C<u>GGGCCAGCTAGTTT</u>-3′ and (B): 5′-CAACAACAGTGTTCGGATGAACTGATGCTAT**GAAGAC**TCCAAA↓AA<u>CCAGCTAGAAAGGGTC</u>-3′.

   These primers contain the recognition site for the restriction enzyme *BbsI/BpiI* (GAAGAC), which cleaves at 6 nt downstream from its recognition site (indicated with an arrow) generating ends that are compatible for ligation into *BbsI/BpiI*-digested psiRNA-hH1GFPzeo. The sequences that align with the specific forward and reverse SOFA-HDV Rz oligonucleotides are underlined.

6. Prepare a PCR mix with cloning primers A and B from **step 5** (2 μM each), the common reverse oligonucleotide from **step 3** or **4** (25 nM), 1× PCR buffer, $MgCl_2$ (1.5 mM), dNTP mix (200 nM each), and Taq DNA polymerase (1 μL/100 μL). Aliquot 100 μL of the PCR mix into labeled PCR tubes, add 1 μL of each specific SOFA-HDV Rz oligonucleotide from **step 2** (2.5 μM) to each tube, and use the following PCR protocol: 94 °C for 2 min followed by 20 cycles (94 °C for 30 s, 50 °C for 30 s, 72 °C for 30 s), and then 72 °C for 2 min.

7. Purify the PCR products using a PCR Purification Kit (Qiagen), and digest the eluate (25 μL) and psiRNA-hH1GFPzeo plasmid (12 μg in 25 μL) overnight with *BbsI/BpiI* (2 μL) and 10× *BbsI/BpiI* buffer (3 μL) at 37 °C. Run the digestion products on an agarose gel, excise the bands, and purify using a Gel Extraction Kit (Qiagen). Ligate each extracted PCR product to the extracted psiRNA-hH1GFPzeo vector using 100 ng of

PCR product and 100 ng of vector in 10 μL ddH_2O with 1× ligation buffer and 1 μL of T4 ligase. Ligate overnight at 14 °C, warm to 37 °C, and transform *E. coli* DH5α subcloning efficiency (40 μL) with each ligation using heat shock (10 min on ice, 30 s at 42 °C, 2 min on ice). Add 350 μL LB-Miller broth and incubate at 37 °C with agitation for 2 h. Plate transformed bacteria onto LB-Miller agar plates containing Zeocin (25 μg/mL) and pre-coated with 30 μL Xgal (2 % in DMSO) and 30 μL IPTG (0.1 M) for blue/white colony selection. Incubate the plates upside down at 37 °C overnight.

8. Amplify white colonies from each plate in 4 mL LB-Miller broth + 1 μL of 100 mg/mL Zeocin stock solution, overnight at 37 °C with agitation. Use a mini-prep kit designed for the purification of transfection-grade DNA to purify the plasmids (*see* **Note 10**).
9. Confirm that psiRNA-hH1 SOFA-HDV Rz constructs are correct by sequencing using the primer 5′-TCTACGGGGT CTGACGC-3′ (*see* **Note 11**).

3.3 Screening of SOFA-HDV Rzs for Activity Against HIV Production

1. Plate HEK 293T cells (ATCC) in up to four 24-well plates, 24 h prior to transfection. Cells should be at approximately 50 % confluency prior to transfection.
2. Prepare HIV molecular clone (example pNL4-3) DNA at a concentration of 10 ng/μL in ddH_2O. Aliquot 7.5 μL into 1.5 mL microcentrifuge tubes, one for each well. Dilute psiRNA-hH1 SOFA-HDV Rz plasmid preparations from Subheading 3.2 to 100 ng/μL in ddH_2O, and add 7.5 μL of each plasmid to the transfection tubes containing HIV DNA in triplicate. For each set of transfections, include the empty psiRNA-hH1 vector and an irrelevant psiRNA-hH1 SOFA HDV Rz (example HBV303, Subheading 3.2, **step 4**) in triplicate.
3. Dilute the DNA in 25 μL of media without serum. In a biosafety level 3 (BL3) laboratory, add 2 μL of TransIT-LT1 to each tube, applied under the surface, making sure not to touch the sides of the tube. Let DNA-TransIT complexes form for 20 min. Apply the DNA-TransIT complexes dropwise over each well in the 24-well plate, gently swirl the plate, and incubate at 37 °C for 48 h.
4. Gently swirl the plates. For each well, transfer 200 μL of supernatant to wells of a 96-well plate in eight rows of 6 wells (48 samples from two 24-well plates correspond to one DEAE paper, **step 8**). The supernatants can be stored at −80 °C or immediately used for the determination of HIV RT activity.
5. Prepare 96-well plates with 25 μL of nonradioactive cocktail in eight rows of six wells, one for each set of supernatants that you would like to measure.

6. Using a multichannel pipette, transfer 5 μL of supernatant to the 25 μL of nonradioactive cocktail. Incubate at room temperature for 15 min, and remove the plate from the BL3 laboratory to a certified radioactive laboratory (*see* **Note 12**).
7. Pipette 25 μL of radioactive cocktail (DTT and TTP added) to each well containing nonradioactive cocktail and supernatant (*see* **Note 13**), and incubate at 37 °C for 2 h.
8. Spot 5 μL of the reaction mixture onto DEAE papers (using every other square, 48 samples), and allow the papers to dry (5 min). Wash the papers five times in 2× SSC buffer for 5 min with agitation. Wash twice for 1 min in 95 % ethanol, and allow the papers to dry (30 min).
9. Seal papers in plastic bags (Perkin Elmer), and measure cpm on a plate reader (Microbeta, Perkin Elmer).
10. For each SOFA-HDV Rz that has activity against HIV production compared to the empty plasmid and irrelevant control (HBV303), generate the inactive version by following **steps 6–9** in Subheading 3.2 using the common reverse oligonucleotide from Subheading 3.2, **step 3**.
11. Repeat **steps 1–9** for each active and inactive Rz including the empty plasmid and irrelevant control. This will provide information on the effects of Rz turnover on the inhibition of HIV production (Fig. 2).

4 Notes

1. In Jalview Version 2.7, data on the identity and percentage of nts other than the consensus nt will be missing if the "show consensus logo" is not selected. A preview of what will be exported can be viewed by scrolling over the histogram at the bottom of the window.
2. Export annotation can also be selected by right clicking on the word "consensus," next to the histogram.
3. Insertions that occur at a low frequency relative to the consensus sequence are common and can be ignored when selecting conserved target sites (we use less than 10 % as the cutoff value). Deleting these rows from the data set makes it easier to find and document potential target sites.
4. The conservation cutoffs can be altered depending on the sequence diversity of the target. If using a Pol III promoter for expression of the Rz, target sites that include four or more As in the RD- or Bs-binding sites should be avoided, as this will result in possible Pol III termination signals (TTTT) within the Rz. For the same reason, the Bs-binding site should not

end with A, as there are three Us between the stabilizer and Bs (Fig. 1a) and the RD-binding site should not end with four Ts, as it will generate a Bl with four Ts [19]. The number of nts between the RD and Bs (called the spacer) can be 1–7 nt (3–5 nt is generally used), and the optimal Bs length is 10 nt [31]. The length of the spacer can be modified to avoid poorly conserved nts in the Bs [19].

5. If the target site from the consensus sequence does not align perfectly with the target site in the HIV viral strain selected, either "no significant similarity found" or the best match with gaps (horizontal line) and mutations (missing vertical line) will be displayed under "alignment." If the discrepancy between the consensus sequence (sbjct) and the query sequence is in the RD- or the Bs-binding sites, eliminate the target site. If the discrepancy is in the $n-2$ or $n-1$ positions, ensure that these are not G ($n-1$) or CC in the query sequence. If the discrepancy is in the spacer (region between the RD- and the Bs-binding sites) continue to **step 19**.
6. Details on the scoring for the Ribosubstrate software have been described [32]. Hammerhead Rz target sites can also be screened using the Ribosubstrate software.
7. The dsDNA generated from these oligonucleotides can also be used to evaluate the activity of SOFA-HDV Rzs in vitro [19].
8. The C76A mutation described here and the C76U or the C76G mutations inactivate the catalytic activity of the SOFA-HDV Rz. They are generated by mutation in the reverse oligonucleotide. The inactive SOFA-HDV Rzs are used to evaluate the antisense activity of specific Rzs [19].
9. For the design of SOFA-HDV Rzs with different targets, ensure that the irrelevant Rz selected does not have activity against that target. The selection of irrelevant Rzs has been described [19].
10. It is very important that the DNA for each SOFA-HDV Rz construct is purified in the same way, as differences in the plasmid preparation may influence the transfection efficiency of the cotransfected HIV molecular clone in the activity evaluation (Subheading 3.3). We use the Purelink HiPure plasmid miniprep kit from Invitrogen.
11. Align the sequencing results to the following sequence to confirm that the Rz is inserted properly and that there are no mutations from the PCR (the vector sequences are in bold): 5′-**TTCTGTATGAGACCACGGTACCTC**GGGCCAGCT AGTTT(Bs)(Bl)CAGGGTCCACCTCCTCGCGGT (RD)GGGCATCCGTTCGCGGATGGCTAAGGACCCT TTCTAGCTGG**TTTTTGGAAAAGCTT**-3′.

12. The nonradioactive cocktail contains NP40 at 1 %. Exposure to 0.5 % (v/v) NP-40 for 1 min is able to inactivate HIV in solution [33]. This step ensures that the virus is fully inactivated before the samples are removed from the BL3 laboratory and avoids using radioactivity in a BL3 laboratory.
13. To avoid radioactive contamination of pipettes and the bench space, this and subsequent steps are done with filter tips. The radioactive cocktail is mixed in a 2 mL microcentrifuge tube (1,250 μL for each 96-well plate) with a 1,000 μL pipette and added to each well of the 96-well plate using a single-channel pipette.

References

1. De Clercq E (2010) Antiretroviral drugs. Curr Opin Pharmacol 10(5):507–515
2. Tsibris AM, Hirsch MS (2010) Antiretroviral therapy in the clinic. J Virol 84(11): 5458–5464
3. Taiwo B, Hicks C, Eron J (2010) Unmet therapeutic needs in the new era of combination antiretroviral therapy for HIV-1. J Antimicrob Chemother 65(6):1100–1107
4. Scherer LJ, Rossi JJ (2011) Ex vivo gene therapy for HIV-1 treatment. Hum Mol Genet 20(R1):R100–R107
5. Burnett JC, Rossi JJ (2012) RNA-based therapeutics: current progress and future prospects. Chem Biol 19(1):60–71
6. Serganov A, Patel DJ (2007) Ribozymes, riboswitches and beyond: regulation of gene expression without proteins. Nat Rev Genet 8(10):776–790
7. Haasnoot J, Berkhout B (2009) Nucleic acids-based therapeutics in the battle against pathogenic viruses. Handb Exp Pharmacol 189:243–263
8. Lainé S, Scarborough RJ, Lévesque D, Didierlaurent L, Soye KJ, Mougel M, Perreault JP, Gatignol A (2011) In vitro and in vivo cleavage of HIV-1 RNA by new SOFA-HDV ribozymes and their potential to inhibit viral replication. RNA Biol 8(2):343–353
9. Reyes-Darias JA, Sánchez-Luque FJ, Berzal-Herranz A (2008) Inhibition of HIV-1 replication by RNA-based strategies. Curr HIV Res 6(6):500–514
10. Taylor JM (2006) Hepatitis delta virus. Virology 344(1):71–76
11. Bergeron LJ, Perreault JP (2005) Target-dependent on/off switch increases ribozyme fidelity. Nucleic Acids Res 33(4): 1240–1248
12. Bergeron LJ, Reymond C, Perreault JP (2005) Functional characterization of the SOFA delta ribozyme. RNA 11(12):1858–1868
13. Asif-Ullah M, Lévesque M, Robichaud G, Perreault JP (2007) Development of ribozyme-based gene-inactivations; the example of the hepatitis delta virus ribozyme. Curr Gene Ther 7(3):205–216
14. Lévesque MV, Lévesque D, Brière FP, Perreault JP (2010) Investigating a new generation of ribozymes in order to target HCV. PLoS One 5(3):e9627
15. Motard J, Rouxel R, Paun A, von Messling V, Bisaillon M, Perreault JP (2011) A novel ribozyme-based prophylaxis inhibits influenza A virus replication and protects from severe disease. PLoS One 6(11):e27327
16. Robichaud GA, Perreault JP, Ouellette RJ (2008) Development of an isoform-specific gene suppression system: the study of the human Pax-5B transcriptional element. Nucleic Acids Res 36(14):4609–4620
17. Ben Aissa M, April MC, Bergeron LJ, Perreault JP, Levesque G (2012) Silencing of amyloid precursor protein expression using a new engineered delta ribozyme. Int J Alzheimers Dis 2012:947147
18. D'Anjou F, Routhier S, Perreault JP, Latil A, Bonnel D, Fournier I, Salzet M, Day R (2011) Molecular validation of PACE4 as a target in prostate cancer. Transl Oncol 4(3):157–172
19. Lévesque MV, Perreault JP (2012) Target-induced SOFA-HDV ribozyme. Methods Mol Biol 848:369–384
20. DeYoung MB, Hampel A (1997) Computer analysis of the conservation and uniqueness of ribozyme-targeted HIV sequences. Methods Mol Biol 74:27–36
21. Lee SK, Dykxhoorn DM, Kumar P, Ranjbar S, Song E, Maliszewski LE, François-Bongarçon V, Goldfeld A, Swamy NM, Lieberman J, Shankar P (2005) Lentiviral delivery of short hairpin RNAs protects CD4 T cells from multiple clades and primary isolates of HIV. Blood 106(3):818–826

22. Waterhouse AM, Procter JB, Martin DM, Clamp M, Barton GJ (2009) Jalview Version 2–a multiple sequence alignment editor and analysis workbench. Bioinformatics 25(9):1189–1191
23. Brummelkamp TR, Bernards R, Agami R (2002) A system for stable expression of short interfering RNAs in mammalian cells. Science 296(5567):550–553
24. Good PD, Krikos AJ, Li SX, Bertrand E, Lee NS, Giver L, Ellington A, Zaia JA, Rossi JJ, Engelke DR (1997) Expression of small, therapeutic RNAs in human cell nuclei. Gene Ther 4(1):45–54
25. Puerta-Fernández E, Barroso-del Jesus A, Romero-López C, Tapia N, Martínez MA, Berzal-Herranz A (2005) Inhibition of HIV-1 replication by RNA targeted against the LTR region. AIDS 19(9):863–870
26. Unwalla HJ, Li H, Li SY, Abad D, Rossi JJ (2008) Use of a U16 snoRNA-containing ribozyme library to identify ribozyme targets in HIV-1. Mol Ther 16(6):1113–1119
27. Chang LJ, Liu X, He J (2005) Lentiviral siRNAs targeting multiple highly conserved RNA sequences of human immunodeficiency virus type 1. Gene Ther 12(14):1133–1144
28. Peden KWC, Martin MA (1995) Virological and molecular genetic techniques for studies of established HIV isolates. In: Karn J (ed) HIV, a practical approach: virology and immunology. IRL Press, Oxford, pp 21–45
29. Clerzius G, Gélinas JF, Daher A, Bonnet M, Meurs EF, Gatignol A (2009) ADAR1 interacts with PKR during human immunodeficiency virus infection of lymphocytes and contributes to viral replication. J Virol 83(19):10119–10128
30. Altschul SF, Madden TL, Schaffer AA, Zhang J, Zhang Z, Miller W, Lipman DJ (1997) Gapped BLAST and PSI-BLAST: a new generation of protein database search programs. Nucleic Acids Res 25(17):3389–3402
31. Lévesque MV, Rouleau SG, Perreault JP (2011) Selection of the most potent specific on/off adaptor-hepatitis delta virus ribozymes for use in gene targeting. Nucleic Acid Ther 21(4):241–252
32. Lucier JF, Bergeron LJ, Brière FP, Ouellette R, Elela SA, Perreault JP (2006) RiboSubstrates: a web application addressing the cleavage specificities of ribozymes in designated genomes. BMC Bioinformatics 7:480
33. Butler PJG (1995) Biological safety when working with HIV. In: Karn J (ed) HIV, a practical approach: virology and immunology. IRL Press, Oxford, pp 1–12

Chapter 4

Directing RNase P-Mediated Cleavage of Target mRNAs by Engineered External Guide Sequences in Cultured Cells

Xiaohong Jiang, Naresh Sunkara, Sangwei Lu, and Fenyong Liu

Abstract

Ribonuclease P (RNase P) complexed with external guide sequence (termed as EGS) represents a novel nucleic acid-based gene interference approach to modulate gene expression. In previous studies, by using an in vitro selection procedure, we have successfully generated EGS variants that are complementary to target mRNAs, and these variants exhibit higher efficiency in directing human RNase P to cleave the target mRNAs than those derived from nature RNAs in vitro. This chapter describes the procedure of using engineered EGSs for in vitro trans-cleavage of target viral mRNAs in cultured cells. Detailed information is focused on (1) generation and in vitro cleavage assay of the customized EGS variants and (2) stable expression of EGS and evaluation of its activity in inhibition of viral gene expression and growth in cultured cells. These methods should provide general guidelines for using engineered EGS to direct RNase P-mediated cleavage of target mRNAs in cultured cells.

Key words Engineered external guide sequence, Human RNase P, In vitro evaluation, Human cytomegalovirus (HCMV), Gene targeting

1 Introduction

Nucleic acid-based gene interference strategies represent powerful tools to unravel functions of genes and promising therapeutic agents for human diseases [1]. As one of the most abundant, stable, and efficient enzymes in cells, ribonuclease P (RNase P) is responsible for the 5′ end maturation of all tRNAs by catalyzing a hydrolysis reaction to remove a 5′ leader sequence from tRNA precursors (ptRNA) [2] (Fig. 1a). The extensive studies on RNase P recognition revealed that it recognizes the structure rather than the sequence of the substrates [3, 4]. According to this unique feature, any complex of two RNA molecules that resembles a ptRNA can be cleaved by RNase P (Fig. 1b). Thus, in principle, RNase P can be recruited to cleave an mRNA if the mRNA substrate forms a hybrid complex with a custom-designed sequence (external guide sequence or EGS) to resemble a ptRNA molecule [5, 6] (Fig. 1c).

Daniel Lafontaine and Audrey Dubé (eds.), *Therapeutic Applications of Ribozymes and Riboswitches: Methods and Protocols*, vol. 1103, DOI 10.1007/978-1-62703-730-3_4, © Springer Science+Business Media New York 2014

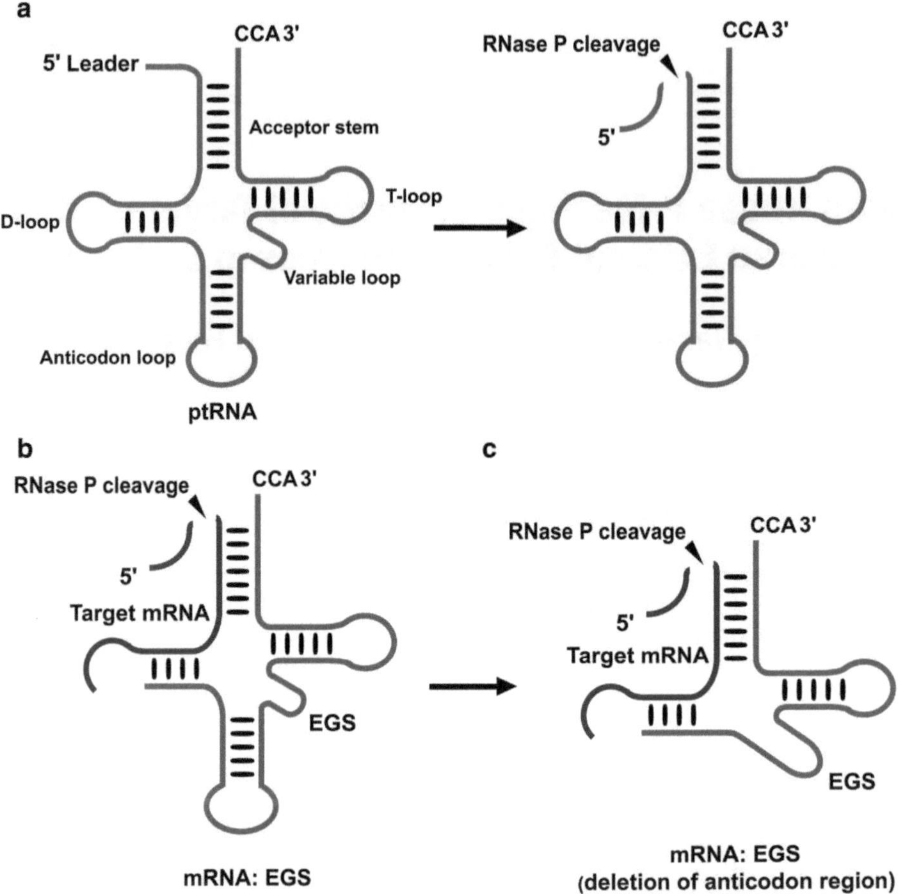

Fig. 1 Substrates for ribonuclease P (RNase P). (**a**) Schematic representation of a natural substrate (precursor tRNA, ptRNA) and can be cleaved by RNase P. (**b**) A hybridized complex of a target mRNA and an external guide sequence (EGS) that resembles the structure of a ptRNA. (**c**) Results from (**b**) achieved deleting the anticodon domain of the EGS, which is indispensable for EGS-targeting activity [18]. The site of cleavage by RNase P is indicated with an *arrowhead*. The sequence of the target mRNA around the targeting site is shown in *purple*, and the EGS sequence is shown in *green*

Unlike other nucleic acid-based interference approaches such as antisense oligonucleotides and RNAi [7], EGS-based technology is unique in inducing endogenous RNase P for targeted cleavage. Importantly, the RNase P-mediated cleavage is highly specific and does not generate nonspecific "irrelevant cleavage" that is observed in RNase H-mediated cleavage induced by conventional antisense phosphothioate molecules [8, 9]. EGS RNAs derived from natural tRNA sequences have been shown to be effective in blocking gene expression both in bacteria and mammalian cells [6, 8, 10–13]. Moreover, in vitro selection procedure has been exploited to increase the targeting activity of EGSs in directing human RNase P to cleave an mRNA, which can lead to better efficacies of the EGSs in inducing RNase P-mediated inhibition of the expression of the target mRNA in cultured cells [14, 15]. Therefore, RNase P complexed with EGS represents a novel and promising nucleic acid-based gene interference strategy for specific inhibition of target mRNAs [15].

Human cytomegalovirus (HCMV), a human herpes virus, is a common opportunistic pathogen and the leading cause of congenital infections associated with mental retardation in newborns and can cause serious clinical manifestations in immunocompromised individuals, including AIDS patients and transplant recipients [16]. The emergence of drug-resistant strains of HCMV has posed a need for the development of effective antiviral approaches for the treatment and prevention [17]. In previous studies, an engineered EGS variant, which was used to target mRNA encoding the protease of HCMV, exhibited higher activity in inducing human RNase P to cleave the mRNA in vitro than the EGS derived from a natural tRNA [15]. These results demonstrated the feasibility of using engineered EGS as potential therapeutic agent against HCMV as well as other viruses.

In this chapter, detailed protocols involved in the in vitro generation of efficient EGS against protease (PR) mRNA of HCMV are described. The series of experiments undertaken to down-regulate the expression of the PR mRNA by EGS are (1) generation of EGS and target PR mRNAs in vitro, (2) determination of the efficiency of RNase P-mediated degradation of the target mRNA in the presence of the selected EGSs by in vitro cleavage assay, (3) stable expression of EGS in cultured cells, and (4) evaluation of EGS activity in inhibition of viral gene expression and growth in cultured cells.

2 Materials

2.1 Reagents and Solutions

1. β-Mercaptoethanol.
2. Dulbecco's modified Eagle's medium (DMEM).
3. Nu-serum, fetal bovine serum (FBS).
4. Neomycin.
5. Deoxynucleotide 5′ triphosphates (dNTPs) and nucleotide 5′ triphosphates (NTPs).
6. Diethylpyrocarbonate (DEPC)-treated H_2O.
7. [^{32}P]-labeled nucleotides.
8. 1× Denhardt's solution: 0.1 % Bovine serum albumin (BSA), 0.1 % polyvinylpyrrolidone, 0.1 % Ficoll, 0.1 % SDS, and 200 μg/mL denatured salmon sperm.
9. 20× Standard saline citrate (SSC): 3 M NaCl, 0.3 M trisodium citrate.

10. 10× TBE: 0.89 M Tris–borate, 10 mM EDTA.
11. 2× RNA dye solution: 8 M urea, 20 mM EDTA, 0.25 mg/mL bromophenol blue, 0.25 mg/mL xylene cyanol FF (XCFF).
12. TNT: 10 mM Tris–HCl, pH 7.0, 150 mM NaCl, 0.05 % Tween-20.

2.2 Enzymes, Reaction Buffers, and Kits

1. 10× PCR buffer: 25 mM $MgCl_2$, 10 mM dNTPs.
2. Buffer A (cleavage buffer): 50 mM Tris–HCl, pH 7.4, 10 mM $MgCl_2$, 100 mM NH_4Cl.
3. Buffer B (binding buffer): 50 mM Tris–HCl, pH 7.5, 0.5 mM EDTA, 10 mM NaCl, 10 mM $MgCl_2$, 3 % glycerol, 0.1 % xylene cyanol.
4. Annealing buffer: 10 mM Tris–HCl, pH 7.5, 10 mM KCl.
5. Prehybridization buffer: 6× SSC, 0.05 % sodium pyrophosphate, 2× Denhardt's solution.
6. *Taq* DNA polymerase (Perkin-Elmer).
7. T7 in vitro transcription kit (Promega).
8. RNA isolation kit (SurePrep™ kit, Fisher Bioreagents).
9. Mammalian transfection kit (GIBCO/BRL).
10. ECL Western blot detection kit (GE Healthcare).
11. Random primed Labeling kit (Roche).

2.3 Virus, Cells, Plasmids

1. HCMV (Strain AD169, ATCC).
2. Murine PA317 cells (amphotropic retrovirus packaging cell line).
3. Human foreskin fibroblasts (HFF).
4. Human U373MG cells (ATCC).
5. LXSN (retroviral vector).

3 Methods

3.1 Construction of DNA Sequences Coding for EGS and PR mRNAs

Here we generate active EGS variants that can be used to effectively target mRNA. As described previously [14], by using the in vitro selection procedure to isolate EGS RNA variants, we chose variant 125 for this study. By covalently linking the EGS domain of C125 to the targeting sequences that are complementary to the PR mRNA, we constructed EGS PR-125. Another EGS, PR-SER, which was derived from the natural $tRNA^{Ser}$ sequence, was also constructed in a similar way. The DNA sequences coding for the EGS were synthesized by PCR amplification, using construct plasmid as the template and were cloned under the control of the T7 RNA polymerase promoter.

1. To generate PR-SER, construct pTK112 DNA [12] was used as the template and the 5′ and 3′ primers were oligoPR31 (5′-GGAATTCTAATACGACTCACTATAGGTTAA CGCGCCGGGTGCGGTCTCC-3′) and oligoPR32 (5′-AAG CTTTAAATGCTCTCCGC AGGATTTGAACCTGCGCG CG-3′). To generate PR-C125, construct Ptkc125 DNA was used as the template and the 5′ and 3′ primers were oligoPR41 (5′-GGAATTCTAATACGACTCACT ATAGGTTAACGCG CCGGGCAGCTACTAGC AG-3′) and oligoPR42 (5′-AAGC TTTAAATGCTCTCCGCAGGCTCCGAACT GCTAGTAG-3′). The DNA sequences coding for EGS PR-SER-C and PR-C125-C were derived from those for PR-SER and PR-C125, respectively, and contained point mutations (5′-TTC-3′→AAG) at the three highly conserved positions in the T-loop of these EGSs.
2. The DNA sequence that encodes substrate pr39, which contained a PR mRNA sequence of 39 nucleotides, was constructed by annealing oligonucleotide AF25 (5′-GGAATT CTAATACGACTCACTATAG-3′) with sPR (5′-CGGGATC CGCAGCGCCGGCTGG AGAGCGAGAGGCCGG CCTAT AGTGAGTCGTATTA-3′).
3. For a 100 μL PCR reaction, mix the following: 10 μL of 10× PCR buffer; 6 μL of 25 mM $MgCl_2$; 8 μL each of 10 mM dNTP; 20 pmol of pTK112 DNA templates; 400 pmol each of PR31 and PR32 oligomer; and 2.5 U *Taq* DNA polymerase.
4. PCR amplification program: Denaturing for 2 min at 94 °C; 30 cycles of 94 °C for 1 min, 47 °C for 1 min, and 72 °C for 1 min; final extension for 10 min at 72 °C.
5. The PCR DNA products are separated in 5 % polyacrylamide gels under non-denaturing conditions and are purified and used as the template for the next in vitro transcription step (*see* **Note 1**).

3.2 Generation of EGS and PR mRNAs by In Vitro Transcription

1. The gel-purified PCR DNA products are used as templates for the in vitro transcription of EGSs. A 100 μL in vitro transcription reaction mix is prepared by adding the following: 20 μL of 5× in vitro transcription buffer; 10 μL of 100 mM DTT; 5 μL each of 10 mM ATP, GTP, CTP, and UTP; 7 μL of PCR-generated DNA templates (~10 μg DNA); 1.5 μL of RNasin RNase inhibitor; 1.5 μL of T7 RNA polymerase.
2. Incubate at 37 °C for at least 16 h, stop the reaction by adding 100 μL of denaturing dye, and load on 8 % denaturing gels containing 7 M urea.
3. Visualize the full-length RNA products by autoradiography, and extract RNA (*see* **Note 2**) from the excised gel slice by the crush-soak method using DEPC-treated water (*see* **Note 3**).

4. The yield of the in vitro RNA synthesis is determined either by measuring the concentration of the RNA using a spectrophotometer or by quantitation of the total radioactivity by using a STORM 840 PhosphoImager.
5. Synthesize labeled PR mRNA as described above, except using 30 μCi α-[^{32}P]-GTP and 1 mM GTP instead of 10 mM GTP in the reaction.

3.3 In Vitro Cleavage Assay for Substrate pr39

1. Human RNase P was prepared from HeLa cellular extracts as described previously [18].
2. The EGSs and [^{32}P]-labeled pr39 were incubated with human RNase P at 37 °C in a volume of 10 μL for 45 min in buffer A.
3. Cleavage products were separated in denaturing gels and analyzed with a STORM840 PhosphorImager (GE Healthcare).

3.4 Measurement of K_d for the Target RNA by EGSs

To determine the stability of EGS-mRNA complex, equilibrium dissociation constant (K_d) for the target RNA by EGS variants was measured.

1. Prepare a 5 % polyacrylamide gel with a 30:1 weight ratio of acrylamide to *N*,*N'*-methylene *bis*-acrylamide in 36 mM Tris base/64 mM HEPES, pH 7.5, 10 mM $MgCl_2$, and 0.1 mM EDTA. Run the gel in the same buffer at constant power, and maintain the temperature of the gel at 37 °C.
2. Dilute EGS to varying concentrations, ranging from 0.1 to 1,000 nM.
3. Mix 10 μL EGS dilutions at 2× final concentration with 6 μL ^{32}P-labeled target mRNA (1,000 cpm and 0.1 nM). Heat at 80 °C for 3 min, and add 4 μL of 5× binding buffer.
4. Incubate the samples at 37 °C for 10 min (to reach an equilibrium). Load samples onto the pre-warmed gel and run at a constant temperature of 37 °C immediately (*see* **Note 4**).
5. Dry the gel and quantitate free target RNA and bound RNA at each concentration of EGS on a PhosphorImager. The dissociation constant of an EGS is determined by inspection of the gel midpoint, where RNA$[P_{free}]=[P_{bound}]$. $K_d=[E_{total\ at\ midpoint}]-1/2[P_{total}]$ (here $[P]$ and $[E]$ symbolize the target RNA and the EGS, respectively).

3.5 Kinetic Analyses of the Cleavage Reactions with EGSs

Multiple turnover kinetic analyses to determine the values Michaelis constant (K_m) and the maximum velocity (V_{max}) of the enzymatic reactions were carried out.

1. Mix varying amounts of ^{32}P-labeled target mRNA with an excess amount of EGS RNA in buffer A, and calculate the concentration of target RNA–EGS complex in each mixture based on the dissociation constants (K_d).

2. Perform standard RNase P reactions for 0.5, 1, 2, 5, and 10 min.
3. Extrapolate an initial rate of RNase P reaction for each concentration of RNA–EGS complex.
4. Determine the values of K_m and V_{max} using Lineweaver–Burk plots (double-reciprocal plots) [14].

3.6 Stable Expression of EGS RNAs in Human Cells

The DNA sequence coding for PR-C125 and PR-SER EGS was sub-cloned into retroviral vector LXSN and was placed under the control of the small nuclear U6 RNA promoter [6, 19, 20].

1. To construct cell lines that express variant EGS RNAs, amphotropic packaging PA317 cells [21] were cultured in 6-well plates 1 day before transfection. When cell confluence was about 80 % (about 16 h after cell plating), transfect cells with 10–20 μg following retroviral vector DNAs (LXSN-PR-SER, LXSN-PR-SER-C, LXSN-PR-C125, LXSN-PR-C125-C) by using the mammalian transfection kit (GIBCO/BRL).
2. 48 h post transfection, collect culture supernatants (about 3 mL) and infect U373MG cells.
3. 1.5 mL of the collected retroviral stock was used to infect U373MG cells. The infected cells were incubated for 4–12 h with occasional shaking (*see* **Note 5**), and then the inoculums were replaced with fresh DMEM supplemented with 10 % FBS.
4. 48–72 h post transfection, cells were incubated in culture medium that contained 600 μg/mL of neomycin.
5. Neomycin-resistant cells were selected and cloned in the presence of neomycin for 2 weeks. Cells infected with retrovirus were subsequently split sparsely over ten cultured flasks and placed under neomycin to select for cloned EGS-expressing cell lines (*see* **Note 6**). Aliquot and freeze the selected cells for long-term storage in liquid nitrogen or for use in further studies.

3.7 Northern Analysis of the Expression of EGS RNAs in Cultured Cells

1. Wash the infected cells with PBS; both nuclear and cytoplasmic RNA fractions were isolated using SurePrep™ kit (Fisher Bioreagents). RNA concentration was tested by Bio-spectrophotometer (*see* **Note 7**).
2. 10 μg of RNAs are loaded onto a 2.5 % formaldehyde agarose gel with 45 mL of formaldehyde and 50 mL of 5× Northern buffer. RNAs are separated by running the gel at a constant voltage.
3. After washes of the gel with deionized water, transfer the RNAs onto a nitrocellulose membrane.

4. The transfer sandwich is set up in a large glass dish filled with 1 L 20× SSC solution from the bottom up: a long glass plate, a half sheet of Whatman 3 mm paper, three pieces of Whatman 3 mm paper (larger than the gel), the gel, a saran wrap cut with a window to expose gel, a piece of nitrocellulose membrane wetted with RNase-free water (larger than the gel), five sheets of Whatman 3 mm (the same size as the membrane), a stack of paper towels of about 4 cm at height, a glass, and a weight (1 kg). At each step, air bubbles are trapped by gently rolling a test tube over the last layer added. After overnight transferring, the membrane is rinsed three times with deionized water for 10 min. Then the membrane is placed on top of a piece of Whatman 3 mm paper and baked in an oven at 80 °C for 1.5 h.
5. The nitrocellulose membrane is pre-hybridized for 4 h and hybridized for 16 h with the ^{32}P-radiolabeled DNA probes at 65 °C in the hybridization buffer.
6. Wash the membrane with 2× SSC, 1× SSC, and 0.5× SSC (containing 0.1 % SDS).
7. Expose the membrane and analyze with a STORM840 PhosphorImager.

3.8 Determination of RNase P-Mediated Inhibition of Viral Gene Expression and Growth in Cultured Cells

3.8.1 Viral Infection

1. 1×10^6 Cells were either mock-infected or infected with HCMV at a multiplicity of infection (m.o.i.) of 0.5–5 in an inoculum of 1.5 mL of DMEM supplemented with 1 % fetal calf serum.
2. The inoculum was replaced with DMEM supplemented with 10 % (v/v) FBS after 2-h incubation with cells.
3. The infected cells were incubated for 4–72 h before harvesting for isolation of viral mRNA or protein.
4. To measure the levels of viral immediate-early (IE) transcripts, some of the cells were also treated with 100 μg/mL cycloheximide prior to and during infection.

3.8.2 Northern Blot Analysis

1. Wash the infected cells with PBS; RNA fractions were isolated as described in Subheading 3.7 (**step 1**).
2. For detection of viral mRNAs, the RNA fractions were separated in 1 % agarose gels that contained formaldehyde, transferred to a nitrocellulose membrane, hybridized with the ^{32}P-radiolabeled DNA probes that the HCMV DNA sequences, and analyzed with STORM840 PhosphorImager.

3.8.3 Western Blot Analysis

Western analyses are performed to determine the expression level of HCMV PR proteins.

1. Wash the cells twice with 5 mL PBS, and spin down cells with 3,000 × *g* at 4 °C for 5 min.

2. Suspend cell pellets in 50–100 μL cold PBS, followed by adding the same volume of 2× disruption buffer.
3. Vortex the mixture for 1 min, followed by three times sonication on ice. Each lasts for 20–30 s.
4. Boil the sample for 5 min before loading on SDS polyacrylamide gel.
5. Load 50 μg of proteins onto either 7.5 or 9 % SDS polyacrylamide gel cross-linked with *N,N′*-methylenebisacrylamide with a stacking layer of 4.5 % acrylamide/bisacrylamide. Separate the proteins by running the gels at a constant power setting.
6. The proteins are transferred onto a nitrocellulose membrane (*see* **Note 8**), using an electrophoretic transfer apparatus with a constant current of 150 mA for 2 h.
7. Incubate the nitrocellulose membrane with TNT buffer plus 2.5 % skim milk for 1 h in an orbital shaker (blocking step).
8. After blocking, incubate the membrane with primary antibody at a dilution of 1:500 in TNT supplemented with 1.25 % of skim milk for 1 h.
9. Wash the membrane with TNT buffer three times and 5 min for each wash.
10. Incubate the membrane with secondary antibody diluted at 1:1,000 dilutions for 1 h, followed by three washes of the membrane with TNT buffers and 5 min for each wash.
11. Incubate with ECL substrates for 1 min at RT, expose membrane, and develop film.

3.8.4 Analysis of Viral Growth by Plaque Assay

1. 5×10^5 of EGS-expressing cells are infected with HCMV at an m.o.i. of 0.5–2.
2. Incubate the cells with DMEM for 1.5 h, wash with PBS, followed by adding 0.5 mL fresh DMEM supplemented with 10 % Nu-serum.
3. The cells and medium are harvested at 1, 2, 3, 4, 5, 6, and 7 days post infection.
4. Viral stocks are prepared by adding an equal volume of 10 % (v/v) skim milk, followed by sonication.
5. Prepare tenfold serial dilution of the viral stock in 2 mL of DMEM for each dilution.
6. Infect 1×10^5 HFF cells in a 6-well plate with 1 mL of viral dilutions, and incubate for 2 h.
7. Wash the cells with DMEM, and overlay the cells with fresh 1 % agarose and DMEM containing 10 % Nu-serum in a 1:1 ratio.

8. Count the number of viral plaques 10–14 days after infection.
9. Plaque-forming unit (PFU) is determined by the highest viral dilution that yields plaque.

3.8.5 Determination of the Level of Intracellular HCMV Genome

To determine the antiviral mechanism resulting from the EGS-directed cleavage, we carried out a series of experiments to investigate whether EGS-based inhibition of PR expression affects viral genomic DNA replication as well as viral DNA encapsidation. The encapsidated viral DNAs would be resistant to DNaseI digestion, whereas those that are not packaged in the capsid would be susceptible.

1. The level of intracellular viral genomic DNA was determined by PCR detection of the sequence of immediate-early IE1 sequence, using the human β-actin sequence as the internal control. The 5′ and 3′ primers for detecting the IE1 sequence were CMV3 (5′-CCAAGCGGCCTCTGATAACCAAGCC-3′) and CMV4 (5′-CAGCACCATCCTCC TCTTCCTCTGG-3′), respectively. The 5′ and 3′ primers used to amplify the β-actin sequence were actin5 (5′-TGACGGGGTCACCCA CACTGT GCCCATCTA-3′) and actin3 (5′-CTAGAAGCATTGCGGT GGCAGATGG AGGG-3′), respectively.
2. 5×10^5 cells grown on 6-well plates were mock-infected or infected with HCMV.
3. After 1.5-h incubation at 37 °C, the inoculum was removed and the cells were further incubated and harvested at 72–96 h post infection.
4. Total and encapsidated (DNase I-treated) DNAs were isolated essentially as described [22] and used as the PCR DNA templates.
5. The PCR amplification consisted of 20 cycles (*see* **Note 9**) with denaturation at 94 °C for 1 min, followed by primer annealing at 47 °C for 1 min and extension at 72 °C for 1 min. The last cycle was again an extension at 72 °C for 10 min.
6. The amplified HCMV DNA (481 bp) and actin sequence (610 bp) were separated on 4 % non-denaturing polyacrylamide gels.

4 Notes

1. Although there are several methods available for oligonucleotide purification, gel purification is suggested here, since the unused primers can be removed in this process.
2. It is recommended to warm up the RNA samples at 37 °C for at least 5 min before loading on the gel. To avoid urea precipitation, flush the well ahead of loading samples.

3. DEPC-treated H_2O is recommended for the entire gel-running process to avoid mRNA degradation.
4. Maintaining the gel at 37 °C is critical to obtain accurate values of K_d; thus, pre-running the gel in the same buffer at constant power is suggested in this step.
5. To increase the efficiency of retroviral infection in targeted cells, medium supplemented with 8 μg/mL polybrene is recommended. Alternatively, multiple rounds of infection can also be carried out to improve the efficiency.
6. More than ten clones are selected in this step, and most of them were found to express high level of ribozymes.
7. RNA fractions are extracted from nucleus and cytoplasm and tested by Northern blot separately, since the EGS RNA expressed by U6 promoter is primarily localized in the nucleus.
8. The nitrocellulose membrane should be soaked in distilled water for 2 min, followed by placing into the transfer buffer, and allowed to soak for 5 min. Transfer of protein from the gel to the membrane should be performed at 4 °C.
9. PCR cycles and other conditions were optimized to assure that the amplification is within the liner range. We also generated a standard (dilution) curve by amplifying different dilutions of the template DNA. The plot of counts for both HCMV and β-actin versus dilutions of DNA did not reach a plateau for the saturation curve under certain conditions, indicating that quantitation of viral DNA could be accomplished.

Acknowledgments

We are grateful to Yong Bai and Paul Rider for technical assistance and invaluable suggestions. X. J. was a recipient of a China Graduate Student Scholarship from the Chinese Ministry of Education. This research has been supported by grants from NIH (RO1-AI041927, RO1-AI091536, and RO1-DE014842).

References

1. Lundblad EW, Altman S (2010) Inhibition of gene expression by RNase P. New Biotechnol 27:212–221
2. Kazantsev AV, Pace NR (2006) Bacterial RNase P: a new view of an ancient enzyme. Nat Rev Microbiol 4:729–740
3. Gopalan V, Vioque A, Altman S (2002) RNase P: variations and uses. J Biol Chem 277:6759–6762
4. Marvin MC, Engelke DR (2009) Broadening the mission of an RNA enzyme. J Cell Biochem 108:1244–1251
5. Forster AC, Altman S (1990) External guide sequences for an RNA enzyme. Science 249: 783–786
6. Yuan Y, Hwang ES, Altman S (1992) Targeted cleavage of mRNA by human RNase P. Proc Natl Acad Sci U S A 89:8006–8010
7. Castanotto D, Rossi JJ (2009) The promises and pitfalls of RNA-interference-based therapeutics. Nature 457:426–433
8. Zhu J, Trang P, Kim K, Zhou T, Deng H, Liu F (2004) Effective inhibition of Rta expression

and lytic replication of Kaposi's sarcoma-associated herpesvirus by human RNase P. Proc Natl Acad Sci U S A 101:9073–9078

9. Ma M, Benimetskaya L, Lebedeva I, Dignam J, Takle G, Stein CA (2000) Intracellular mRNA cleavage induced through activation of RNase P by nuclease-resistant external guide sequences. Nat Biotechnol 18:58–61
10. Guerrier-Takada C, Li Y, Altman S (1995) Artificial regulation of gene expression in Escherichia coli by RNase P. Proc Natl Acad Sci U S A 92:11115–11119
11. Plehn-Dujowich D, Altman S (1998) Effective inhibition of influenza virus production in cultured cells by external guide sequences and ribonuclease P. Proc Natl Acad Sci U S A 95: 7327–7332
12. Kawa D, Wang J, Yuan Y, Liu F (1998) Inhibition of viral gene expression by human ribonuclease P. RNA 4:1397–1406
13. Kraus G, Geffin R, Spruill G, Young AK, Seivright R et al (2002) Cross-clade inhibition of HIV-1 replication and cytopathology by using RNase P-associated external guide sequences. Proc Natl Acad Sci U S A 99: 3406–3411
14. Zhou T, Kim J, Kilani AF, Kim K, Dunn W et al (2002) In vitro selection of external guide sequences for directing RNase P-mediated inhibition of viral gene expression. J Biol Chem 277:30112–30120
15. Jiang X, Bai Y, Rider P, Kim K, Zhang CY et al (2011) Engineered external guide sequences effectively block viral gene expression and replication in cultured cells. J Biol Chem 286: 322–330
16. Mocarski ES, Shenk T, Pass RF (2007) Cytomegalovirus. In: Knipe DM, Howley PM, Griffin DE, Martin MA, Lamb RA, Roizman B, Straus SE (eds) Fields virology. Lippincott-Williams and Wilkins, Philadelphia, PA, pp 2701–2772
17. Schreiber A, Harter G, Schubert A, Bunjes D, Mertens T, Michel D (2009) Antiviral treatment of cytomegalovirus infection and resistant strains. Expet Opin Pharmacother 10: 191–209
18. Yuan Y, Altman S (1994) Selection of guide sequences that direct efficient cleavage of mRNA by human ribonuclease P. Science 263:1269–1273
19. Liu F, Altman S (1995) Inhibition of viral gene expression by the catalytic RNA subunit of RNase P from Escherichia coli. Gene Dev 9:471–480
20. Bertrand E, Castanotto D, Zhou C, Carbonnelle C, Lee NS et al (1997) The expression cassette determines the functional activity of ribozymes in mammalian cells by controlling their intracellular localization. RNA 3:75–88
21. Miller AD, Rosman GJ (1989) Improved retroviral vectors for gene transfer and expression. BioTechniques 7:980–982, 4–6, 9–90
22. Matusick-Kumar L, Hurlburt W, Weinheimer SP, Newcomb WW, Brown JC, Gao M (1994) Phenotype of the herpes simplex virus type 1 protease substrate ICP35 mutant virus. J Virol 68:5384–5394

Chapter 5

Design and Analysis of Hammerhead Ribozyme Activity Against an Artificial Gene Target

James R. Carter, Pruksa Nawtaisong, Velmurugan Balaraman, and Malcolm J. Fraser Jr.

Abstract

In vitro cleavage assays are routinely conducted to properly assess the catalytic activity of hammerhead ribozymes (HHR) against target RNA molecules like dengue virus RNA. These experiments are performed for initial assessment of HHR catalysis in a cell-free system and have been simplified by the substitution of agarose gel electrophoresis for SDS-PAGE. Substituting mobility assays enables the analysis of ribozymes in a more rapid fashion without radioisotopes. Here we describe the in vitro transcription of an HHR and corresponding target from T7-promoted plasmids into RNA molecules leading to the analysis of HHR activity against the RNA target by in vitro cleavage assays.

Key words Dengue, Flavivirus, Ribozyme, Catalysis, Electrophoresis, RNA, In vitro transcription

1 Introduction

Huge strides have been made in the development of antiviral strategies. One such approach involves the use of ribozymes, catalytic RNAs, with hammerhead ribozymes (HHRs) at the forefront [1–6]. There are a number of advantages to using HHR as a gene ablation tool. The catalytic activities of HHRs are independent of host cell's machinery [7]. Secondly, HHRs can simultaneously cleave a target RNA at more than one site by covalently linking several HHRs to form maxizymes [8].

HHRs cleave their target RNAs on the 3′ end of the uracil located within an NUH nucleotide consensus (where N = any nucleotide, U = uracil, and H = any nucleotide but G) in a magnesium ion-dependent manner, provided there are sequence homologies between the ribozyme arms and the corresponding target sequence [9] (Fig. 1).

Initial assessment of HHR activity is most effectively executed through in vitro cleavage assays. The in vitro cleavage reactions we

Daniel Lafontaine and Audrey Dubé (eds.), *Therapeutic Applications of Ribozymes and Riboswitches: Methods and Protocols*, vol. 1103, DOI 10.1007/978-1-62703-730-3_5, © Springer Science+Business Media New York 2014

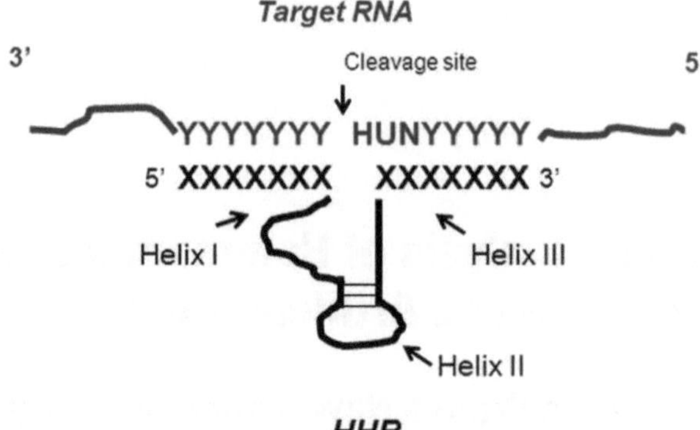

Fig. 1 Design and construction of a typical hammerhead ribozyme (HHR). HHRs are small ribonucleic-based RNAs that cleave target RNA in a sequence-specific manner. Catalysis is initiated by base pairing of the 5′ helix I and 3′ helix III arms of the HHR to complementary 3′ and 5′ bases on the target RNA to form helices I and III, respectively. The catalytic core of the HHR, helix II, is responsible for cleavage at a 5′–NUH–3′ triplet site on the target RNA, where N can be any 4 nucleotide and H can be A, C, or U. Target RNA is shown in gray with "Y" depicting each nucleotide residue to be bound by the HHR. The HHR is in black with "X" depicting each nucleotide residue of the 5′ and 3′ arms

have used are adopted from a protocol developed by Shao et al. [10]. These assays are performed in two general steps: (1) In vitro transcription of HHR and the corresponding artificial target, each from a separate T7 promoter and (2) a general in vitro cleavage reaction to assess the ability of these ribozymes to catalyze the cleavage of the intended substrates.

The success of in vitro cleavage assays depends on the precise determination of experimental conditions necessary for optimal HHR activity. Successful determination of the optimal in vitro cleavage conditions is achieved by varying the amounts of HHRs used, the concentration of $MgCl_2$ necessary for HHR catalysis, and the time necessary to achieve optimal catalysis. We previously produced T7-promoted anti-dengue virus (DENV) HHRs and corresponding DENV target regions and showed through in vitro cleavage analysis that several of our HHRs were capable of catalyzing the cleavage of DENV RNA substrates [1]. The protocol presented here can be tailored for use with any HHR/target RNA pair.

2 Materials

2.1 In Vitro Analysis of HHR Cleavage Activity

The success of in vitro transcription and in vitro cleavage assays is greatly increased if performed under RNase-free conditions. To ensure this, wipe down all gel chambers and pipettors with an RNase-eliminating solvent. Sterilize all glassware and solutions prior to use unless indicated otherwise. All water used in making the described buffers should be certified RNase free by the manufacturer or at least millipure, diethyl pyrocarbonate (DEPC) treated, and autoclaved prior to use. Store all reagents at room temperature unless indicated otherwise. Closely follow all waste disposal regulations when disposing of waste materials. All concentrations given in parentheses are the final concentrations.

1. 1 M Tris–HCl (pH 8.0): Mix 121.1 g of Tris–HCl base with 700 mL of millipure H_2O by stirring. Add 42 mL concentrated HCl to reach pH of 8.0. Fine adjust to the desired pH (8.0) with concentrated HCl if needed (*see* **Note 1**). Add millipure H_2O until final volume is 1 L and autoclave.
2. 5 M NaCl stock: Weigh 292.2 g NaCl and add to 800 mL millipure water. Mix until dissolved using a stir bar and stir plate, add millipure H_2O to 1 L, and autoclave.
3. 1 M NaCl: In an autoclaved graduated cylinder place 10 mL 5 M NaCl with 40 mL DEPC-treated H_2O. Mix well, and place in an autoclaved bottle.
4. 1.5 M $MgCl_2$ stock solution: Weigh 304.95 g $MgCl_2 \cdot 6H_2O$ and add to 800 mL millipure water. Mix until dissolved using a stir bar and stir plate. Add millipure H_2O to 1 L and autoclave.
5. RNase inhibitor: "RNaseout" purchased from Life Technologies. Only 40 U (1 μL) per reaction is needed.
6. HHR: The T7-promoted forward and antisense forms should be obtained from a gene or a primer synthesis company as DNA, annealed at 95 °C for 5 min, immediately cooled on ice, and then in vitro transcribed from the T7 promoter (*see* Subheading 3.2) prior to use in the described in vitro cleavage reactions.
7. Artificial target: This is a mimic of your gene or gene segment of choice and is in vitro transcribed from a T7 promoter (*see* Subheading 2.1, **item 1**).
8. In vitro transcription kit: We use the *Megascriptshort* and *Megascript* kits from Life Technologies (*see* **Note 2**).
9. Agarose gel extraction kit: We use the *Promega SV Gel* and *PCR Clean-Up System*.
10. GuTC extraction buffer: In lieu of GuTC buffer, "Trizol," from Life Technologies can be used. Be sure to add the prescribed amount of LiCl to the GuTC extraction buffer.
11. RNA gel loading buffer: 95 mL formamide, 18 mM EDTA (pH 7.5), 0.025 % SDS, 0.05 g bromophenol blue, and 0.05 g xylene cyanol. Add enough DEPC-treated water to make the total volume 100 mL and then autoclave after mixing.

12. RNase-free ethidium bromide (10 mg/mL): Dissolve 1 g of ethidium bromide with 100 mL DEPC-treated dH_2O. Mix by stirring, and store in the dark at 4 °C.
13. 10× MOPS running buffer: Add 41.85 g 4-morpholino-propanesulfonic acid (MOPS-free acid) and 6.80 g sodium acetate–$3H_2O$ to 800 mL DEPC-treated dH_2O and stir until completely dissolved. Add 20 mL of a DEPC-treated 0.5 M Na_2EDTA solution and adjust pH to 7.0 with 10 M sodium hydroxide or glacial acetic acid. Adjust volume to 1 L with DEPC-treated dH_2O (*see* **Note 3**).

3 Methods

3.1 Selection of HHR Target Sites on DENV-2 Genome

Align all sequences of the gene of interest using ClustalX [11, 12]. Select HHR target sites by scanning for conserved sequences that contain the nucleic acid triplets GTC, GTA, ATC, or CTC, since HHR-mediated cleavage occurs more efficiently when these triplets are targeted. At the minimum a target sequence containing an NUH triplet (*see* Subheading 1) is required. For example, our primary criterion for the selection of a DENV target site was that this target site must be present in all 29 strains of DENV-2. Another important criterion for selecting a suitable site for HHR cleavage is the length of conserved flanking arms which determine the level of specificity of the ribozyme. The length of each arm should range from 5 to 10 bp [1]. Once target sites are selected, design the ribozyme arms such that the 5′ arm of the HHR will be a complementary base pair to the 3′ end of the target site flanking the NUH triplet and the 3′ arm of the HHR will be a complementary base pair to the 5′ end of the target site flanking the NUH triplet. Ensure that the lengths of the binding arms are not longer than the targeted conserved region.

3.2 In Vitro Transcription

3.2.1 Linearization of the Ribozyme Plasmid and DENV Template

1. In separate reaction tubes place 2.5 μg ribozyme and 1.5 μg DENV target with 1 μL of the restriction endonuclease (REN) of interest (*see* **Note 4** and Fig. 2).
2. Allow these REN reactions to proceed for 4 h at a temperature that is optimal for REN activity to ensure that all plasmids in the reaction tube are linearized (*see* **Note 5**).
3. Treat the linearized plasmids with 1 μL Antarctic phosphatase (New England Biolabs) for 15 min at 37 °C to prevent self-ligation of the digested ends. This will prevent self-ligation of the digested plasmid.
4. Heat inactivate the Antarctic phosphatase and REN by incubating the reaction tubes at 65 °C for 20 min.

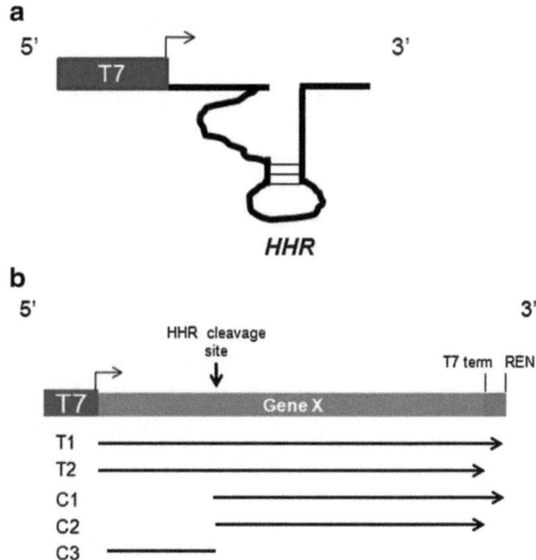

Fig. 2 (**a**) Diagram of a typical T7-promoted ribozyme construct, T7-HHR. We typically insert T7-HHR into a Bluescript expression vector (Invitrogen) prior to in vitro transcription. (**b**) Schematic representation of the predicted in vitro-produced transcripts (T1 and T2) from a T7-promoted hypothetical gene (Gene X) following in vitro transcription and cleavage products resulting from HHR activation (C1, C2, and C3). T7 term = T7 termination sequence. HHR cleavage site is indicated by *arrows*

5. Separate bands by 1 % agarose gel electrophoresis, excise the band of interest using a scalpel, and gel purify this fragment using any agarose gel extraction kit.

3.2.2 In Vitro Transcription

In vitro transcribe the HHR and target using the *Megashortscript* (for HHR RNA production) and *Megascript* (for target RNA production) kits that each contains the T7 polymerase (Life Technologies, USA). Below is an in vitro transcription kit protocol as we use it in the lab.

1. Immediately following gel extraction, place 1 μg of the ribozyme and target in separate 1.5 mL Eppendorf tubes.
2. To each tube add 2 μL of each triphosphated nucleotide, 2 μL reaction buffer, and 2 μL enzyme mix (*see* **Note 6**). Add nuclease-free or DEPC-treated water.
3. Incubate the in vitro transcription reactions containing the target plasmid for 4 h at 37 °C. The in vitro transcription reactions containing the ribozyme require a longer incubation of 16 h at 37 °C (*see* **Note 7**). DNase digestion is not necessary since the quantity of digested plasmid is in a much lower concentration relative to RNA, but DNase treatment can be performed if desired.

3.2.3 RNA Isolation

1. To each sample tube add DEPC-treated water for a final volume of 100 μL.
2. Add an equal volume of GuTC extraction buffer, mix very well, and incubate the samples at room temperature for 5 min.
3. Add 100 μL of phenol, mix very well, and incubate at room temperature for 5 min. If you have chosen to use Trizol for this step, the addition of phenol is not necessary since it is included.
4. Add 100 μL of chloroform–isoamyl alcohol (*see* **Note 8**), close cap, and mix manually by vigorously shaking tubes to create an emulsion. Do not vortex. Spin at 15,294 × *g* in a table microcentrifuge, at 4 °C, for 15 min (*see* **Note 9**). Transfer the top aqueous to a fresh, prechilled 2 mL microcentrifuge tube. The top aqueous phase contains all the isolated RNA molecules.
5. Add 100 μL chloroform–isoamyl alcohol (24:1) to the transferred aqueous phase. Close cap and shake vigorously to create an emulsion. Spin at 15,294 × *g* in a table microcentrifuge, at 4 °C, for 15 min. Once again transfer the top aqueous phase to a fresh, prechilled 2 mL microcentrifuge tube (*see* **Note 10**).
6. Add 2 volumes of 100 % ethanol and 0.2 μg of glycogen to each sample, and incubate samples overnight at −20 °C (*see* **Note 11**).
7. The following day spin samples at 15,294 × *g* in a table microcentrifuge, at 4 °C, for 30 min. Remove as much liquid as possible without disturbing the pellet. Air-dry the RNA pellet at room temperature ensuring that all liquid has dissipated (*see* **Note 12**).
8. Resuspend the RNA pellet in a suitable amount of DEPC-treated water or nuclease-free water from your favorite manufacturer. Determine the concentration spectrophotometrically.

3.3 Ribozyme In Vitro Cleavage Reaction

1. Design separate experiments to vary (a) the amount of in vitro-transcribed ribozymes from 0 to 50 pM; (b) the concentration of $MgCl_2$ from 5 to 50 mM (from a 1 M stock), in a solution containing 50 mM Tris–HCl, pH 8.0 and 40 U (1 μL) of RNase inhibitor; and (c) the incubation time from 0 to 30 min in 5-min increments. The total volume of these reactions should be no more than 10 μL.
2. Place 10 pM of in vitro-transcribed target and the corresponding in vitro-transcribed HHR (0–50 pM) in the same 0.5 mL thin-walled PCR tube containing 50 mM Tris–HCl, pH 8.0.
3. Heat denature the components by placing the PCR tubes in the thermocycler at 85 °C for 3 min, and immediately place on ice (*see* **Note 13**).
4. Initiate the cleavage reaction by adding 0.1 μL of 1 M $MgCl_2$ (final concentration of 10 mM) and 1 μL (40 U) of RNase

inhibitors. Mix by gently pipetting samples, and incubate at 37 °C for 0–30 min.

5. Stop the HHR cleavage reaction by adding 30 μL (i.e., 3× volume) of the gel loading buffer.

3.4 2 % Agarose/Formaldehyde/MOPS Gel Electrophoresis

1. Rinse the gel box, gel combs, and gel caster you will use for RNA electrophoresis with copious amounts of millipure water, making sure to remove any salt deposits from prior electrophoresis experiments. Air-dry components: a 50 °C incubator can be used to facilitate the drying process. Wipe components with an RNase-eliminating solvent. Overnight soaking of the gel caster and comb in the RNase-eliminating solvent is recommended but not essential.
2. In a 250 mL Erlenmeyer flask melt 2 g agarose in 70 mL RNase-free ddH_2O and cool down to 60 °C.
3. In a fume hood, add 20 mL 37 % formaldehyde and 10 mL 10× MOPS (*see* **Note 14**).
4. Cast the gel in the fume hood. Place gel comb in gel caster and place gel comb/caster assembly in gel box. Pour the melted 2 % agarose/formaldehyde/MOPS into the gel caster, cover, and allow the gel to cool and polymerize.
5. Heat the in vitro cleavage samples to 65–70 °C for 5 min, and load 30 μL into each well of the cooled/polymerized 2 % agarose/formaldehyde/MOPS gel (*see* **Note 15**).
6. Run the gel in 1× MOPS buffer for 1 h at 105 V (*see* **Note 16**).
7. Rinse the gel with millipure water and place in a receptacle suitable for staining and destaining. We have found that used pipette tip boxes are ideal for this. Stain the gel for 1 h in DEPC-water containing 10 μg/mL ethidium bromide (*see* **Note 17**). Destain the gel in millipure water for a minimum of 5 min. Change the water and destain for another 10 min (*see* **Note 18**). Photograph the gel or destain more if necessary. It may be also necessary to resolve the bands more. *See* Nawtaisong et al. [1] for an example of a finished gel depicting the HHR cleavage product resulting from an in vitro cleavage assay.

4 Notes

1. Many electronic pH meters are unable to accurately determine the pH of concentrated Tris–HCl solutions. Be sure to use an appropriate probe. pH is temperature dependent: ~0.03 pH units per 1 °C increase. Make sure that the Tris–HCl solution is at room temperature before making final pH adjustments.

2. *MEGAscriptshort* in vitro transcription kits are used to produce RNA transcripts less than 500 nucleotides in length, such as HHRs which are approximately 40 nucleotides in length. *MEGAscript* kits are typically used to produce RNA transcripts that are larger than 500 bases, such as the DENV-2 target sequence we produced [1].
3. Always protect MOPS from light by wrapping container in aluminum foil, and store 10× MOPS at 4 °C. If the solution turns yellow, make anew.
4. We insert the vesicular stomatitis virus transcription termination signal (5′-TTTTTTTCATA-3′; [13]) immediately downstream of the gene or the gene segment we wish to transcribe but immediately upstream of the restriction site used to cleave the vector prior to in vitro transcription. Be sure to insert this sequence for efficient transcription of your HHR target.
5. If digestion of the parental DNA vector prior to in vitro transcription is not performed circular DNA plasmid templates will generate extremely long, heterogeneous RNA transcripts. To insure proper digestion, linearized DNA templates should be examined on an agarose gel to confirm complete digestion. Although any restriction enzyme may be used, one should avoid restriction enzymes that leave 3′ overhanging ends as this can lead to low-level transcription [14]. Be sure to choose a restriction site that is digested by an REN that can be inactivated by heat. Most RENs can be completely inactivated at 65 °C for 20 min.
6. Once thawed, store the ribonucleotides on ice while assembling the transcription reaction. The RNA polymerase enzyme mix should be placed on ice immediately following removal from the freezer since it is stored in glycerol and will not freeze at −20 °C. Keep the 10× reaction buffer at room temperature while assembling the reaction. Mix these components in a single master mix, large enough to accommodate all samples to be transcribed and pipette from that vessel into the sample tubes containing the linearized plasmid. This will decrease your chances for RNase contamination.
7. For transcripts shorter than 500 nt, a longer incubation time of up to 16 h is advantageous since more transcription initiation events are required to synthesize a given mass amount of RNA compared to transcription of longer transcripts.
8. Take care when using chloroform as overexposure can result in skin and eye irritation, mutagenic for mammalian somatic cells, and may be toxic to kidneys, liver, and heart. Repeated or prolonged exposure to the substance can produce damage to target organs.

9. In our experience spinning RNA at a maximum of 15,294 × *g* greatly increases yield since centrifugation at speeds greater than this can lead to the shearing of RNA transcripts.
10. Keep samples on ice or at 4 °C as much as possible. This will enhance RNA yield.
11. Overnight incubation of RNA precipitations at −20 °C greatly facilitates precipitation of short (>500 nt) transcripts. LiCl is generally the precipitant of choice for RNA. However this method is contraindicated for RNA molecules smaller than 300 nucleotides.
12. Do not wash pellet in 70 % EtOH since glycogen is soluble in water and even the water contained in the 70 % EtOH can prematurely solubilize the pellet. Guard against overdrying the RNA pellet as this will make resuspension very difficult.
13. Heating newly transcribed RNA to high temperatures (>80 °C) followed by rapid cooling enhances HHR–target interaction and the re-formation of secondary structures essential for HHR activity. Performance of this step is a must.
14. Always use a fume hood when pouring or decanting formaldehyde. EB is a potential carcinogen. Always wear appropriate protection when handling formaldehyde or EB.
15. Heating samples at this stage in the presence of formide will denature RNA [15], enabling their migration on a MOPS gel while providing protection from RNases [16].
16. It is best to recirculate the running buffer. If recirculation is not available, stop the gel after 30 min and manually mix the buffer by pipetting from one side to the other.
17. The time given is just a baseline. The amount of time necessary to stain gels with ethidium bromide depends on the amount of RNA in the gel, the concentration of the gel itself, and the desired concentration of EB used. Formaldehyde gels are notoriously difficult to stain with ethidium bromide, since the dye appears to bind to the gel matrix and cause high backgrounds. The background can be minimized by limiting the exposure to the staining solution to less than 5 min and destaining in water. Alternatively, the gel can be soaked in water for 10–15 min prior to staining to remove the formaldehyde and allow the RNA to renature, thus increasing its ability to bind ethidium bromide.
18. The time given is just a baseline. The amount of time necessary to stain gels with ethidium bromide depends on the amount of RNA in the gel, the concentration of the gel itself, and the desired concentration of EB used.

Acknowledgement

This work was supported by NIH grant AI097554 to MJF.

References

1. Nawtaisong P, Keith J, Fraser T, Balaraman V, Kolokoltsov A, Davey RA, Higgs S, Mohammed A, Rongsriyam Y, Komalamisra N, Fraser MJ Jr (2009) Effective suppression of Dengue fever virus in mosquito cell cultures using retroviral transduction of hammerhead ribozymes targeting the viral genome. Virol J 6:73
2. Sun LQ, Wang L, Gerlach WL, Symonds G (1995) Target sequence-specific inhibition of HIV-1 replication by ribozymes directed to tat RNA. Nucleic Acids Res 23:2909–2913
3. Lieber A, He CY, Polyak SJ, Gretch DR, Barr D, Kay MA (1996) Elimination of hepatitis C virus RNA in infected human hepatocytes by adenovirus-mediated expression of ribozymes. J Virol 70:8782–8791
4. Rossi JJ, Elkins D, Zaia JA, Sullivan S (1992) Ribozymes as anti-HIV-1 therapeutic agents: principles, applications, and problems. AIDS Res Hum Retroviruses 8:183–189
5. Jackson WH Jr, Moscoso H, Nechtman JF, Galileo DS, Garver FA, Lanclos KD (1998) Inhibition of HIV-1 replication by an anti-tat hammerhead ribozyme. Biochem Biophys Res Commun 245:81–84
6. von Weizsacker F, Blum HE, Wands JR (1992) Cleavage of hepatitis B virus RNA by three ribozymes transcribed from a single DNA template. Biochem Biophys Res Commun 189:743–748
7. Chachulska AM (1992) Ribozymes–catalytic RNA molecules. Postepy Biochem 38:64–74
8. Iyo M, Kawasaki H, Taira K (2004) Maxizyme technology. Methods Mol Biol 252:257–265
9. Blount KF, Uhlenbeck OC (2002) The hammerhead ribozyme. Biochem Soc Trans 30:1119–1122
10. Shao Y, Wu S, Chan CY, Klapper JR, Schneider E, Ding Y (2007) A structural analysis of in vitro catalytic activities of hammerhead ribozymes. BMC Bioinformatics 8:469
11. Chenna R, Sugawara H, Koike T, Lopez R, Gibson TJ, Higgins DG, Thompson JD (2003) Multiple sequence alignment with the Clustal series of programs. Nucleic Acids Res 31: 3497–3500
12. Thompson JD, Gibson TJ, Plewniak F, Jeanmougin F, Higgins DG (1997) The CLUSTAL_X windows interface: flexible strategies for multiple sequence alignment aided by quality analysis tools. Nucleic Acids Res 25: 4876–4882
13. Rose NF, Roberts A, Buonocore L, Rose JK (2000) Glycoprotein exchange vectors based on vesicular stomatitis virus allow effective boosting and generation of neutralizing antibodies to a primary isolate of human immunodeficiency virus type 1. J Virol 74:10903–10910
14. Schenborn ET, Mierendorf RC Jr (1985) A novel transcription property of SP6 and T7 RNA polymerases: dependence on template structure. Nucleic Acids Res 13:6223–6236
15. Chomczynski P (1992) Solubilization in formamide protects RNA from degradation. Nucleic Acids Res 20:3791–3792
16. Pinder JC, Staynov DZ, Gratzer WB (1974) Properties of RNA in formamide. Biochemistry 13:5367–5373

Chapter 6

Knockdown Strategies for the Study of Proprotein Convertases and Proliferation in Prostate Cancer Cells

François D'Anjou, Frédéric Couture, Roxane Desjardins, and Robert Day

Abstract

Gene silencing strategies targeting mRNA are suitable methods to validate the functions of specific genes. In this chapter, we sought to compare two knockdown strategies for the study of proprotein convertases and proliferation in prostate cancer cells. We used both SOFA-HDV ribozyme and lentiviral-mediated shRNA delivery system to reduce PACE4 mRNA levels and validate its implication in the proliferation of DU145 prostate cancer cells. The cellular effects of PACE4 knockdown were assessed (1) in vitro using two tetrazolium salts (MTT and XTT assays) and (2) in vivo using a tumor xenograft approach in immunodeficient mice (Nu/Nu). Our results confirm the unique role of the proprotein convertase PACE4 in prostate cancer cell proliferation while demonstrating advantages and disadvantages of each approach. Achieving target validation in an effective manner is critical, as further development using a drug development approach is highly laborious and requires enormous resources.

Key words HDV-ribozyme, shRNA, Lentivirus, MTT, XTT, Xenografts, Proprotein convertases, Prostate cancer

1 Introduction

In the last decade, an improved definition of modified cellular functions essential for cancer development and growth has been elaborated [1]. Among the six distinctive and complementary characteristics, sustained proliferative capability remains central to the entire process. In this chapter, we describe two complementary cell viability and proliferation assays to rapidly evaluate, in vitro and in vivo, the implication of target genes, such as the proprotein convertases, in cancer cell growth using specific gene silencing approaches. The first method measures the metabolic activity of cultured cells using tetrazolium salts such as MTT or XTT compounds [2]. The second method is based on the use of a mouse model bearing tumor xenografts of stable knockdown cell lines to evaluate the contribution of specific genes to tumor growth [3].

Daniel Lafontaine and Audrey Dubé (eds.), *Therapeutic Applications of Ribozymes and Riboswitches: Methods and Protocols*, vol. 1103, DOI 10.1007/978-1-62703-730-3_6, © Springer Science+Business Media New York 2014

Based on these approaches, we have recently reported the important role of the endoprotease PACE4 in prostate cancer cell proliferation [4]. PACE4 is one of the nine known proprotein convertase (PC), a mammalian enzyme family implicated in the post-translational processing of multiple secretory proteins by limited endoproteolysis [5]. This widespread activation process allows cells to generate single or multiple products from a single precursor. PC cleavages result in a diversity of bioactive products, zymogen activation, and sometimes inactivation of key proteins, including various receptors, growth factors, matrix-associated enzymes and components, and peptide hormone precursors [5, 6]. In addition to their endogenous physiological functions, a number of studies have established the involvement of PCs in pathologies due to their actions on exogenous molecules such as viral-coat proteins and bacterial toxins [7, 8]. Disruptions in PCs, either by overexpression or through mutations, can lead to important disorders such as cancer, arthritis, and dyslipidemia, which has increased interest in PCs as promising pharmacological targets [7].

As the development of selective enzymatic inhibitors for PACE4 or any other PC remains a great challenge [7, 9] and requires considerable efforts, it is first important to establish enough evidence to justify these efforts. We therefore sought to rapidly investigate the contribution of PACE4 in prostate cancer cell growth through molecular targeting. Indeed, various specific silencing technologies, such as catalytic RNAs [4, 10], antisense oligonucleotides [11], or RNA interference [12, 13], have been used in the past to reveal the redundant and shared functions of each PC. In comparison with overexpression strategies, widely used in the study of PCs [6, 14], those knockdown/knockout approaches of endogenously expressed PCs will more accurately reflect normal cellular/physiological conditions.

We employed a unique mRNA silencing approach based on the use of target-induced SOFA-HDV ribozyme [15, 16]. Using the well-established DU145 model epithelial-like cell line (derived from a human metastatic carcinoma of the prostate [17]), we have shown that specific and significant reduction of PACE4 mRNA expression leads to reduced in vitro cell proliferation capabilities that became dramatic in immunodeficient mice (Nu/Nu) bearing low PACE4 DU145 cell tumor xenografts. This study was done with a clonal cell line stably expressing the SOFA-HDV ribozyme. Although very effective, this approach is time consuming at two main levels: (1) for the design of the catalytic RNA itself and (2) for the establishment and screening of derived clonal cell lines. Thus, as a more rapid alternative silencing approach, we used a polyclonal cell population expressing short-hairpin RNAs (shRNAs) targeting PACE4 mRNA (Table 1).

Table 1
Comparison of SOFA-HDV ribozyme and shRNA knockdown strategies

SOFA-HDV ribozyme (for review *see* [25, 26])		shRNA (for reviews *see* [27, 28])	
Pros	**Cons**	**Pros**	**Cons**
High specificity	Lower activity	High activity	Specificity might be lower
Validation with in vitro cleavage assays	Less suitable for polyclonal transfection	Performance high enough for polyclonal transfection	No in vitro validation
High cellular stability	Design and cloning is time consuming	Libraries commercially available	Necessitates additional controls
		Well-established viral delivery systems	

In this chapter, we describe how silencing methods can be used to study the implication of discrete genes on the proliferation of cancer cell lines both in vitro and in vivo to provide essential proof of concept for a target gene. As the design of SOFA-HDV ribozymes and lentiviral-mediated shRNA delivery system have been extensively described elsewhere [4, 18–20], we will put emphasis on the comparative aspects of the proliferation assays used to monitor the effects of PACE4 silencing on DU145 proliferation both in vitro (MTT and XTT assays) and in vivo (tumor xenografts in Nu/Nu mice).

2 Materials

The cell lines DU145 and HT-1080 were purchased at the ATCC (www.atcc.org; HTB-81 and CCL-121, respectively). All plasticware used (cell culture dishes and plates, disposable pipets, conical tubes) was from BD Biosciences.

2.1 Clonal Cell Lines Expressing SOFA-HDV Ribozyme

1. PACE4 SOFA-HDV ribozyme design and cloning were described earlier [4].
2. RPMI 1640, 1× (350-000-CL), Fetal Bovine Serum (FBS) Premium (090-150), Phosphate-Buffered Saline (PBS; 311-010-QL), and Hygromycin B solution (450-141-xl) are from Wisent Bioproducts.
3. 0.05 % trypsin solution: Dilute 4 mL of 2.5 % trypsin solution (Gibco; 15090-046) in 200 mL of versene 1× buffer (137 mM NaCl, 2.7 mM KCl, 8.1 mM Na_2HPO_4, and 1.15 mM

KH_2PO_4, 0.5 mM EDTA) and sterilize with a 0.22 μM steritop filter unit (Millipore; SCGPT05RE).

4. Opti-MEM reduced serum medium (Gibco; 31985-070) and Lipofectamine 2000 transfection reagent (Invitrogen; 11668-019).

2.2 Polyclonal Cell Lines Expressing shRNA

1. HEK293FT (Invitrogen; R700-07), Dulbecco's Modified Eagle Medium (DMEM; Wisent Bioproducts; 319-010-CL), FBS Premium (Wisent Bioproducts; 090-150).
2. pLKO.1-puro with shRNA construct targeting human PACE4 (SHCLNG-NM_002570, clone TRCN0000075250), pLKO.1-TRC control (SHC001), and pLKO.1-scramble shRNA (SHC002) are from Sigma-Aldrich.
3. ViraPower™ Packaging Mix (Invitrogen; K4975-00), Lipofectamine 2000 (Invitrogen; 11668-019), Opti-MEM (Gibco; 31985-070), Acrodisc® 25 mm syringe filter units 0.45 μm (Pall Corporation; PN 4184).
4. For virus titration, EMEM (Wisent Bioproducts; 320-005-CL), FBS, polybrene (hexadimethrine bromide; Sigma-Aldrich; H9268), puromycin (Wisent Bioproducts; 400-160-EM), PBS, methylene blue, methanol.

2.3 In Vitro Cell Proliferation Assay Using MTT/XTT Time-Course Assay

1. Thiazolyl blue tetrazolium bromide (MTT): Dissolve MTT powder (Sigma-Aldrich; M2128-5G) at a final concentration of 5 mg/mL in PBS by heating at 60 °C. Sterilize by filtering on a 0.22 μM Millex-GP filter unit (Millipore; SLGP033RS). The solution can be stored under light protection at 4 °C for weeks and is stable for up to 6 months when stored below 0 °C.
2. MTT revealing solution: Mix isopropanol with HCl 1 N at a 24:1 ratio.
3. XTT Cell Proliferation Kit II (Roche; 11 465 015 001).
4. Microplate reader (Molecular Devices; SpectraMax 190).

2.4 In Vivo Subcutaneous Tumor Xenografts

1. 28–42-day-old male athymic nude mice (Nu/Nu) are from Charles River Laboratories.
2. Insulin seringes 28G 1/2 or bigger (BD; 329461).
3. Isoflurane USP (Abbot Laboratories; DIN 02032384).

3 Methods

The general principles for the design, in vitro characterization, and cloning of selective SOFA-HDV ribozymes have been extensively described elsewhere [4, 10, 15, 16, 18, 19]. In the case of shRNAs, of five shRNAs targeting the same mRNA [20], we have selected the best candidates for further studies (*see* **Note 1**). In this chapter, we compare the effect of two knockdown strategies targeting PACE4 mRNA on in vitro and in vivo proliferation rates of the human prostate cancer cell line DU145.

3.1 Knockdown of PACE4 in Stable Cell Lines

We chose to target endogenous PACE4 mRNA in DU145 prostate cancer cell lines as we found that this particular proprotein convertase was exclusively overexpressed in human surgical specimens [4]. Thus, we sought to study its potential contribution to sustain the proliferation of this highly tumorigenic cancer cell line.

3.1.1 The SOFA-HDV Ribozyme Strategy

1. An expression vector containing the $tRNA^{Val}$ promoter and the hygromycin resistance gene was used to clone PACE4 selective ribozyme [4].
2. Transfection of DU145 cells is usually done at around 75 % confluence in 100-mm dish using the Lipofectamine 2000 transfection reagent. Cells are maintained in RPMI 1640 growth medium supplemented with 5 % FBS at 37 °C/5 % CO_2. The transfection mixture is prepared by mixing 10 μg of plasmid with Opti-MEM growth medium to a final volume of 100 μL. A transfection control without plasmid should be included in the experiment. In a second tube, 20 μL of transfection reagent is combined to 80 μL of Opti-MEM at room temperature for 5 min. The mixture containing plasmid DNA is then slowly incorporated into the transfection reagent mixture. After an incubation period of 45 min at room temperature, 800 μL of Opti-MEM is slowly added, and the transfection mixture is carefully distributed to the DU145 cell in a 100-mm dish previously washed twice with 5 mL PBS and containing 3 mL of Opti-MEM. After 5-h incubation in cell incubator, discard the transfection mixture and replace with RPMI 1640/5 % FBS.
3. The day after transfection, the cells are passaged at 1:10 and 1:5 split, each with two dishes, in complete growth media.
4. The next day, the selection agent hygromycin is added to a final concentration of 200 μg/mL. This concentration should be determined for each cell line used (*see* **Note 2**). Growth medium with selection agent should be replaced every 48 h. Using the transfection control as a reference, all untransfected cells should be dead within 5 days.
5. The surviving clones are allowed to grow until they form colonies big enough to be picked up using a pipet tip with 5 μL of PBS and transferred into a 96-well culture plate for further growth.
6. Freeze the stable cell clonal population as soon as possible. In addition, produce and keep to −80 °C cell pellets for further characterization.

3.1.2 The shRNA Strategy

As we sought for a stable knockdown of the targeted mRNA, we chose a strategy where shRNAs were cloned into the pLKO.1-puro vector [21]. This vector allows the transient or the stable (using the selection agent puromycin) expression of shRNAs as well as the production of lentiviral particles. Lentiviral-based particles allow efficient infection and integration of shRNA-containing vectors into a large spectrum of dividing and nondividing cells.

1. Lentiviral particle production is done by co-transfecting HEK293FT cells with the three packaging vectors (vsvg, pLp1, pLp2) and the shRNA expression vector pLKO.1-puro (Fig. 1) [22, 23]. Seed 5×10^6 freshly unfrozen (*see* **Note 3**) HEK293FT cells in 100 mm cell culture petri dishes with DMEM supplemented with 10 % FBS and incubate for 24 h at 37 °C/5 % CO_2.
2. Prepare the transfection mixture by diluting 6 μg of each plasmid (24 μg total DNA) with 1.5 mL of Opti-MEM growth medium. Various control lentiviral particles may be used such as empty pLKO.1-puro vector (i.e., without shRNA insert) or containing a nontarget shRNA. Separately, dilute 48 μL of transfection reagent with 1.5 mL of Opti-MEM and incubate for 5 min at room temperature. Gently transfer the DNA mixture to the transfection reagent and incubate for 30 min at room temperature. Discard the HEK293FT cell growth medium, wash twice with PBS, and add 5 mL of Opti-MEM to each dish. Add the 3 mL DNA–Lipofectamine2000 mixture carefully to the dish and incubate for 5 h in CO_2 incubator. Discard the transfection mixture and replace with DMEM/10 % FBS to start lentiviral particle production.
3. After a 48-h production period (*see* **Note 4**), collect the cell growth medium, filter using a 0.45 μM syringe unit to remove any cell debris, aliquot in multiple tubes (*see* **Note 5**), and store at −80 °C.
4. As the amount of lentiviral particles used to perform the infection may influence the final result (*see* **Note 6**), it is important to perform a titration for each virus production. We usually seed 40,000 of HT-1080 cells in 6-well plates in EMEM/10 % FBS (*see* **Note 7**) and incubate for 24 h at 37 °C/5 % CO_2.
5. Dilute the lentiviral preparation by tenfold serial dilutions, starting at 1:100 to 1:10^6 using complete growth medium to a final volume of 1 mL. Add polybrene to each tube to a final concentration of 6 μg/μL (*see* **Note 8**). Apply 1 mL of each dilution on cells, except for a control well where only complete medium should be added. After an overnight incubation at 37 °C/5 % CO_2, discard media containing virus dilutions and replace with complete media. The next day, add the selection agent puromycin to a final concentration of 10 μg/mL (*see* **Note 2**). Keep culturing the cells for 10–12 days by replacing

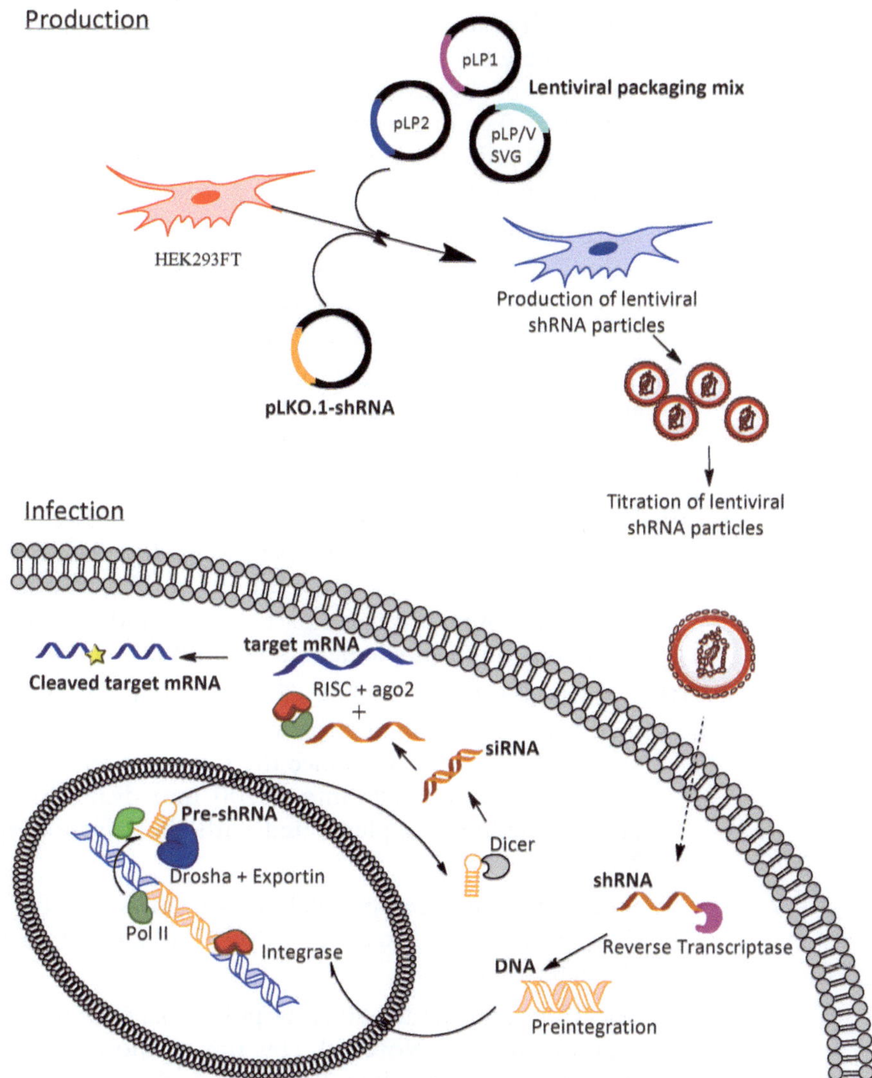

Fig. 1 *Production of shRNA lentiviral particles.* Lentiviruses are produced by co-transfecting the shRNA-containing expression vector pLKO.1 with the packaging plasmids (pLP1, pLP2, and pLP/VSVG) into the producing cell line HEK293FT. Upon titration, those replicative-incompetent lentiviral particles are used to deliver genetic material into a wide range of cell line. The infected cells are selected and maintained with the selective agent puromycin. Upon characterization (e.g., quantitative PCR, Western blots), knockdown cell lines issued from intracellular shRNA expression can be used for further proliferation studies

growth media on a 2-day basis. All cells in the control well should be dead after 5 days, while growing colonies should be seen in the others.

6. Once the colonies have reached a size big enough to be seen visually (*see* **Note 9**), discard the growth media, and wash carefully twice with PBS. Stain the colonies for 10 min with the methylene blue solution. Remove staining solution and gently

wash with water until washing water is no longer stained in blue. Where possible, count the separate colonies to determine the viral titer by dividing the number of colony-forming unit (CFU) by the corresponding dilution factor, e.g., 50 colonies in the 10^{-5} dilution = 5×10^6 CFU/mL. Calculate the mean with every counted well.

7. Once the viral titer is determined, the infection of DU145 cells is done at around 50 % confluence in 6-well plate. In addition to the shRNA targeting PACE4, nontarget shRNA, empty vector, and/or wild-type cells may be used as controls.
8. Thaw the lentiviral particle aliquots rapidly using a water bath at 37 °C. Mix the corresponding amount of viral stocks with RPMI 1640 medium with FBS to a final volume of 1 mL containing 4 μg/mL polybrene.
9. Discard the complete growth media of DU145 cells, and carefully transfer the viral mixtures in wells. Incubate overnight at 37 °C/5 % CO_2 for the virus absorption. The next day, remove the media containing virus particles and add to 2 mL of fresh growth medium supplemented with FBS.
10. Incubate for 2 days at 37 °C/5 % CO_2, and replace cell growth media with fresh complete growth medium supplemented with puromycin (2 μg/mL). Once the surviving cells reach confluence, transfer the cells into a 100 mm dish with complete growth medium supplemented with the selection agent (*see* **Note 10**).
11. Freeze the stable cell population as soon as possible. In addition, produce and keep to −80 °C cell pellets for further characterization.
12. Select the appropriate clonal or polyclonal cell lines for further experiments (*see* **Note 11**) by the method of your choice (western/northern blots, RT-qPCR).

3.2 In Vitro Cell Proliferation Assays

Assessing the importance of the target gene for the sustained growth of cancer cells can be carried out using two different colorimetric dyes: the MTT and XTT tetrazolium salts. Both assays are based on the same general principle where reduction by mitochondrial/membrane enzymes generates a colorimetric signal directly proportional to the number of cells [24]. These two assays are nonradioactive, reproducible, sensitive, and suitable for high throughput. As a major difference, the XTT assay produces a soluble formazan compound, while the derivative produced in the MTT assay is non-soluble and thereby necessitates a media removal/solubilization step that may influence its accuracy and reproducibility. Table 2 summarizes the advantages and drawbacks for both methods.

Table 2
Comparisons of tetrazolium salt MTT and XTT methods to assess cellular proliferation in vitro

MTT		XTT	
Pros	**Cons**	**Pros**	**Cons**
Lower cost	Less sensitive	Higher sensitivity	Requires a reagent mixing/heating step before each use
No background	Requires medium removal and formazan crystal solubilization	Higher dynamic range	Very susceptible to variation in incubation time
	Not suitable for low adhesive/suspension cells	High-throughput compatible	Requires background subtraction as a blank

1. Seed DU145 cells at 2,000 cells/well for each knockdown and control cell lines in 4 different 96-well plates with complete growth media (*see* **Notes 12** and **13**). If XTT technique is chosen, fill up four additional wells with only the complete medium (blank). Prepare one plate for each day-point scheduled and incubate at 37 °C/5 % CO_2.
2. Each day, take out a plate and add either 25 μL of a sterile MTT solution (5 mg/mL) to each well or 50 μL of a freshly heated and prepared XTT-electron coupling reagent (ECR) mix (*see* **Note 14**).
3. Incubate at 37 °C in 5 % CO_2 incubator for precisely 4 h (*see* **Note 15**).
4. Depending on the technique used:
   (a) XTT-treated plates are read directly at 490 nm with a reference at 690 nm using a microplate reader. Remove the absorbance of blank wells as FBS in media may cause a background reduction.
   (b) For MTT-treated plate, carefully discard cell media taking care of leaving the cell attached to the plate by gently pipetting back as much liquid as possible. Perform the formazan solubilization step by adding 100 μL of isopropanol:HCl 1 N (1:24) solution to each well, and gently shake the plate until uniform purple coloration appears. Read the absorbance at 570 and 650 nm (reference wavelength).
5. Draw the growth curves by setting the absorbance values of each cell line at the first time-point (24 h) as a reference (Fig. 2).

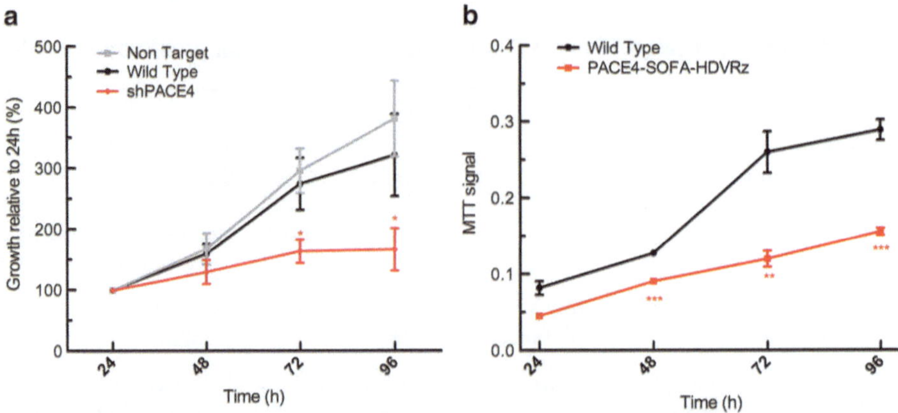

Fig. 2 In vitro proliferation assays. (**a**) XTT and (**b**) MTT time-course assay pointing out the implication of PACE4 on DU145 cell proliferation. Both tetrazolium salts and knockdown techniques were used to assess the reproducibility of the observed phenotype

3.3 In Vivo Subcutaneous Tumor Xenografts

In vivo tumor xenograft studies are a much more representative way to assess the proliferative capabilities of the knockdown cells as in these circumstances, the cells have to demonstrate their tumorigenic properties in a poorer growth environment than the FBS-supplemented culture media. For example, tumor cells will have to adhere to an unfamiliar extracellular matrix, generate their own auto- and paracrine factors to reinitiate proliferation, and recruit mouse vascularization in order to receive sufficient amount of nutrients. Gene knockdown affecting any of these parameters might be observable only in in vivo conditions as growth in controlled conditions with constant supply of fresh culture medium might mask one of those multiple facets of proliferative properties.

1. Using 150 mm petri dishes, scale up each cell line (*see* **Notes 16–18**).
2. Once cells have grown, collect and count all the cell lines in the same day.
3. Pellet the cells needed in an amount exceeding the total number of tumors scheduled. Carefully resuspend in PBS (*see* **Note 19**) to the desired concentration (e.g., 2×10^7 cells/mL for DU145). As cell pellets might constitute themselves a significant volume, carefully adjust the final volumes to avoid any difference between groups.
4. Anesthetize the mice using an isoflurane:oxygen mixture while homogenizing the cell suspensions frequently to avoid sedimentation.
5. Using disposable insulin syringes 28G or bigger (as their dead volume is negligible), slowly inject 100 μL of cell suspension

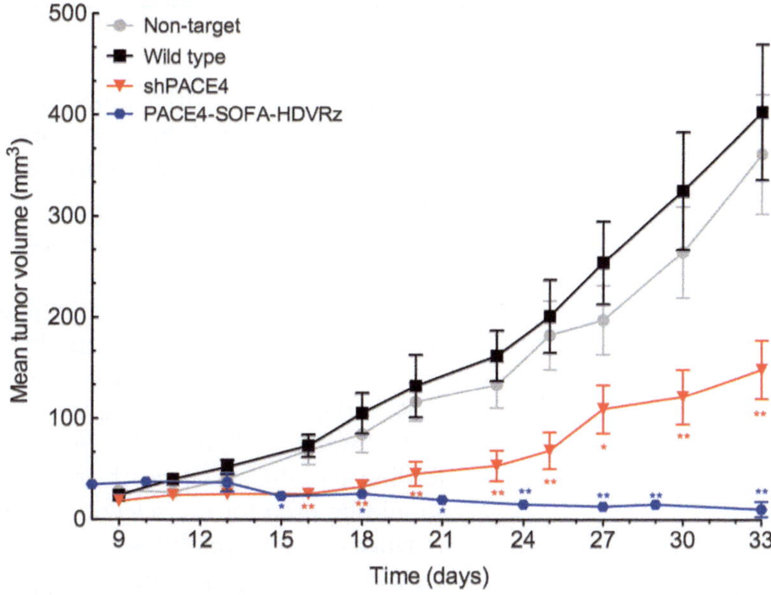

Fig. 3 In vivo tumor xenograft growth curves. Plot of tumor xenograft volumes showing that both polyclonal shRNA (*red*)- and monoclonal SOFA-HDVRz (*blue*)-mediated knockdowns of PACE4 lead to slower growing tumors relative to control cells (wild-type, *black*; nontarget shRNA, *grey*). The monoclonal cell line expressing the PACE4-SOFA-HDVRz leads however to more consistent and uniform tumor volumes, whereas the polyclonal cell lines have a trend towards growth reestablishment, which could be attributable to the heterogeneity of the cell population

(e.g., 2×10^6 cells/injection for DU145) subcutaneously on the shoulders and flanks of each mouse.

6. Allow at least 7 days to form solid and measurable tumors (*see* **Note 20**).
7. Every 2–3 days, measure the length (L) and width (W) of each tumor with a digital caliper. To calculate the tumor volumes, use the following formula: $V = (L \times W^2) \times \pi/6$. Use the L value as the longest measurement.
8. Report the mean volumes for each group on a growth curve (Fig. 3; *see* **Note 21**).
9. When the tumors have reached the maximum volume prescribed by your approved protocol, sacrifice the mice (*see* **Note 22**).
10. Tumors may be excised and weighed, frozen or fixed for further characterization such as western blotting or immunohistochemistry after tumor paraffin embedment [4].

4 Notes

1. Manufacturers now offer multiple pre-cloned sets of shRNAs targeting over 22,500 human genes. The individual clones are available in frozen bacteria, purified plasmid DNA, or lentiviral transduction particle formats.
2. To determine the optimal selection agent concentration for a specific cell line, treat untransfected cells with a series of diluted selection agent (0–200 and 0–20 μg/μL for hygromycine and puromycine, respectively). Plate the cells to a density reaching 75–80 % confluence the next day. Culture for 10–14 days, replace the media containing the selection agent every 3 days, and examine viability daily. Select the lowest concentration leading to 100 % cell death in 3–5 days. This concentration should be used for the selection of transfected cells and maintenance of clonal cell lines.
3. Using cells having less than 20 passages ensures a more efficient lentiviral production.
4. Dead cells floating in growth media may be found during viral particle production.
5. Storing virus in small aliquots (e.g., 500–700 μL) to avoid repeated freeze/thaw cycles ensures maximum performance.
6. For lentiviral infection, the amount of virus per cells is named "multiplicity of infection" (MOI). It is noteworthy to consider the MOI during transduction as it can directly influence the knockdown results (interferon response, absence of knockdown, high mortality rate during selection, etc.). For this reason, it is important to determine the titer of every virus lots. Additionally, performing this step ensures a quality control of the virus produced.
7. For this step, any cell line may be used. For comparison purposes, we use in our laboratory the HT-1080 cells as a standard to determine our viral titers. As usual, it is important to first determine the optimal concentration of the selection agent used (*see* **Note 2**).
8. Polybrene is a cationic polymeric compound that improves retroviral infection by neutralizing the cell surface charges.
9. To facilitate their counting, colonies should be stained before they start fusing with adjacent ones.
10. At this point, if clonal cell lines are generated, dilute the infected cell 1:10, 1:50, and 1:100 in 100 mm dishes to obtain distinct colonies. In our case, we will work with a polyclonal cell population.
11. Usually, expression levels of the targeted gene should be at least 70 % lower than those measured in control cells (Fig. 4a).

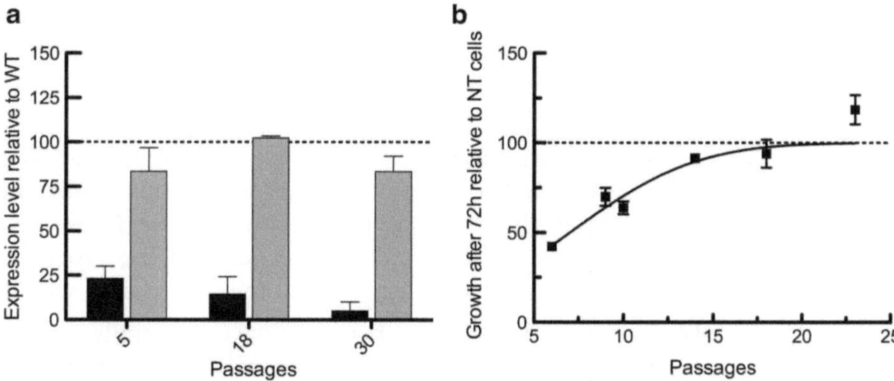

Fig. 4 Knockdown and phenotype stabilities. (**a**) Quantification of PACE4 mRNA level in DU145/shPACE4 (*black*) and nontarget shRNA DU145 (*grey*) relative to wild-type (WT) DU145 (100 %). As observed, the knockdown levels of the polyclonal cell population are stable along with passages (1:5 splitting every 3–4 days). (**b**) Comparison of the DU145/shPACE4 proliferation rates relative to the nontarget (NT) shRNA DU145 cells after 72 h (assessed by XTT assay). As illustrated, the lower proliferation phenotype of the PACE4 knockdown cells is almost completely lost after 15 passages, although PACE4 mRNA levels are still reduced (panel **a**). Thus, the polyclonal cell population seems to adapt its proliferation rates by other mechanisms than those implicating PACE4 expression

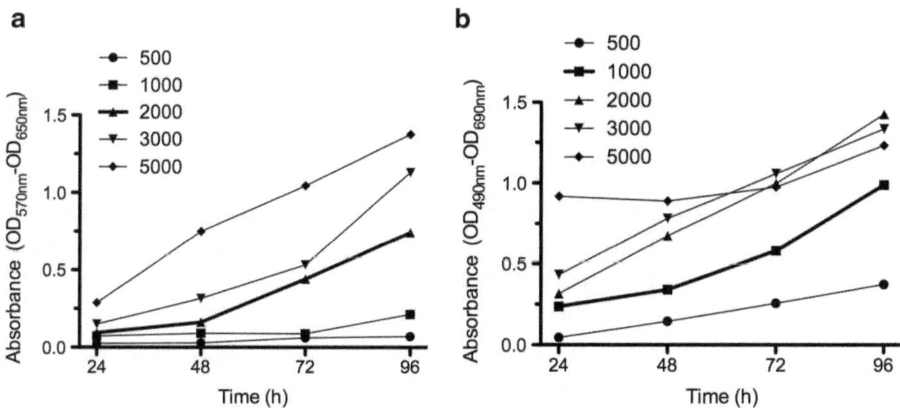

Fig. 5 Comparison of MTT and XTT tetrazolium salt methods. (**a**) MTT and (**b**) XTT time-course assays with different starting number of DU145 cells were performed to set up the optimal conditions for these assays. Data sets in bold show the optimal conditions selected for further assays. These conditions result in growth curves with absorbance units between 0.1 and 1, which is the linear range of the assay. As stated in Table 2, the XTT method is more sensitive than the MTT assay

12. For this assay, use cells with the lowest number of passages, as long-term culturing might affect cell phenotypes, even if knockdown remains unchanged (Fig. 4b).
13. Cell density has to be low enough to avoid formation of a monolayer after 96 h but high enough to have a signal after 24 h. For new cell lines, a setup experiment should be performed to determine the optimal number of cells to seed (Fig. 5).

14. The XTT working solution has to be prepared daily. First, heat both solutions (XTT and ECR) at 37 °C while protecting from light. Mix together in a 50:1 ratio and immediately apply 50 μL per well. Discard any remaining solution.
15. For cells with lower metabolic activity, the incubation time can be extended to obtain a better signal; this should be established during the setup experiments (*see* **Note 13**). Once determined, it is mandatory to perform the experiment with the same incubation time for all the time points scheduled.
16. To maximize the experiment, we generally implant four tumors/animal. The number of cells per tumor ranges from 2 to 5×10^6 cells depending on the growth capabilities. For DU145, 2×10^6 cells/tumor are injected. To reach statistical significance, 16–20 tumors are generally required.
17. Sub-confluent cells (i.e., in exponentially growing phase) are used to improve tumor formation rates and shorten latency phase once implanted.
18. As knockdown cells might grow slower than controls, it is important to adjust the number of petri dishes to ensure that the correct amount of each cell line is available on the day of injection.
19. DU145 are usually implanted in PBS or culture medium. However, other cell lines (e.g., LNCaP) might require solubilized basement membrane preparation (e.g., BD Matrigel) in a 50:50 (Matrigel:PBS) ratio to a final injection volume of 200 μL.
20. Depending on the cells, solid tumors may form after more than 20 days. Monitoring the tumor aspects routinely during this period is thereby critical. Tumors become easy to measure once they have reached about 3×3 mm.
21. Comparing tumor growth curves is a good way to compare knockdown cell lines to controls. To avoid bias, only measurable tumors should be included in the graph, as those unable to form tumor could alter curve interpretation. However, tumors that regress during the experiment should be included even if they reach unmeasurable sizes as they are relevant for comparison. Thus, the implantation rate for a particular cell line should, if available, be considered at the beginning of the experiment to ensure that a sufficient number of tumors will be available for measurements and statistical analysis.
22. As tumors might not form clear spherical masses, the caliper measurements may not reflect the real tumor volumes. In such cases, weighting tumors at the end of the experiment may increase confidence in the potential disparities observed between groups.

Acknowledgments

This work is funded by grants from the Canadian Institutes of Health Research (CIHR) and the Ministère du Développement Économique de l'Innovation et de l'Exportation (MDEIE) to RD. RD is a member of the Centre de Recherche Clinique Etienne-Le Bel (Sherbrooke, QC, Canada). FC holds a Graduate Scholarship from the Fonds de la Recherche Santé Québec (FRSQ).

References

1. Hanahan D, Weinberg RA (2011) Hallmarks of cancer: the next generation. Cell 144: 646–674
2. Berridge MV, Herst PM, Tan AS (2005) Tetrazolium dyes as tools in cell biology: new insights into their cellular reduction. Biotechnol Annu Rev 11:127–152
3. Cespedes MV, Casanova I, Parreno M, Mangues R (2006) Mouse models in oncogenesis and cancer therapy. Clin Transl Oncol 8: 318–329
4. D'Anjou F, Routhier S, Perreault JP, Latil A, Bonnel D, Fournier I, Salzet M, Day R (2011) Molecular validation of PACE4 as a target in prostate cancer. Transl Oncol 4:157–172
5. Seidah NG, Prat A (2012) The biology and therapeutic targeting of the proprotein convertases. Nat Rev Drug Discov 11:367–383
6. Seidah NG, Chretien M (1999) Proprotein and prohormone convertases: a family of subtilases generating diverse bioactive polypeptides. Brain Res 848:45 62
7. Couture F, D'Anjou F, Day R (2011) On the cutting edge of proprotein convertase pharmacology: from molecular concepts to clinical applications. Biomol Concepts 2:421–438
8. Becker GL, Lu Y, Hardes K, Strehlow B, Levesque C, Lindberg I, Sandvig K, Bakowsky U, Day R, Garten W, Steinmetzer T (2012) Highly potent inhibitors of the proprotein convertase furin as potential drugs for the treatment of infectious diseases. J Biol Chem 287(26):21992–22003
9. Fugere M, Day R (2005) Cutting back on proprotein convertases: the latest approaches to pharmacological inhibition. Trends Pharmacol Sci 26:294–301
10. D'Anjou F, Bergeron LJ, Larbi NB, Fournier I, Salzet M, Perreault JP, Day R (2004) Silencing of SPC2 expression using an engineered delta ribozyme in the mouse betaTC-3 endocrine cell line. J Biol Chem 279:14232–14239
11. Gupta N, Fisker N, Asselin MC, Lindholm M, Rosenbohm C, Orum H, Elmen J, Seidah NG, Straarup EM (2010) A locked nucleic acid antisense oligonucleotide (LNA) silences PCSK9 and enhances LDLR expression in vitro and in vivo. PLoS One 5:e10682
12. Scamuffa N, Siegfried G, Bontemps Y, Ma L, Basak A, Cherel G, Calvo F, Seidah NG, Khatib AM (2008) Selective inhibition of proprotein convertases represses the metastatic potential of human colorectal tumor cells. J Clin Invest 118:352–363
13. Yuasa K, Masuda T, Yoshikawa C, Nagahama M, Matsuda Y, Tsuji A (2009) Subtilisin-like proprotein convertase PACE4 is required for skeletal muscle differentiation. J Biochem 146:407–415
14. Khatib AM, Siegfried G, Chretien M, Metrakos P, Seidah NG (2002) Proprotein convertases in tumor progression and malignancy: novel targets in cancer therapy. Am J Pathol 160:1921–1935
15. Bergeron LJ, Reymond C, Perreault JP (2005) Functional characterization of the SOFA delta ribozyme. RNA 11:1858–1868
16. Bergeron LJ, Perreault JP (2005) Target-dependent on/off switch increases ribozyme fidelity. Nucleic Acids Res 33:1240–1248
17. Stone KR, Mickey DD, Wunderli H, Mickey GH, Paulson DF (1978) Isolation of a human prostate carcinoma cell line (DU 145). Int J Cancer 21:274–281
18. Levesque MV, Perreault JP (2012) Target-induced SOFA-HDV ribozyme. Methods Mol Biol 848:369–384
19. Levesque MV, Rouleau SG, Perreault JP (2011) Selection of the most potent specific on/off adaptor-hepatitis delta virus ribozymes for use in gene targeting. Nucleic Acid Ther 21:241–252
20. Moffat J, Grueneberg DA, Yang X, Kim SY, Kloepfer AM, Hinkle G, Piqani B, Eisenhaure TM, Luo B, Grenier JK, Carpenter AE, Foo SY, Stewart SA, Stockwell BR, Hacohen N, Hahn WC, Lander ES, Sabatini DM, Root DE (2006) A lentiviral RNAi library for human and mouse genes applied to an arrayed viral high-content screen. Cell 124:1283–1298

21. Stewart SA, Dykxhoorn DM, Palliser D, Mizuno H, Yu EY, An DS, Sabatini DM, Chen IS, Hahn WC, Sharp PA, Weinberg RA, Novina CD (2003) Lentivirus-delivered stable gene silencing by RNAi in primary cells. RNA 9:493–501
22. Zufferey R, Nagy D, Mandel RJ, Naldini L, Trono D (1997) Multiply attenuated lentiviral vector achieves efficient gene delivery in vivo. Nat Biotechnol 15:871–875
23. Zufferey R, Dull T, Mandel RJ, Bukovsky A, Quiroz D, Naldini L, Trono D (1998) Self-inactivating lentivirus vector for safe and efficient in vivo gene delivery. J Virol 72:9873–9880
24. Roehm NW, Rodgers GH, Hatfield SM, Glasebrook AL (1991) An improved colorimetric assay for cell proliferation and viability utilizing the tetrazolium salt XTT. J Immunol Methods 142:257–265
25. Asif-Ullah M, Levesque M, Robichaud G, Perreault JP (2007) Development of ribozyme-based gene-inactivations; the example of the hepatitis delta virus ribozyme. Curr Gene Ther 7:205–216
26. Akashi H, Matsumoto S, Taira K (2005) Gene discovery by ribozyme and siRNA libraries. Nat Rev Mol Cell Biol 6:413–422
27. Paddison PJ (2008) RNA interference in mammalian cell systems. Curr Top Microbiol Immunol 320:1–19
28. Sliva K, Schnierle BS (2010) Selective gene silencing by viral delivery of short hairpin RNA. Virol J 7:248

Chapter 7

Use of Tumor-Targeting *Trans*-Splicing Ribozyme for Cancer Treatment

Seong-Wook Lee and Jin-Sook Jeong

Abstract

One of the major concerns with regard to successful cancer gene therapy is to enhance both efficacy and safety. Gene targeting may represent an attractive tool to combat cancer cells without damage to normal cells. Here, we introduce a tumor-targeting approach with the *Tetrahymena* group I intron-based *trans*-splicing ribozyme, which cleaves target RNA and *trans*-ligate an exon tagged at the end of the ribozyme onto the downstream U nucleotide of the cleaved target RNA. We develop a specific *trans*-splicing ribozyme that can target and reprogram human cytoskeleton-associate protein 2 (hCKAP2)-encoding RNA to trigger therapeutic transgene herpes simplex virus thymidine kinase (HSVtk) selectively in cancer cells that express the RNA. Adenoviral vectors encoding the hCKAP2-specific *trans*-splicing ribozyme are constructed for in vivo delivery into either subcutaneous tumor xenograft or orthotopically multifocal hepatocarcinoma. We present analyses of the efficacy of the recombinant adenoviral vectors in terms of cancer retardation, target RNA and cell specificity, and in vivo toxicity.

Key words Ribozyme, *Trans*-splicing, Group I Intron, RNA replacement, Adenovirus, Cancer gene therapy

1 Introduction

Gene targeting may represent an attractive approach to the development of strategy for the efficient, specific, and safe treatment of cancer [1]. Three major approaches have been employed to achieve that end: transcriptional or posttranscriptional targeting, transductional targeting, and utilization of cancer-associated cellular pathways [2]. However, the ideal of cancer cell-specific targeting without compromise of anticancer effectiveness remains a significant challenge [3]. In this study, we introduce a novel tumor-targeting gene therapy tool based on group I intron-derived *trans*-splicing ribozyme, and present an evaluation of the inhibitory effects of a recombinant adenovirus encoding the ribozyme on tumor growth.

Daniel Lafontaine and Audrey Dubé (eds.), *Therapeutic Applications of Ribozymes and Riboswitches: Methods and Protocols*, vol. 1103, DOI 10.1007/978-1-62703-730-3_7,

Self-splicing group I intron from *Tetrahymena thermophila* harbors catalytic *trans*-splicing activity and can mediate cleavage of a target RNA and *trans*-ligation of an exon attached at the 3′ end of the intron onto the downstream U nucleotide (nt) of the cleaved target RNA not only in the test tube but also in bacteria and mammalian cells [4–6]. Therefore, the group I intron-based *trans*-splicing ribozyme can be designed to replace a specific disease-associated or causative RNA with transgene transcript conferring a therapeutic effect, selectively in cells expressing the target RNA. The *trans*-splicing ribozyme thus provide an attractive gene therapy tool against diverse human diseases, including genetic or infectious diseases [7–14].

Moreover, we have demonstrated that the *trans*-splicing ribozyme could be a potent anticancer agent through cancer-specific transcript targeting and RNA replacement to trigger anticancer gene activity (Fig. 1). We validated the potential of this approach by developing *trans*-splicing ribozymes specifically targeting and replacing tumor-specific transcripts including RNA of human telomerase reverse transcriptase (hTERT), CEA, AFP, AIMP2-DX2, and hCKAP2 to induce transgene activity specifically and selectively in the target RNA-expressing cancer cells [15–22]. A major advantage of RNA replacement approach is that by attacking tumor-specific transcripts, targeted expression of the therapeutic gene product is generated with a simultaneous reduction in the expression of the targeted RNAs, resulting in additive or synergistic anticancer effects. Of note, hTERT- and hCKAP2-specific ribozymes specifically and efficiently retard tumor growth not only in tissue cultures but also in animals with tumor xenografts with minimal hepatotoxicity [16–18, 22]. Here, we present the experimental procedures to construct hCKAP2-targeting *trans*-splicing ribozyme and evaluate the effects of the ribozyme as anticancer agent.

2 Materials

2.1 hCKAP2 Ribozyme Design

1. RNA mapping library construction: *Trans*-splicing ribozyme library is generated by PCR amplification of plasmid pT7L-21, which encodes a slightly shortened version of the *Tetrahymena* group I intron [5], with a 5′ primer containing a randomized sequence at positions corresponding to the ribozyme's internal guide sequence (IGS) (5′-GGG GGG ATC CTA ATA CGA CTC ACT ATA GNN NNN AAA AGT TAT CAG GCA TGC ACC-3′, where N represents equal amounts of the 4 nt G, A, C, or T) and 3′ primer specific for 3′ exon tag sequences present in the pT7L-21 plasmid (5′-AGT AGT CTT ACT GCA GGG GCC TCT TCG CTA TTA CG-3′). The resulting cDNA library is in vitro transcribed [40 mM Tris–HCl pH 7.5, 5 mM $MgCl_2$, 10 mM DTT, 4 mM spermidine, 80 U RNase inhibitor

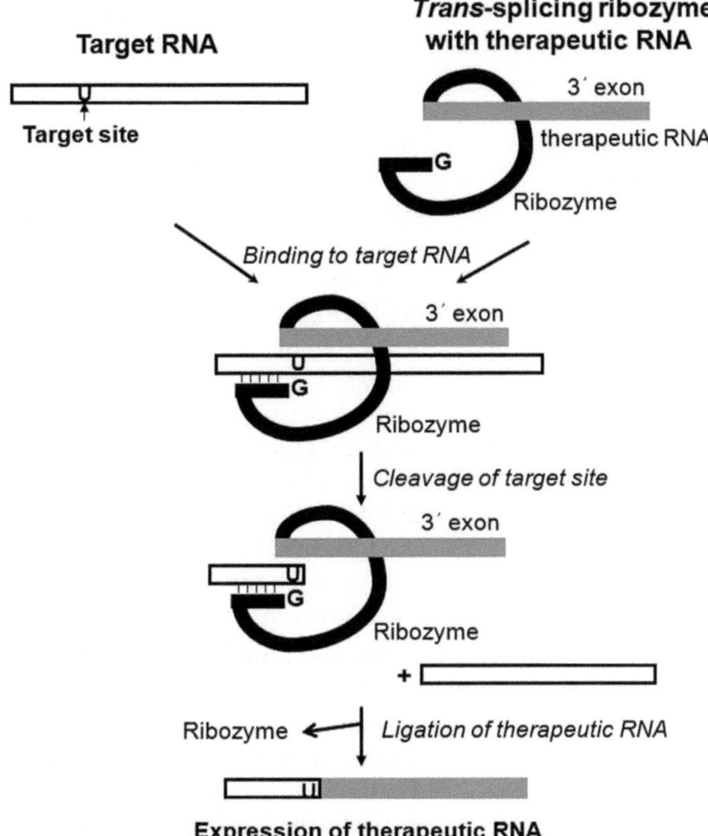

Fig. 1 Scheme for ribozyme-triggered selective expression of therapeutic RNA by targeted *trans*-splicing of cancer-specific RNA. A ribozyme can recognize the target RNA at any accessible uridine residue by base pairing to the sequence through its internal guide sequence. The ribozyme then removes the sequence downstream of the target site and replaces it with a 3′ exon that encodes a therapeutic RNA sequence

(Kosco), 1.5 mM NTP] using T7 RNA polymerase to construct the RNA mapping library (*see* **Note 1**).

2. hCKAP2 RNA construction: hCKAP2 RNA for substrate is generated by PCR amplification with plasmid mixtures of pEGFP-C2-hCKAP2 and hCKAP2-RACE-5′UTR that encode open reading frame and 5′ UTR to part of 3′exon sequence of hCKAP2 RNA, respectively, with primers (5′-CCG GAA TTC TAA TAC GAC TCA CTA TAG GGA GAC GGC AGC CGC GCC-3′, 5′-GCG GGA TCC ATT TTT GTT TTA AGT TTC AGC-3′) (*see* **Note 2**). The resulting of full-length hCKAP2 cDNA is in vitro transcribed [40 mM Tris–HCl pH 7.5, 1.5 mM $MgCl_2$, 10 mM DTT, 4 mM spermidine, 80 U RNase inhibitor (Kosco), 0.5 mM NTP] using T7 RNA polymerase for the substrate RNA.

3. *Trans*-splicing reaction buffer in test tube: 50 mM HEPES (pH 7.0), 150 mM NaCl, and 5 mM $MgCl_2$.

2.2 Adenovirus Vector Construction and Adenovirus Production

1. Vector construction: Expression vectors encoding for hCKAP2-specific *trans*-splicing ribozyme with herpes simplex virus thymidine kinase (HSVtk) gene at its 3′ exon under the control of cytomegalovirus immediate early (CMV) or phosphoenolpyruvate carboxykinase (PEPCK) promoter were constructed to generate recombinant adenoviral vector. To this end, CMV-hCKAP2Rib-TK or PEPCK-hCKAP2Rib-TK expression cassette in pCDNA3.1(+) or pPEPCK-LCR, respectively, is digested with *Spe*I and *Bst*BI restriction enzymes and then ligated with the linearized pAVQ shuttle vector (Qbiogene) with T4 DNA ligase. The cloned shuttle vector is digested with *Pme*I and co-transformed into bacteria (BJ5183) with an E1/E3-deleted adenoviral type5 backbone genome (pAdenoVator ΔE1/E3, Qbiogene). The resulting recombinant adenoviral vectors encoding the ribozyme are generated by in vivo homologous recombination procedure in the bacteria (Fig. 2).
2. Adenovirus amplification: The recombinant adenoviral vectors are linearized with *Pac*I and then transfected into HEK 293 helper cell, which is maintained in DMEM (high glucose with L-glutamine) with 10 % fetal bovine serum (FBS) and 1 % antibiotic/antimycotic (AA). The transfection into HEK 293 cell is performed using lipofectamine 2000 in Opti-MEM buffer (Invitrogen). The regenerated recombinant adenoviruses are purified through three rounds of plaque purification and amplified. To this end, when more than 90 % of transfected cells are detached from tissue culture plate, the cells and medium are harvested and freezed/thawed three times, and one the half of crude viral extract is infected into HEK 293 cells. The amplified adenoviruses are then purified and concentrated by CsCl gradient ultracentrifugation and saved in storage buffer after dialysis [10 % glycerol, 10 mM Tris–HCl (pH 7.5), 1 mM $MgCl_2$] (*see* **Note 3**). Titers of the recombinant adenovirus were then determined by Tissue Culture Infectious Dose for 50 % (TCID50) analysis.

2.3 In Vitro Activity of Adenoviral Vectors Encoding the hCKAP2-Specific Trans-Splicing Ribozyme

1. Cell proliferation assay: The cells in 96-well plates are infected with appropriate multiplicities of infection (MOI) of recombinant adenoviruses. 1 day post infection, ganciclovir (GCV, Roche) is added and the cells are further incubated. Cell viability is assessed by using tetrazolium compound MTS [3-(4,5-dimethylthiazol-2-yl)-5-(3-carboxymethoxyphenyl)-2-(4-sulfophenyl)-2*H*-tetrazolium, inner salt] (Promega). MTS is converted to colored formazan by NADH and NADPH and thus the reduced products are directly proportional to the

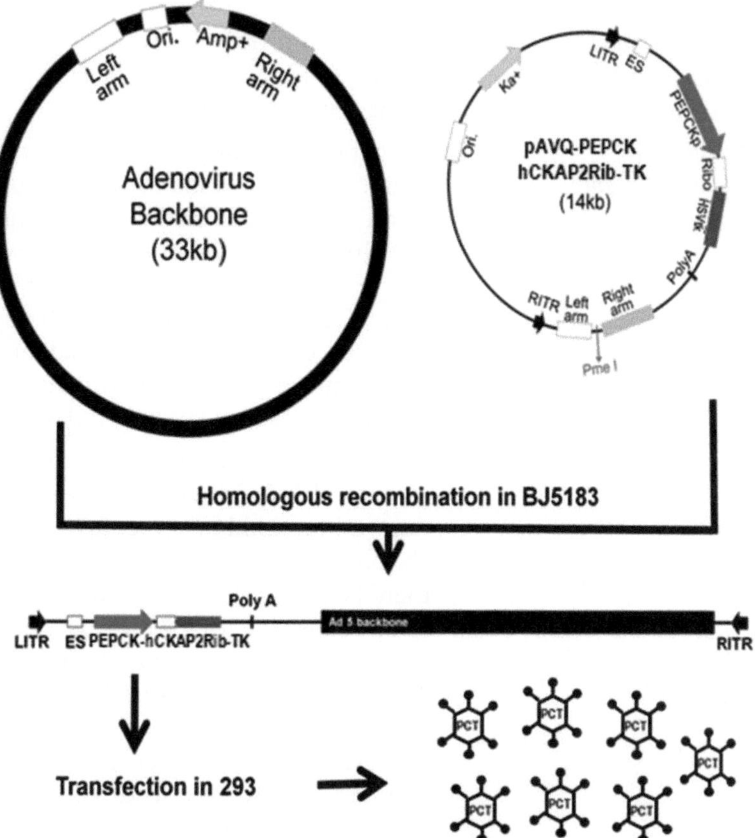

Fig. 2 Scheme for the generation of recombinant adenoviral vectors encoding the ribozyme. pAVQ-PEPCK-hCKAP2Rib-TK is linearized and cotransformed into bacteria (BJ5183) with E1/E3-deleted adenovirus backbone construct. The resulting recombinant adenoviral vectors are generated by in vivo homologous recombination procedure in the bacteria. The adenoviral vectors are linearized with *PacI* and transfected into HEK 293 helper cell. The regenerated recombinant adenoviruses (PCT) are purified, amplified, and concentrated

number of viable cells. The colored formazan is quantitated by determining absorbance at 490 nm.

2. PCR for RNA analysis: The cells are seeded in 100-mm plates, incubated for overnight, and then infected with the recombinant adenoviruses. Total RNAs are extracted from the cells infected with adenoviruses, isolated with Trizol (Invitrogen) supplemented with 20 mM EDTA, and reverse transcribed in the presence of 10 mM L-argininamide. L-Argininamide is used to quench the splicing reaction during the reverse-transcription reaction. Ribozyme RNA levels, *trans*-spliced RNA products, and the reduction level of the hCKAP2 RNA in the ribozyme-encoding adenovirus infected cells are assessed by PCR amplification of the cDNAs.

2.4 In Vivo Anticancer Activity of Adenoviral Vectors Encoding the hCKAP2-Specific Trans-Splicing Ribozyme

1. 4- to 5-week-old male Balb/cAnNCrj-nu/nu nude mice are used through this study. The animals are kept under specific pathogen-free conditions, acclimated to laboratory conditions for at least 1 week before use.
2. Subcutaneous tumor xenograft model: A mouse xenograft tumor model is established by subcutaneous inoculation of colon cancer HCT116 cells into the flanks of male nude mice (*see* **Note 4**). Tumor nodules achieve a size of 6–9 mm in diameter approximately 3 weeks after inoculation. Preestablished tumors are directly injected with appropriate dose of adenovirus. One day after the initial adenoviral injection, the animals are intraperitoneally treated daily with GCV (in PBS). Tumor sizes are measured every 2–3 day using digital caliper.
3. Multifocal hepatocarcinoma (HCC) model: A mouse model of multifocal HCC is established by splenic subcapsular inoculation of human HCC (Hep3B) cells. To this end, the left flank abdominal wall is linearly incised to visualize the spleen, and then Hep3B cells in 100 μL of PBS are injected under the spleen capsule with a 29-gauge needle. Adenoviruses prepared in PBS are injected through tail vein with a 29-gauge needle, followed by intraperitoneal administration of GCV (in PBS). On the next day after last GCV treatment, blood sampling is collected, the whole liver lobes are removed, measured, photographed, and serially sectioned, and the tumor weight is analyzed (*see* **Note 5**).

3 Methods

In this study, we describe the specific RNA-targeting *trans*-splicing ribozyme as a new cancer-targeting agent, which directs the expression of therapeutic genes selectively in the target RNA-expressing cancer cells through the targeted RNA replacement. As a working model, we introduce the methods for the development and assessment of hCKAP2-specific *trans*-splicing ribozyme as anticancer regimen.

Although any uridine residue in a target RNA can be theoretically targeted by *Tetrahymena trans*-splicing ribozyme through modification of nucleotide composition of 6-nt-long IGS on the ribozyme, only a limited number of uridines on the target RNA in cells will be accessible to the ribozyme due to the complex tertiary structure of the substrate RNA and/or plausible intracellular interaction with proteins [9]. Therefore, an RNA mapping strategy should be initially employed that is based on a *trans*-splicing ribozyme library and RNA tagging to determine which regions of the target transcript are most accessible to the ribozymes [6, 7]. The group I *trans*-spicing ribozymes with only a 6-nt-long IGS are relatively inefficient in their activity and specificity when expressed in mammalian cells [6]. Therefore, in order to construct ribozyme-expressing vectors, we modify a ribozyme, which targets the most accessible uridine on the target RNA, to contain an extension of the P1 helix, an addition of the P10 helix with/without an antisense sequence that is complementary to the downstream region of the targeted uridine of the target RNA.

The modified *trans*-spicing ribozyme derivative is further modified to specifically harbor the suicide HSVtk gene as its 3′ exon. HSVtk is shown to convert the protoxic nucleoside analog GCV into a highly genotoxic phosphorylated GCV that acts as a chain terminator of DNA synthesis during cellular replication, thereby selectively inducing cell death only in dividing cells such as are found in tumors. Therefore, when exposed to *trans*-splicing ribozyme with HSVtk and GCV prodrug, target RNA-expressing cancer cells are selectively regressed by specific replacement of target transcript with HSVtk transgene RNA. Additionally, both the ribozyme-exposed cells and neighboring adjacent cells can be killed in the presence of GCV due to the bystander effect, amplifying the impact of the ribozyme activity. For effective delivery into tumor tissues embedded in the animal xenograft model, the cDNA harboring the *trans*-spicing ribozyme tagged with HSVtk is inserted into replication-defective adenovirus [16, 22].

3.1 hCKAP2 Ribozyme Design

1. For in vitro mapping, the ribozyme library with randomized IGS (40 nM) is incubated with full-length hCKAP2 transcripts (200 nM) in a *trans*-splicing reaction buffer with guanosine (0.1 mM) at 37 °C for 3 h. For intracellular mapping, the ribozyme library (1 pmol) is cotransfected with the full-length hCKAP2 RNA (1 pmol) into non-hCKAP2-expressing NIH3T3 cells using 3 μL of DMRIE-C reagent (Invitrogen). The resulting *trans*-splicing products are reverse transcribed with primer specific for 3′ exon tagged at the 3′ end of the ribozyme (5′-ATG TGC TGC AAG GCG ATT-3′) and amplified with a 3′ primer specific for the ribozyme's 3′ exon (5′-TGT AAA ACG ACG GCC AGT G-3′) and a 5′ primer specific for the target hCKAP2 RNA (5′-CCG GAA TTC TAA TAC GAC TCA CTA TAG GGA GAC GGC AGC CGC GCC-3′). The amplified *trans*-spliced cDNA is cloned into pUC19 vector and sequenced to identify sequences around the spliced Us of the substrate RNA.
2. The accessible sites identified through the mapping analyses were validated by comparing the in vitro and intracellular *trans*-splicing activities of several ribozymes targeting different uridines in the hCKAP2 transcript. To this end, different ribozymes targeting uridines at positions 30, 73, 84, 115, 135, or 147 in the hCKAP2 RNA (Rib30-, Rib73-, Rib84-, Rib-115,

Rib-135, and Rib147-3′tag ribozymes, respectively) are constructed through exchange of original IGS of *Tetrahymena* group I intron with 5′-GAGCCG in Rib30, 5′-GGAACT in Rib73, 5′-GCCGCC in Rib84, 5′-GGCCTC in Rib115, 5′-GCGGCC in Rib135, or 5′-GGGTCC in Rib147. The deleted form in the catalytic center of each ribozyme is used as an inactive control. Each specific ribozyme is incubated with hCKAP2 RNA in a splicing buffer with guanosine and the amount of resulting *trans*-splicing products amplified by PCR is compared. Alternatively, CMV promoter-driven expression vector encoding each ribozyme with luciferase as 3′ exon is transfected into hCKAP2-expressing 293 cells and the reporter gene induction by the ribozyme is compared.

3. Enhanced ribozyme constructs are designed because group I *trans*-splicing ribozymes with only a 6-nt-long IGS are inactive with regards to specificity and efficacy when expressed in mammalian cells [6]. Each ribozyme is modified to contain an extension of the P1 helix and an addition of the 7-nt-long P10 helix with or without 100- or 300-nt-long antisense sequence complimentary to the downstream region of each targeted uridine of the hCKAP2 RNA [22]. To this end, hCKAP2Rib-F.Luc is constructed by modification of the Rib by insertion of synthesized complementary oligonucleotides containing an extended P1 plus a 7 nt-long P10 helix upstream to the ribozyme's IGS and introduction of firefly luciferase (F.Luc) cDNA as a 3′ exon of the ribozyme. The DNA fragment consisting of Rib sequence with the extended IGS and F.Luc cDNA is inserted between the *Hin*dIII and *Not*I sites of pcDNA3.1(+) encoding transgene under CMV promoter (Clontech) to create CMV-hCKAP2Rib-F.Luc. The CMV-hCKAP2RibAS100-F.Luc or CMV-hCKAP2RibAS300-F.Luc is constructed through the insertion of a 100- or 300-nt-long antisense sequence complementary to the downstream region of the targeted uridine of the hCKAP2 RNA into the *Hin*dIII site of CMV-hCKAP2Rib-F.Luc. CMV-hCKAP2Rib-TK and CMV-hCKAP2RibAS100-TK are created by replacement of F.Luc cDNA present in CMV-hCKAP2Rib-F.Luc and CMV-hCKAP2RibAS100-Fluc, respectively, with HSVtk cDNA. CMV-F.Luc and CMV-TK encoding the cDNA of F.Luc and the HSVtk gene, respectively, under the control of the CMV promoter are constructed as controls. PEPCK-hCKAP2Rib-TK encoding hCKAP2-specific ribozyme under the liver-specific PEPCK promoter is constructed by insertion of hCKAP2Rib-TK into pPEPCK-LCR (kindly gifted by Dr. K. Oka from Bayler College of Medicine).

3.2 Generation of Recombinant Adenoviruses Encoding Trans-Splicing Ribozyme

1. CMV-hCKAP2Rib-TK or PEPCK-hCKAP2Rib-TK expression cassette is digested with *Spe*I and *Bst*BI restriction enzymes and subcloned into pAVQ shuttle vector. The resulting pAVQ-CMV-hCKAP2Rib-TK or pAVQ-PEPCK-hCKAP2Rib-TK is digested with *Pme*I and transformed into bacteria (BJ5183 strain) with pAdenoVator ΔE1/E3 backbone vector that has the most of adenovirus genome with the deletion of E1 and E3. The transformants are selected with kanamycin and screened for the identification of clones with the recombinant adenoviral vectors encoding the hCKAP2-specific ribozyme that results from in vivo homologous recombination.
2. The recombinant adenoviral vectors are digested with *Pac*I and transfected into HEK 293 helper cell using lipofectamine 2000. One day before transfection, HEK293 cells are transferred into 6 cm plates containing DMEM media supplemented with 10 % FBS and 1 % AA. Eight hours after transfection, the cells are washed twice with DMEM/10 % FBS/1 % AA and overlaid with a 10 mL of 1:1 mixture of 2 % SeaPlaque agarose and 2 × MEM mix. Viral plaques that appear 7–10 days after transfection are picked and resuspended in 500 μL of DMEM. Adenoviruses are released from cells through 3 times of freeze/thaw and stored at −70 °C. Adenoviruses isolated by plaque assays are amplified in HEK293 cells and purified by double cesium chloride gradient ultracentrifugation. We designated the adenovirus products expressing hCKAP2-specific ribozymes with HSVtk under CMV or PEPCK promoter as Ad-CMV-hCKAP2Rib-TK (CCT) or Ad-PEPCK-hCKAP2Rib-TK (PCT), respectively (Fig. 2). As a control, we employed adenoviruses with HSVtk (Ad-CMV-TK or Ad-PEPCK-TK) driven by the CMV or PEPCK promoter, respectively, and Ad-Mock, which harbors only the adenoviral backbone.

3.3 In Vitro Activity of Adenoviral Vectors Encoding the hCKAP2-Specific Trans-Splicing Ribozyme

1. Cell proliferation is evaluated using MTS, with CellTiter 96 AQueous one solution reagent (Promega). The cells are seeded at 5×10^3 cells per well in 96-well plates and incubated overnight at 37 °C. Triplicate wells are then infected with varying MOI of recombinant adenoviruses. One day post infection, GCV (100 μM) is added and the cells are further incubated for 5 days. Alternatively, cells are infected with adenovirus (100 MOI), and varying doses of GCV are treated for 5 days 1 day post infection. After adenovirus and GCV treatment, one solution reagent is added into each well of 96-well plate and incubated for 2 h at 37 °C. Cell viability is then assessed by determination of absorbance at 490 nm. Cell viability after GCV treatment is quantitated as the fraction of the absorbance at cells without GCV treatment and then represented as the percentage relative to that of the mock-infected cells.

2. For the analysis of the ribozyme RNA levels, total RNAs are extracted from cells infected with recombinant adenoviruses, isolated with Trizol (Invitrogen) supplemented with 20 mM EDTA, and reverse transcribed with an oligo(dT) primer in the presence of 10 mM L-argininamide. The cDNA is then amplified with primers specific for 3′ exon HSVtk (5′-GCG AAC ATC TAC ACC ACA CA-3′ and 5′-AGT TAG CCT CCC CCA TCT C-3′). As the internal control, cDNAs are amplified with GAPDH specific primers (5′-TGA CAT CAA GAA GGT GGT GA-3′ and 5′-TCC ACC ACC CTG TTG CTG TA-3′). For the *trans*-spliced RNA products in cells, the total RNA is reverse transcribed with a primer specific for HSVtk (5′-CGG GAT CCT CAG TTA GCC TCC CCC AT-3′) in the presence of 10 mM L-argininamide, and the resulting cDNA is amplified with a 5′ primer specific to the 5′ end of the hCKAP2 RNA (5′-GGG AGA CGG CAG CCG CGC C-3′) and with the 3′ primer specific to the nested HSVtk sequence (5′-GTT ATC TGG GCG CTT GTC AA-3′), cloned, and sequenced. To assess the reduction level of the hCKAP2 RNA in the ribozyme-encoding adenovirus infected cells, hCKAP2 cDNA is amplified through real-time PCR from total RNA acquired from cells at 2 days post infection using primers (5′-CCC GGA ATT CTA ATA CGA CTC ACT ATA GGG AGA CGG CAG CCG CGC C-3′ and 5′-GCG GGA TCC ATT TTT GTT TTA AGT TTC AGC-3′). All reagents with the exception of Taq polymerase (Takara) are purchased from the SYBR-Green core reagent kit (Molecular Probes). The conditions for the PCRs are at 95 °C for 30 s, 58 °C for 30 s, and 72 °C for 30 s, for 30 cycles. For the standard curve in the reaction mix, 18S RNA is used. The threshold levels acquired from the hCKAP2 RNA are adjusted to those obtained in the 18S RNA reaction to revise for minor variations in cDNA loading.

3.4 In Vivo Anticancer Activity of Adenoviral Vectors Encoding the hCKAP2-Specific Trans-Splicing Ribozyme

1. A mouse xenograft tumor model is established by subcutaneous inoculation of HCT116 (5×10^6 cells) into the flanks of male nude mice. When tumor nodules reach a size of 6–9 mm in diameter, they are randomly assigned to treatment groups. Preestablished tumors are then directly injected with 1×10^7 plaque forming units (pfu) of adenoviruses. The injections are repeated twice more at 2 day intervals. One day after the initial adenoviral injection, the animals are intraperitoneally injected daily with 50 mg/kg GCV (in PBS) for 10 days. Tumor sizes are measured across two diameters of the implanted tumor every 2–3 day using digital caliper. Tumor volume is calculated as follows: (largest diameter × smallest diameter^2)/2.
2. Multifocal HCC model: A mouse model of multifocal HCC is established by splenic subcapsular inoculation of human HCC

(Hep3B) cells. A linear incision of the left flank abdominal wall is made to visualize the spleen, and 3×10^6 Hep3B cells in 100 μL of PBS are injected under the spleen capsule with a 29-gauge needle. After injection of cells, the injection site is pressed with an aseptic cotton sponge for several minutes to prevent further leakage, and the abdominal wall is then sutured with silk. The nude mice displayed multiple tiny tumor nodules along the liver margin, easily detectable by gross inspection, on the 11th or 12th day. On 12th day, adenoviruses prepared in 100 μL of PBS (0.25×10^9 or 1×10^9 pfu) are injected into the mice through tail vein, followed by intraperitoneal administration of GCV (50 mg/kg) twice per day for 10 days. On the next day after last GCV treatment, blood sampling through heart for liver enzyme [glutamic-oxaloacetic transaminase (GOT) and glutamic-pyruvic transaminase (GPT)] analysis is processed, and all mice are euthanized, the whole liver lobes are removed, measured, photographed, and serially sectioned in 2–3 mm thickness. Entire liver slices from each mouse are fixed in 10 % neutralized buffered formalin and processed for paraffin embedding. Tissue sections of 4- to 6-μm thickness are stained with hematoxylin and eosin (H&E) for morphologic examination. The microscopic images are scanned under the virtual microscope (Aperio Technologica). The tumor fraction is calculated from the program of Aperio Imagescope v10.2.2.2319, and then the tumor weight is estimated by multiplication of total liver mass and measured tumor fraction.

3. Statistical analysis is performed with Statistical Analysis System software (SAS Institute). The between-group differences are analyzed through assessment of variance (ANOVA). In the case of highly skewed distribution of measurements and small sample sizes, nonparametric statistical tests (Kruskal–Wallis test for overall comparison and Wilcoxon's rank-sum test for pairwise comparison) are used. All data are expressed as the average ± standard deviation. Differences are considered to be statistically significant at $p < 0.05$.

4 Notes

1. With the in vitro transcription condition, two major RNA bands can be synthesized (longer transcript, unspliced ribozyme-3′ exon chimeric transcript; shorter one, self-spliced ribozyme RNA). For the mapping study, transcribed RNAs are separated on a 5 % acrylamide gel with urea. Longer transcripts are then eluted using elution buffer (0.6 M ammonium acetate, 1 mM EDTA, 0.2 % SDS) at 37 °C for 3 h, extracted with phenol, and concentrated with ethanol.

2. For the proper RNA conformation, hCKAP2 RNA is preheated 37 °C for 5 min in the *trans*-splicing reaction buffer with GTP. Ribozyme RNA is heated at 50 °C for 5 min in the reaction buffer, incubated at 37 °C for 2 min, and then added into the preheated substrate RNA.
3. HEK 293 cells are cultured with 80 % confluency before infection. The cells come apart 2–3 days post infection depending on the concentration of adenoviral vectors. Sometimes, cells become so acidic without any evident cytopathic effect. Then, the cells are splitted and fed with fresh DMEM. The bulk of produced adenovirus is dialyzed using the dialysis bag of 30 kDa MW cut-off (PIERCE).
4. When cells are harvested from culture plate, minimum amount of trypsin/EDTA is utilized for avoiding the damage to cells. Cells for xenograft formation are counted using 0.8 mM trypan blue solution in PBS. In contrast dead cells stain blue due to trypan blue uptake, viable cells exclude trypan blue. The inoculation area of the mice is sterilized using ethanol.
5. Halothane is used for anesthesia during operation mouse, since inhalation of anesthesia is superior to other injectable forms of anesthesia in safety and efficacy.

Acknowledgement

This work was supported by grants from National Research Foundation of Korea funded by the Ministry of Science, ICT and Future Planning (2011-0002169, 2012R1A1A2A10039116, 2012M3A9B6055200) and a grant from National R&D Program for Cancer Control, Korean Ministry of Health & Welfare (0720520).

References

1. McCormick F (2001) Cancer gene therapy: fringe or cutting edge? Nat Rev Cancer 1:130–141
2. Wu L, Johnson M, Sato M (2003) Transcriptionally targeted gene therapy to detect and treat cancer. Trends Mol Med 9: 421–429
3. Tarner IH, Muller-Ladner U, Fathman CG (2004) Targeted gene therapy: frontiers in the development of 'smart drugs'. Trends Biotechnol 22:304–310
4. Been MD, Cech TR (1986) One binding site determines sequence specificity of Tetrahymena pre-rRNA self-splicing, *trans*-splicing, and RNA enzyme activity. Cell 47:207–216
5. Sullenger BA, Cech TR (1994) Ribozyme-mediated repair of defective mRNA by targeted, *trans*-splicing. Nature 371:619–622
6. Jones JT, Lee SW, Sullenger BA (1996) Tagging ribozyme reaction sites to follow *trans*-splicing in mammalian cells. Nat Med 2:643–648
7. Lan N, Howrey RP, Lee SW, Smith CA, Sullenger BA (1998) Ribozyme-mediated repair of sickle beta-globin mRNAs in erythrocyte precursors. Science 280:1593–1596
8. Phylactou LA, Darrah C, Wood MJ (1998) Ribozyme-mediated *trans*-splicing of a trinucleotide repeat. Nat Genet 18:378–381
9. Lan N, Rooney BL, Lee SW, Howrey RP, Smith CA, Sullenger BA (2000) Enhancing RNA repair efficiency by combining *trans*-splicing ribozymes that recognize different accessible sites on a target RNA. Mol Ther 2:245–255

10. Watanabe T, Sullenger BA (2000) Induction of wild-type p53 activity in human cancer cells by ribozymes that repair mutant p53 transcripts. Proc Natl Acad Sci U S A 97:8490–8494
11. Rogers CS, Vanoye CG, Sullenger BA, George AL Jr (2002) Functional repair of a mutant chloride channel using a *trans*-splicing ribozyme. J Clin Invest 110:1783–1789
12. Ryu KJ, Kim JH, Lee SW (2003) Ribozyme-mediated selective induction of new gene activity in hepatitis C virus internal ribosome entry site-expressing cells by targeted *trans*-splicing. Mol Ther 7:386–395
13. Shin KS, Sullenger BA, Lee SW (2004) Ribozyme-mediated induction of apoptosis in human cancer cells by targeted repair of mutant p53 RNA. Mol Ther 10:365–372
14. Kastanos E, Hjiantoniou E, Phylactou LA (2004) Restoration of protein synthesis in pancreatic cancer cells by *trans*-splicing ribozymes. Biochem Biophys Res Commun 322:930–934
15. Kwon BS, Jung HS, Song MS, Cho KS, Kim SC, Kimm K, Jeong JS, Kim IH, Lee SW (2005) Specific regression of human cancer cells by ribozyme-mediated targeted replacement of tumor-specific transcript. Mol Ther 12:824–834
16. Hong SH, Jeong JS, Lee YJ, Jung HI, Cho KS, Kim CM, Kwon BS, Sullenger BA, Lee SW, Kim IH (2008) In vivo reprogramming of hTERT by *trans*-splicing ribozyme to target tumor cells. Mol Ther 16:74–80
17. Jeong JS, Lee SW, Hong SH, Lee YJ, Jung HI, Cho KS, Seo HH, Lee SJ, Park S, Song MS, Kim CM, Kim IH (2008) Antitumor effects of systemically delivered adenovirus harboring *trans*-splicing ribozyme in intrahepatic colon cancer mouse model. Clin Cancer Res 14: 281–290
18. Song MS, Jeong JS, Ban G, Lee JH, Won YS, Cho KS, Kim IH, Lee SW (2009) Validation of tissue-specific promoter-driven tumor-targeting *trans*-splicing ribozyme system as a multifunctional cancer gene therapy device in vivo. Cancer Gene Ther 16:113–125
19. Jung HS, Lee SW (2006) Ribozyme-mediated selective killing of cancer cells expressing carcinoembryonic antigen RNA by targeted *trans*-splicing. Biochem Biophys Res Commun 349:556–563
20. Won YS, Lee SW (2007) Targeted retardation of hepatocarcinoma cells by specific replacement of alpha-fetoprotein RNA. J Biotechnol 129:614–619
21. Won YS, Lee SW (2012) Selective regression of cancer cells expressing a splicing variant of AIMP2 through targeted RNA replacement by *trans*-splicing ribozyme. J Biotechnol 158: 44–49
22. Ban G, Jeong JS, Kim A, Kim SJ, Han SY, Kim IH, Lee SW (2011) Selective and efficient retardation of cancers expressing cytoskeleton-associated protein 2 by targeted RNA replacement. Int J Cancer 129:1018–1029

Chapter 8

Characterization of Hairpin Ribozyme Reactions

Preeti Bajaj and Christian Hammann

Abstract

Hairpin ribozymes are small RNA catalytic motifs naturally found in the satellite RNAs of tobacco ringspot virus (TRsV), chicory yellow mottle virus (CYMoV), and arabis mosaic virus (ArMV). The catalytic activity of the hairpin ribozyme extends to both cleavage and ligation reactions. Here we describe methods for the kinetic analysis of the self-cleavage reaction under transcription reaction conditions. We also describe methods for the generation of DNA templates for subsequent in vitro transcription reaction of hairpin ribozymes. This is followed by a description of the preparation of the suitable RNA molecules for ligation reaction and their kinetic analysis.

Key words Catalytic RNA, Kinetic analysis, Plant virus satellite RNA, Self-cleavage, Ligation

1 Introduction

The hairpin ribozyme belongs to the family of the small nucleolytic ribozymes [1] and it requires no external cofactors for its catalytic activity. The catalytic center is formed by the intimate connection between formally unpaired nucleotides in the bulges of stem A and B [2, 3]. Its formation is brought about by an accelerated folding due to the formation of the four-way helical junction [4]. Further also stems C and D are arranged in parallel (Fig. 1). Both kinetic and structural data revealed the importance of four helices to be present in these ribozymes forming a four-way helical junction [5–7]. The hairpin ribozyme is exclusively found in the minus strand of the three plant viral satellite RNAs [8–11], which can accompany their cognate viruses thereby aggravating the symptoms of the virus [8–10, 12, 13].

The catalytic center of the predominantly studied sTRsV hairpin ribozyme and of sArMV is organized around a four-way helical junction. We have shown that sCYMoV features a five-way helical junction instead [14]. Mutational analysis indicates that the fifth arm does not influence the kinetic parameters of the sCYMoV hairpin ribozyme in vitro reactions and thereby seems an appendix to

Daniel Lafontaine and Audrey Dubé (eds.), *Therapeutic Applications of Ribozymes and Riboswitches: Methods and Protocols*, vol. 1103, DOI 10.1007/978-1-62703-730-3_8, © Springer Science+Business Media New York 2014

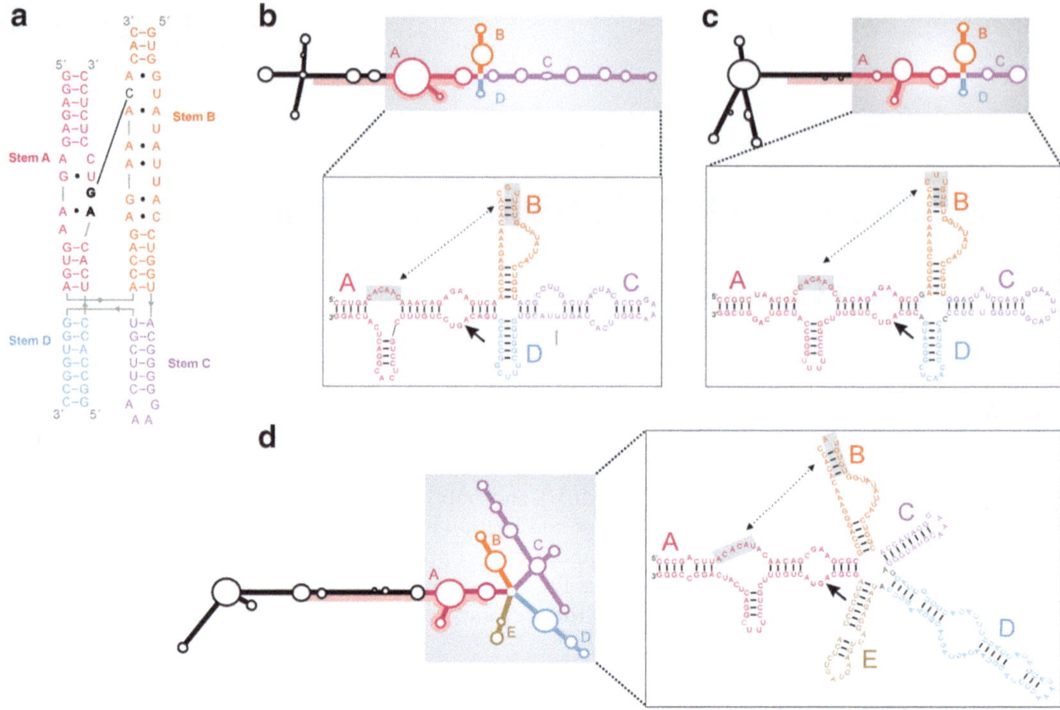

Fig. 1 Hairpin ribozymes. (**a**) Secondary structure with the cleavage site in *grey embossed* letters. Helical stems *A*, *B*, *C*, and *D* are color-coded (throughout all panels the same coloring is used) and nucleic acid strand connectivities are shown by *grey lines*. Non-Watson–Crick interactions between the loops in stem *A* and *B* are indicated by *black dots*. The Watson–Crick base pair between the G + 1 nucleotide in loop A and the essential nucleotide C25 in loop B is indicated by a *black line*. Interactions are based on the crystal structure by Ferré-d'Amaré and coworkers [7]. In the secondary structure predictions of the complete minus strands of the three viral satellite RNAs that harbor hairpin ribozymes (**b**–**d**), these loop–loop interactions are omitted for clarity. The predicted structures of the tobacco ringspot virus satellite RNA (sTRsV) and of the arabis mosaic virus satellite RNA (sArMV) are shown on the *top* of panels (**b**) and (**c**), respectively. The helical stems *A*, *B*, *C*, and *D* form four-way helical junctions, around which the hairpin ribozymes in these RNAs are organized. Parental molecules were termed C_{2B} and C_{1B}, respectively. The stems are denoted and identically color-coded both in the *lower* and in the *upper* part of either panel. The *lower* part shows the sequences of the constructs used in this study. *Black arrows* denote the position of the hairpin ribozyme self-cleavage. The *pink* stretch starting in stem *A* in the *upper* part of either panel shows the position of the hammerhead ribozyme encoded in the plus strand of the respective satellite RNA. *Boxed* in *grey* are sequences in stem *A* and the apical loop B that feature Watson–Crick base complementarity (indicated by *double-headed arrows*). (**d**) The structure of the chicory yellow mottle virus satellite RNA (sCYMoV) is shown in the *left* part. In this RNA, the hairpin ribozyme is organized around a five-way helical junction featuring an extra stem *E*. The *right* part shows the sequence of the construct used in this study as sCYMoV wild type sequence that is termed D_{2B}. All other descriptions are as for (**a**) and (**b**). Secondary structure predictions were calculated using Mfold [25]. Figure and legend modified from ref. 14. Reprinted with kind permission

that junction in the other ribozymes. Additionally, all three natural sequences feature a three-way helical junction outside the catalytic core in stem A, with Watson–Crick complementarity to loop nucleotides in stem B (Fig. 1). Kinetic analyses of the cleavage and

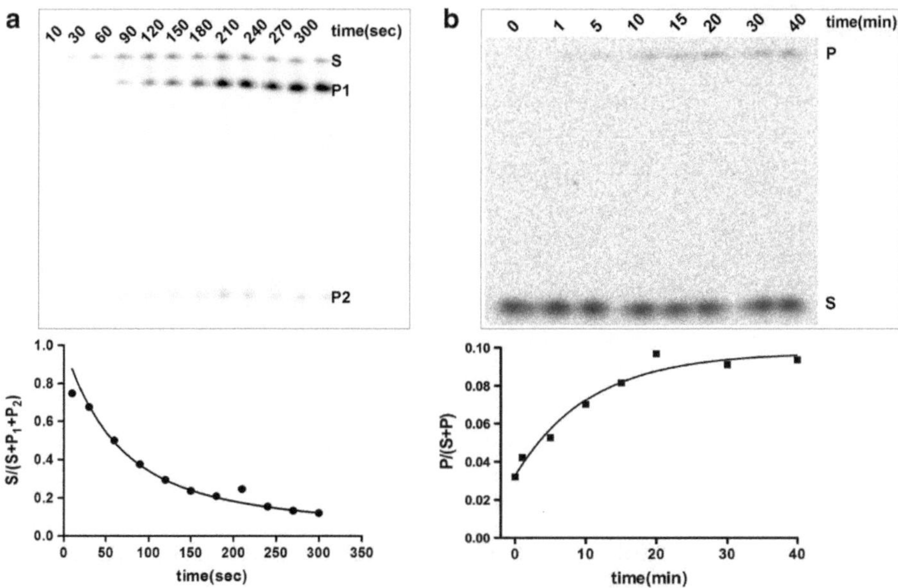

Fig. 2 Examples for the kinetic analyses of cleavage and ligation reactions. Exemplarily, co-transcriptional cleavage analysis (**a**) is shown with the original data (*top*) and the resulting progress curve (*bottom*). (**b**) Analysis of the ligation reaction (**b**) is shown with a gel picture (*top*) and the resulting progress curve (*bottom*). Figure and legend modified from ref. 14. Reprinted with kind permission

ligation reaction of the several variants of sTRsV and sCYMoV hairpin ribozymes in vitro show that the presence of this junction interferes with their reactions, particularly the ligation. Thus, when applying the hairpin ribozyme for gene regulatory purposes in *trans*, care has to be taken to include or exclude those structural elements of the natural sequences that influence the cleavage or ligation reaction.

Usually, self-cleavage of the small endonucleolytic ribozymes like the hairpin can be readily analyzed kinetically by using gel-purified full-length *cis*-cleaving RNA species [15–19]. For the majority of the hairpin ribozyme species, however, we observed a tendency to self-cleave during elution from excised gel slices, prohibiting any kinetic or structural analysis. As a convenient alternative, a transcription-coupled *cis*-cleavage assay [20] can be used. In this setup, the appearance of both the full-length transcript and the two self-cleavage products are monitored over a period of time. If the rate of transcription is constant, and if it does not influence the self-cleavage reaction of the ribozyme, the relative signal intensities of substrate and the products can be used to determine the apparent rate constant of the RNA catalysis k_{cis} (Fig. 2). Because of similar experimental approaches, a part of the protocols provided here have also been published in another chapter of this series [15]. This reference thus also concerns all subsequent text.

2 Materials

All material and solutions should be prepared wearing gloves to avoid contamination with RNases, using deionized ultrapure water and analytical grade reagents (*see* **Note 1**).

2.1 Recursive PCR

1. DNA oligonucleotides (desalted) can be ordered from any suitable suppliers. Stock solutions with a concentration of 100 μM are prepared that can be diluted 20- or 100-fold to yield working concentrations of 5 or 1 μM for overlapping nucleotides. Stock solutions of the outer primers need to be used undiluted.
2. In polymerase chain reaction (PCR), Taq-polymerase is used together with the 10× Buffer containing 10–20 mM $MgCl_2$, following the instructions of the supplier.
3. Solution of dNTP: 2 mM of each dATP, dCTP, dGTP, and dTTP.
4. 10× TBE Buffer: Dissolve 890 mmol Tris base and 890 mmol boric acid with 40 mL 0.5 M EDTA (pH 8.0) in 1 L H_2O. After autoclaving for 20 min, 100 mL can be diluted tenfold to yield 1 L of the working solution 1× TBE.
5. A stock solution of 10 mg/mL ethidium bromide.
6. Any suitable DNA size standard with fragment sizes in the range 100 bp–1 kb, possibly also below 100 bp, depending on fragment sizes, from a supplier of choice.
7. 6× DNA loading dye: 10 mM Tris–HCl, pH 8.0, 60 mM Na_2EDTA, 60 % (w/v) glycerol, 0.03 % (w/v) bromophenol blue, and 0.03 % (w/v) xylene cyanol.
8. Any cloning vector lacking the T7 promoter sequence can be used to clone fragments of recursive PCRs. We normally used pJET1/Blunt that Fermentas (now Thermo Fischer Scientific) has prepared upon request.
9. Competent *Escherichia coli* cells are prepared using standard procedures or purchased commercially. LB medium and the antibiotic ampicillin were used.
10. To record ethidium bromide stained gels, a gel documentation system with UV table is required.
11. PCI mixture: phenol:chloroform:isoamyl alcohol (25:24:1) purchased from any supplier or mixed in the lab. Work with PCI mixture require appropriate protective measures and should be performed under a fume hood.
12. A solution of 3 M NaOAc (pH 5.3) and ethanol, at a concentration of 100 and 70 % (v/v).

2.2 In Vitro Transcription

1. When working with radiolabeled material, appropriate protective equipment is required.
2. The 10× transcription buffer is prepared by mixing ingredients to yield 400 mM Tris–HCl (pH 8.0), 200 mM $MgCl_2$, 20 mM spermidine, and 0.1 % Triton X-100; ACG solution (2.5 mM of each ATP, CTP, and GTP), UTP solution (0.5 mM), RNase inhibitor (10 U/μL), traces of (α-^{32}P) UTP and T7 RNA polymerase (20 U/μL), 0.5 M EDTA (pH 8.0).
3. A mixture of 2.5 mM each rNTP.
4. A DNA template with the desired hairpin ribozyme sequence.
5. For the preparation of gel filtration columns, mix 20 g of Sephadex G-50 fine in a bottle with 200 mL 1× TE containing 10 mM Tris–HCl and 1 mM EDTA (pH 8.0) and boil the mixture in a microwave for 30 s. Upon setting of the sephadex bed in the bottle, pour off the buffer, and replace with fresh buffer. After bringing it again to the boil, the material is stored after cooling at 4 °C.

2.3 Gel Purification of RNA Species

1. For polyacrylamide gels, a vertical gel electrophoresis apparatus with a power supply generating a voltage of 600 V is required.
2. A solution of 10 % polyacrylamide:bis-acrylamide (19:1) with 7 M urea and 1× TBE.
3. A solution 20 % (w/v) of ammoniumpersulfate (APS) in water.
4. A Denaturing RNA loading buffer containing 95 % formamide, 50 mM EDTA (pH 8.0), 0.03 % (w/v) bromophenol blue, and 0.03 % (w/v) xylene cyanol.
5. A Gel extraction solution with 40 % formamide (v/v), 0.7 % (w/v) SDS in 1× TE.
6. A 1 μg/mL solution of ethidium bromide in 1× TBE.

2.4 Transcription-Coupled Cis-Cleavage Assay

1. Ingredients for the 10× transcription buffer are 400 mM Tris–HCl (pH 8.0), 20 mM spermidine, and 0.1 % Triton X-100; ACG solution (2.5 mM of each ATP, CTP and GTP), UTP solution (0.5 mM), RNase inhibitor (10 U/μL), traces of (α-^{32}P) UTP and T7 RNA polymerase (20 U/μL), 0.5 M EDTA (pH 8.0).
2. At least 12 reaction tubes with 10 μL denaturing loading dye buffer (*see* Subheading 2.3, **item 4**), numbered consecutively and placed on ice.

2.5 Kinetic Analysis of the In Vitro Ligation Reactions

1. Reaction Buffer (2×) containing 50 mM Tris–HCl (pH 7.0) and 100 mM NaCl.
2. Start solution (2×) containing 50 mM Tris–HCl (pH 7.0), 100 mM NaCl and 20 mM $MgCl_2$.

3. At least eight reaction tubes with 10 μL denaturing loading dye buffer (*see* Subheading 2.3, **item 4**), consecutively numbered and placed on ice.

3 Methods

3.1 Recursive PCR

1. To allow for subsequent transcription, the DNA template must contain the T7 RNA polymerase promoter sequence (TAATACGACTCACTATA), to which the sequence GGG, GCG, or GGC is added depending upon the structural features of the resulting RNA (*see* **Note 2**). The addition of these triplets ensures high transcription yields [21]. After either triplet, the sequence of the hairpin ribozyme motif under investigation is added. The resulting DNA template is therefore 20 nucleotides (nt) longer than the original ribozyme sequence.
2. The DNA template must be split up in an even number of DNA oligonucleotides that partially overlap. It is crucial that all overlapping sequences for one DNA template have a similar predicted melting temperature (Tm) within 1 °C. A wide range of programs accessible via the Internet can determine Tm values (*see* **Note 3**). The size of the template determines a number of DNA oligonucleotides that are suitable to cover the entire sequence. Currently, the efficient synthesis rate for oligonucleotides can easily yield 50–60 nt. On the basis of this and an average overlapping sequence of about 20 nt (depending on the sequence), the number of required DNA molecules (N) can be estimated according to Eq. 1 in which L is the length of the template (*see* **Note 4**):

$$L = 30nt \times N + 20nt \quad (1)$$

Additionally to these DNA oligonucleotides, two outer primers flanking the complete sequence are required that will amplify the template sequence eventually (Fig. 3). The sequences of these are defined by the requirement to have a similar Tm value, as the overlapping sequences of the other DNA oligonucleotides have.

3. For PCR, set up a mixture containing 1 μL of each overlapping DNA oligonucleotide (5 μM, *see* **Note 5**), 1 μL of each outer primer (100 μM), 2.5 μL 10× PCR Buffer, 2.5 μL dNTP solution, and 0.5 μL Taq-Polymerase in a total reaction volume of 25 μL H_2O in a PCR tube. The PCR is started in a thermocycler with the following parameters.

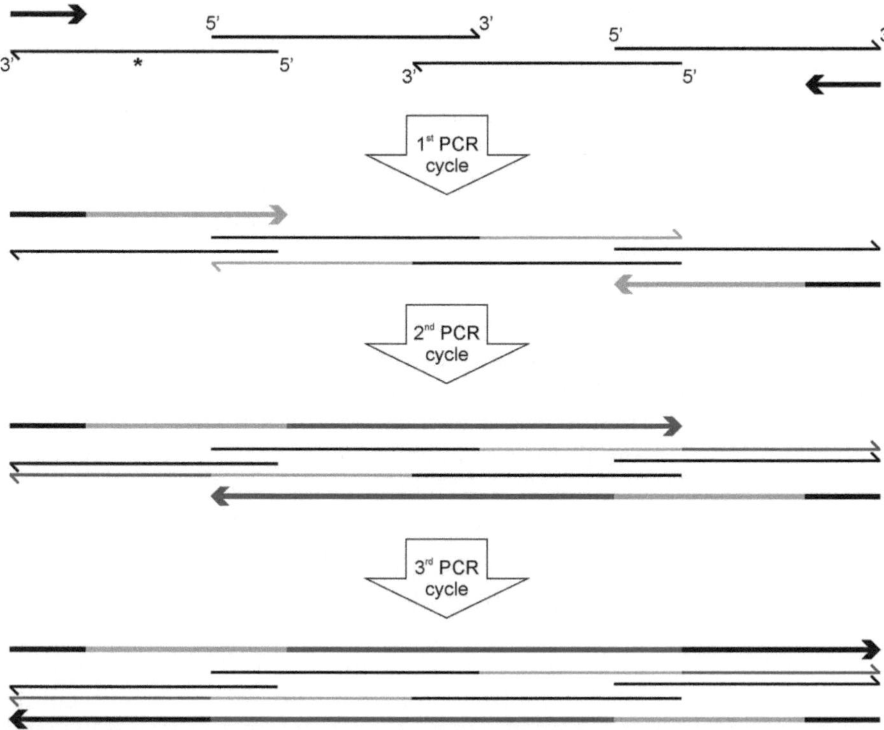

Fig. 3 Principle of recursive PCRs. An even number of overlapping DNA oligonucleotides (*thin arrows*) is designed in a way that the overlapping stretches have the same melting temperature (Tm). To these, two additional outer oligonucleotides with identical Tm to that of the overlapping stretches are added in excess (*thick black arrows*). In a first PCR round, the oligonucleotides are extended as indicated (*light grey lines*). In the second round, DNA strands get further extended (*dark grey lines*). In the third round and thereafter, only the outer oligonucleotides are extended to cover the entire sequence. The DNA oligonucleotide marked with an *asterisk* can be used in subsequent in vitro transcription reactions to prevent self-cleavage of the ribozyme. Reproduced with kind permission from Springer Science + Business Media: Methods in Molecular Biology 848, Ribozymes, Characterization of hammerhead ribozyme reactions, 2012, p. 5–20, A. Kalweit et al., Fig. 4

First step	1×	3 min at 95 °C
Second step	30×	20 s at 95 °C 20 s at Tm of the used DNA oligonucleotides 20 s at 72 °C
Third step	1×	5 min at 72 °C

4. Dissolve 1.5 g agarose in 100 mL 1× TBE in a microwave oven to prepare a 1.5 % standard agarose gel. After the solution has cooled to about 60 °C, 5 μL of ethidium bromide solution are added and the gel casted in a horizontal gel chamber, into which a comb is inserted. After the gel has set, overlay with 1×

TBE. Mix 10 μL of each PCR reaction with 2 μL loading dye and load next to DNA standard covering the size range of the expected PCR product. Connect to a power supply and perform gel electrophoresis at a field strength of 10 V/cm.

5. Make sure to wear appropriate protective gear against the radiation, when analyzing the result by visualization on a UV table. Document the result by a gel documentation system. If no clear product of the expected size is observed, the reaction can be repeated using a reduced amount of overlapping DNA oligonucleotides (*see* **Note 5**). As soon as a clear band is obtained, use any cloning vector lacking the T7 promoter to clone the PCR product. Follow the manufacturer's instructions for the cloning vector to ligate the PCR fragment in the cloning vector. Transform in competent *E. coli* cells and on the next day, perform Plasmid mini-preparations according to the lab's standard protocols, or as described [22].
6. For analytical digestion of the DNA, use 5–10 prepared plasmids to identify clones with inserts of the expected sizes. Sequence 2–3 plasmids at any company offering commercial sequencing. Upon identification of a clone with the correct DNA template, prepare a 100 mL overnight culture of the respective *E. coli* strain and prepare the plasmid DNA. This cell culture volume is expected to yield 100 μL solution of 1 μg/μL plasmid DNA. Use a restriction enzyme that cleaves near the last nucleotides encoding the ribozyme sequence (*see* **Note 6**) and linearize 10 μg of the plasmid preparation. Analyze the completeness of the digest, by checking 1/100 (v/v) on a 1 % agarose gel and compare the migration with that of the same amount of undigested plasmid together with a conventional DNA size marker in the kb range. Proceed as above (**step 4**) and analyze the gel using a gel documentation system. Upon complete digest (*see* **Note** 7), purify the linear DNA by phenol extraction. To this end, mix equal amounts of the linear DNA solution and PCI in a 1.5 mL reaction tube, vortex for 30 s, and separate phases at room temperature in a table centrifuge at maximum speed for 5 min. Remove the upper aqueous phase carefully and transfer to a fresh 1.5 mL reaction tube. Add to the linear plasmid solution 1/10 (v/v) 3 M NaOAc, pH 5.3 and 3 (v/v) parts 100 % ethanol and incubate for 20 min at −20 °C. Place the tube in a cooling table centrifuge and start the run at maximum speed for 30 min at 4 °C. Remove the supernatant, add 1 mL of 70 % ethanol and repeat centrifugation for 10 min. After removal of the liquid again, dry the pellet by placing the open reaction tube in a heating block set to 50 °C and redissolve subsequently in 100 μL H_2O. Determine the DNA concentration of the linear fragment in a UV spectrometer.

3.2 In Vitro Transcription

1. For generation of hairpin ribozyme transcripts, linearized plasmid DNA serves as template that is generated by recursive PCR. Since the in vitro transcription is performed in the presence of (α-^{32}P) UTP, have appropriate protective measures against radioactivity in place in the laboratory and implemented for all work described in the subsequent sections. Mix by pipetting the following ingredients in a total volume of 30 μL: 600 ng linearized plasmid DNA, 3 μL of 10× transcription buffer, 7.5 μL ACG solution, 3 μL UTP solution, 1 μL RNase inhibitor, 2 μL (α-^{32}P) UTP. Start the reaction by the addition of 2 μL T7 RNA Polymerase. Incubate at 37 °C for 1 h.
2. One of the cleavage products is required in a non-radiolabeled version for Kinetic analysis of the in vitro ligation reaction. To generate this fragment, perform a second in vitro transcription reaction as stated above (Subheading 3.2, **step 1**), however in the absence of (α-^{32}P) UTP, and at a uniform concentration rNTP concentration (*see* **Note 8**).
3. Stop all reactions after 1 h by the addition of 30 μL 2× denaturing loading dye.
4. Heat denature the samples at 95 °C for 2 min, snap cool on ice for 2 min and run the samples on denaturing PAGE as discussed below (Subheading 3.3, **steps 1–9**).

3.3 Gel Purification of RNA Species to Study Ligation Reactions

Hairpin ribozymes have the ability to self-cleave during transcription reaction. The transcription-coupled *cis*-cleavage assay (Subheading 3.4) does not require elution of uncleaved transcript and can be analyzed directly under radiolabeled conditions by Phosphoimager analysis. However, the analysis of the ligation reaction requires isolation of products of self-cleavage of the hairpin ribozyme performed in two independent transcription reactions one of which was radiolabeled and the other non-radiolabeled (*see* **Note 9**).

1. Prepare polyacrylamide gels with desired percentage of (10–20 %) by dissolving urea (7 M), acrylamide/bis-acrylamide (19:1) in a final concentration of 1× TBE, followed by filtration.
2. The two glass plates, one normal and other with a notch, are cleaned and sealed from three sides with the help of spacers, held together with the help of clamps.
3. To 60 mL of polyacrylamide solution in a beaker, 300 μL of 20 % (w/v) APS (*see* **Note 10**) and 60 μL of TEMED are added and mixed gently. This gel solution is immediately poured into glass plates from the top, avoiding any bubble formation in the gel.

4. A tethered comb is then inserted from the top to make wells in the gel. The gel is allowed to polymerize at room temperature for 10–15 min. After the gel is formed, the spacer at the bottom of the glass plates is removed.
5. The whole assembly is mounted in a vertical electrophoresis apparatus having an upper reservoir and a lower reservoir containing 1× TBE as the running buffer.
6. The comb is gently removed from the gel and the wells are washed gently with running buffer to remove unpolymerized gel. The upper reservoir of the electrophoresis apparatus is covered with the lid, keeping the level of the buffer above the wells in the gel.
7. The RNA sample to be analyzed is added to the denaturing loading dye, heat denatured at 95 °C for 2 min and snap cooled on ice for 2 min. The denatured RNA samples are then loaded onto the wells along with RNA marker.
8. The electrophoresis apparatus is connected to the power supply and the RNA samples are allowed to separate in the gel from cathode to anode at 25 mA for approximately 2 h.
9. After electrophoresis is finished, the glass plates are disassembled and processed in two different ways depending upon the nature of transcription reaction.
10. For radiolabeled samples, an asymmetric pattern is drawn on those areas of the gel, where no RNA has been loaded by using a mixture of a small amount of [α-^{32}P] UTP, or any other β-emitter, with the colored loading dye (fresh isotope not required). This allows for visual identification of the pattern after phosphoimager exposure. The gel then is wrapped in plastic sheet, sealed on all sides, and kept for exposure to phosphoimager screen in phosphoimager cassette for 5–10 min.
11. The screen is read in the scanner and the desired band is identified as black dots formed by the exposure of radioactivity. Make a 1:1 print out of the gel area and placed under the glass plate with the gel and use the drawn pattern to overlay (*see* **Note 11**).
12. The desired RNA fragment (products of transcription reaction) is cut out of the gel with a sterile scalpel on the UV table and the slice is crushed into small pieces and transferred into sterile reaction tubes containing 300 μL of the respective elution buffer. This mixture is kept shaking overnight at room temperature.
13. The supernatant is transferred to sterile reaction tubes and purified with PCI solution (25:24:1) at 16,000 × *g* in a microcentrifuge for 5 min.

14. The upper aqueous phase is carefully transferred to fresh sterile reaction tubes and precipitated with 1/10 volume of 3 M sodium acetate, pH 4.7, and 0.6 volumes of isopropanol.
15. The samples are incubated at −20 °C for 20 min. The nucleic acid pellet is obtained by centrifugation at 16,000 × *g* for 30 min at 4 °C.
16. The supernatant is discarded and the pellet washed with 70 % ethanol at room temperature for 10 min. Upon removal of the supernatant, pellets are dried and dissolved in 10 μL of sterile water each.
17. For non-radiolabeled samples, the gel is transferred to a tray containing water and ethidium bromide (*see* **Note 12**) and kept for shaking for 2–3 min. Thereafter, the gel is transferred to the UV table to visualize the bands and photographed. The desired band is cut out of the gel with the help of a sterile scalpel and incubated with elution buffer, as described above (Subheading 3.3, **steps 12–16**).
18. The concentration of the non-radiolabeled nucleic acid fragments are determined using UV spectrophotometer.

3.4 Transcription-Coupled Cis-cleavage Assay

In this section we describe a protocol for the analysis of cleavage kinetics of hairpin ribozymes during transcription, in a method called transcription-coupled *cis*-cleavage assay [23].

1. The DNA templates for hairpin ribozymes are generated by recursive PCR and cloned into pJET1/blunt cloning vector are linearized with either *Xho*I or *Xba*I enzyme depending upon orientation in the vector.
2. The radiolabeled transcription reaction was carried out in a 30 μL reaction consisting of 60 ng DNA, 1× Transcription buffer, 5 mM $MgCl_2$ (*see* **Note 13**), 2.5 mM ACG solution, 0.5 mM UTP, 1.0 μL (α-^{32}P) UTP, 0.3 U of 10 U/μL RNasein, diluted with UV-H_2O.
3. All the components except T7 polymerase are mixed in a sterile reaction tube and incubated at 37 °C for 2 min (*see* **Note 14**).
4. Prepare 12 reaction tubes, containing 8 μL of 2× denaturing RNA loading dye and label them.
5. An aliquot of 2 μL is added to first tube with 8 μL of denaturing loading dye as the zero time point.
6. The reaction is initiated by the addition of 1 μL of T7 polymerase. 2 μL aliquots of this reaction is taken out and added to 8 μL of 2× denaturing loading dye starting from 10 s time point.
7. All the subsequent time points are taken at an interval of 30 s till 300 s and added to the respective reaction tube with loading dye.

8. All reaction tubes with the samples stopped at different time points are heat denatured at 95 °C for 2 min and subsequently snap cooled on ice for 2 min.
9. These reactions are then subjected to denaturing PAGE and the gel is exposed to phosphoimager screen for data analysis. The data obtained is fitted to Eq. 2

$$\frac{L}{L+S} = \frac{1}{kt}\left(1 - e^{-kt}\right) \tag{2}$$

wherein L is the concentration of full-length transcript, S is the concentration of cleaved transcript, t is time, and k is the unimolecular rate constant for *cis*-cleavage [23].
10. The equation is applicable for the kinetic analysis of ribozyme self-cleavage reactions under the prerequisite that the transcription reaction itself is constant and not rate limiting.
11. All reactions should be carried out at least in duplicate.

3.5 Kinetic Analysis of the In Vitro Ligation Reaction

The hairpin ribozymes have the ability to cleave during transcription reaction, which is exploited to prepare substrates for the ligation reaction. For this purpose, the substrates for the ligation reaction are prepared in two independent transcription reactions using the same DNA template. One of the reactions is radiolabeled, while other is non-radiolabeled. From the former, the 3′ cleavage product is gel purified, and from the latter the 5′ cleavage product.

1. The non-radiolabeled strand is quantified using spectrophotometer. 200 nM of 5′ substrate strand is used to perform ligation on traces of radiolabeled 3′ substrate, thus the ligation reaction is performed under pseudo-first-order conditions.
2. Prepare 20 μL of 5′ substrate and 3′ substrate strands each in 1× buffer.
3. Both strands are denatured separately at 95 °C for 2 min, renatured on ice for 5 min and incubated at 25 °C for 15 min (*see* **Note 14**).
4. To both solutions, 20 μL of 1× start buffer containing 10 mM $MgCl_2$ (*see* **Note 13**) is added and the mixture is incubated at 25 °C for 5 min.
5. The zero time point is taken by adding 5 μL of radiolabeled 3′ substrate strand to 5 μL of denaturing loading dye.
6. From the 5′ substrate solution, 5 μL are discarded at this time to make the reaction volumes of both the strands even before to start the cleavage reaction.
7. Prepare seven reaction tubes, containing 10 μL of 2× denaturing RNA loading dye and label them.

8. The ligation reaction is initiated by mixing the 5′ and 3′ substrate. An aliquot of 10 μL reaction mixture is taken at each time point and added to 10 μL of denaturing loading dye.
9. The samples are denatured at 95 °C for 2 min, snap cooled on ice for 2 min and subjected to denaturing PAGE for analysis of the ligation reaction.
10. Using nonlinear regression analysis, the rate of ligation was determined by fitting the data to the single exponential Eq. 3

$$\frac{L}{S_1 + L} = F_0 + F_\infty\left(1 - e^{-kt}\right) \tag{3}$$

wherein L is the concentration of ligation product, S_1 the concentration of radiolabeled 3′ substrate, F_0 the fraction ligated at the time zero, and $F\infty$ that at the end of the reaction. Time is symbolized by t and the observed pseudo-first-order rate constant is k [24].
11. All reactions should be carried out at least in duplicate.

4 Notes

1. We do not use diethylpyrocarbonate (DEPC) that can be applied to alkylate and thus inactivate RNases, as we find it dispensable if normal lab standards are maintained.
2. The addition of the triplet might interfere with folding of the RNA into the hairpin ribozyme structure. It therefore is essential to compare the motif's predicted secondary structure with and without the triplet by using Mfold [25], or any other RNA folding program. In case the presence of the optimal GGG changes the folding, GGC or GCG can be used alternatively [21].
3. Because of different algorithms that are employed, only Tm values determined from a given program are comparable with another but not necessarily between programs. One example calculator is available at http://www.basic.northwestern.edu/biotools/oligocalc.html.
4. Since N in Eq. 1 has to be an even number, one can round the calculated value. For example, 100 nt is covered by $N=8/3=2.66$, i.e., two DNA oligonucleotides; for a template sequence of 180 nt, the number comes to $N=16/3=5.33$, i.e., six oligonulceotides will be required.
5. Since the overlapping DNA oligonucleotides are required for the sake of providing the backbone from which the entire sequence is generated, one can dilute them manifold (1:5, 1:10, 1:20, …). This might become important when additional bands appear after PCR that have another size than the

expected product. In case dilution of the overlapping DNA oligonucleotides does not yield a uniform PCR product of expected size, one needs to gel purify the desired PCR product using a gel purification kit. For that purpose, bands are visualized on a UV table, excised with a sterile scalpel, and processed according to the instructions provided by the manufacturer of the gel purification kit.

6. For linearization of plasmid DNA that is to be used for subsequent in vitro run off transcription, there is a preference for restriction enzymes: those producing blunt ends or 5′ overhangs are better suited than those generating 3′ overhangs, as these might lead to extended RNA transcripts.
7. In case the digest remains incomplete, it is advisable to gel purify the linearized plasmid, as described (*see* **Note 5**).
8. In the generation of the cleavage products as substrates for ligation reaction, occasionally additional shorter RNAs are observed, in addition to the full-length RNA and its cleavage products. In this case, gel purification of the desired band is required.
9. The success of the in vitro transcription of radiolabeled RNA can be immediately assessed by measuring the fraction of radionucleotides in the column compared to screw cap reaction tube, using a Geiger-Müller counter.
10. The APS solution must be stored at 4 °C and should be prepared freshly every month.
11. If no “old” radiosource is available, take a small amount of the (^{32}P) UTP.
12. The ethidium bromide solution might be reused if stored in a dark bottle.
13. It is essential to apply Mg^{2+} concentrations in the low mM or better sub-mM range during transcription-coupled *cis-cleavage* assay. This allows progressive monitoring of the cleavage reaction of the hairpin ribozyme. However, the higher mM Mg^{2+} concentration was used in the ligation reaction.
14. Different reaction temperature and time were adjusted depending upon the type of the reaction investigated.

References

1. Fedor MJ (2009) Comparative enzymology and structural biology of RNA self-cleavage. Annu Rev Biophys 38:271–299
2. Ryder SP, Strobel SA (1999) Nucleotide analog interference mapping of the hairpin ribozyme: implications for secondary and tertiary structure formation. J Mol Biol 291:295–311
3. Ryder SP, Strobel SA (2002) Comparative analysis of hairpin ribozyme structures and interference data. Nucleic Acids Res 30:1287–1291
4. Zhao ZY, Wilson TJ, Maxwell K, Lilley DM (2000) The folding of the hairpin ribozyme: dependence on the loops and the junction. RNA 6:1833–1846

5. Murchie AI, Thomson JB, Walter F, Lilley DM (1998) Folding of the hairpin ribozyme in its natural conformation achieves close physical proximity of the loops. Mol Cell 1:873–881
6. Rupert PB, Ferre-D'Amare AR (2001) Crystal structure of a hairpin ribozyme–inhibitor complex with implications for catalysis. Nature 410:780–786
7. Rupert PB, Massey AP, Sigurdsson ST, Ferre-D'Amare AR (2002) Transition state stabilization by a catalytic RNA. Science 298: 1421–1424
8. Buzayan JM, Gerlach WL, Bruening G (1986) Satellite tobacco ringspot virus RNA: a subset of the RNA sequence is sufficient for autolytic processing. Proc Natl Acad Sci U S A 83:8859–8862
9. Buzayan JM, Gerlach WL, Bruening G (1986) Non-enzymatic cleavage and ligation of RNAs complementary to a plant virus satellite RNA. Nature 323:349–353
10. Kaper JM, Tousignant ME, Steger G (1988) Nucleotide sequence predicts circularity and self-cleavage of 300-ribonucleotide satellite of arabis mosaic virus. Biochem Biophys Res Commun 154:318–325
11. Rubino L, Tousignant ME, Steger G, Kaper JM (1990) Nucleotide sequence and structural analysis of two satellite RNAs associated with chicory yellow mottle virus. J Gen Virol 71:1897–1903
12. Kaper JM, Tousignant ME (1984) Endeavour 8:194–200
13. Buzayan JM, Gerlach WL, Bruening G, Keese P, Gould AR (1986) Nucleotide sequence of satellite tobacco ringspot virus RNA and its relationship to multimeric forms. Virology 151:186–199
14. Bajaj P, Steger G, Hammann C (2011) Sequence elements outside the catalytic core of natural hairpin ribozymes modulate the reactions differentially. Biol Chem 392:593–600
15. Kalweit A, Przybilski R, Seehafer C, de la Pena M, Hammann C (2012) Characterization of hammerhead ribozyme reactions. Methods Mol Biol 848:5–20
16. Seehafer C, Kalweit A, Steger G, Gräf S, Hammann C (2011) From alpaca to zebrafish: hammerhead ribozymes wherever you look. RNA 17:21–26
17. Przybilski R, Hammann C (2007) The tolerance to exchanges of the Watson/Crick basepair in the hammerhead ribozyme core is determined by surrounding elements. RNA 13:1625–1630
18. Przybilski R, Hammann C (2007) Idiosyncratic cleavage and ligation activity of individual hammerhead ribozymes and core sequence variants thereof. Biol Chem 388:737–741
19. Przybilski R, Gräf S, Lescoute A, Nellen W, Westhof E, Steger G, Hammann C (2005) Functional hammerhead ribozymes naturally encoded in the genome of arabidopsis thaliana. Plant Cell 17:1877–1885
20. Long DM, Uhlenbeck OC (1994) Kinetic characterization of intramolecular and intermolecular hammerhead RNAs with stem II deletions. Proc Natl Acad Sci U S A 91: 6977–6981
21. Weber U, Gross HJ (1997) In vitro RNAs. In: Lichtenstein C, Nellen W (eds) Antisense technology, a practical approach. Oxford University Press, Oxford, pp 75–92
22. Sambrook J, Fritsch EF, Maniatis T (1989) Molecular cloning, 2nd edn. Cold Spring Harbor Laboratory Press, Cold Spring Harbor, NY
23. Long DM, Uhlenbeck OC (1994) Kinetic characterization of intramolecular and intermolecular hammerhead RNAs with stem II deletions. Proc Natl Acad Sci U S A 91: 6977–6981
24. Stage-Zimmermann TK, Uhlenbeck OC (1998) Hammerhead ribozyme kinetics. RNA 4:875–889
25. Zuker M (2003) Mfold web server for nucleic acid folding and hybridization prediction. Nucleic Acids Res 31:3406–3415

Chapter 9

Finding Instances of Riboswitches and Ribozymes by Homology Search of Structured RNA with Infernal

Amell El Korbi, Jonathan Ouellet, Mohammad Reza Naghdi, and Jonathan Perreault

Abstract

In the genomics era, computational tools are essential to extract information from sequences and annotate them to allow easy access to genes. Fortunately, many of these tools are now part of standard pipelines. As a consequence, a cornucopia of genomic features is available in multiple databases. Nevertheless, as novel genomes are sequenced and new structured RNAs are discovered, homology searches and additional analyses need to be performed. In this chapter, we propose simple ways of finding instances of riboswitches and ribozymes in databases or in unannotated genomes, as well as ways of finding variants that deviate from the typical consensus.

Key words ncRNA, Noncoding RNA, Infernal, Covariation, Homology search, RNA structure, Secondary structure, Riboswitches, Ribozymes

1 Introduction

The diversity of roles attributed to noncoding RNA (ncRNA) has increased at a rapid pace in the last decade. As additional classes of RNAs were discovered and studied, so were their structures [1]. At the same time, sequence databases have grown exponentially, largely due to next-generation sequencing technologies. Public databases such as GenBank [2] or the metagenome-focussed CAMERA [3] database provide incredible opportunities to discover functional structured RNAs with computational screening which have proven extraordinarily useful for many ground-breaking discoveries of new ncRNAs [4–7].

Several computational methods have been used to that end. Some of the most commonly used tools for de novo prediction of ncRNAs include Evofold [8], QRNA [9], RNAz [10–12], CMfinder [13], Dynalign [14], LocARNA [15], Pfold [16], and the Vienna RNA package [17]. A comprehensive list can also be found in this wikipedia page [18].

Daniel Lafontaine and Audrey Dubé (eds.), *Therapeutic Applications of Ribozymes and Riboswitches: Methods and Protocols*, vol. 1103, DOI 10.1007/978-1-62703-730-3_9, © Springer Science+Business Media New York 2014

These tools allowed many research groups to find various ncRNAs: small RNAs that base pair on multiple target mRNAs to inhibit gene expression [19], self-cleaving ribozymes [20, 21], and riboswitches that bind a metabolite with their aptamer domain to change gene expression through their expression platform [22].

In parallel to the blossoming field of ncRNAs, the increasing rate of DNA sequencing requires efficient methods to annotate known RNAs. This becomes as important as ever since some of these RNAs are used as antibiotic targets. Indeed, a few riboswitches are already known to be sensitive to natural antibiotics analogous to their ligand [23–27] and new ligand-analogs are being developed in order to overcome the increasing worldwide resistance of bacteria against antibiotics [28–30]. Therefore, finding all instances of a targeted RNA can help determine the sensitive pathogenic strains as well as potentially sensitive beneficial strains.

1.1 Browsing Rfam: The RNA Families Database

In that regard, the collection of RNA families database (Rfam [31]) is particularly useful. The Sanger Institute performs homology searches with the Infernal software suite [32] for all known RNA families, which includes riboswitches and ribozymes, to update the database approximately once a year. Therefore, the quickest way of finding a riboswitch that could be a potential antibiotic target in any organism is by browsing Rfam. The cutoff scores typically used by Rfam to accept a predicted RNA with a relatively complex structure, such as for riboswitches, are high enough that there are almost no false positives in microbial genomes. On this subject, readers might also be interested to look at other recent publications in Methods in Molecular Biology [33, 34].

1.2 Search for a Motif in New Genomes with Infernal

However, there are some instances where Rfam does not provide the information needed, particularly if a given riboswitch exists in a newly sequenced genome that has not been screened by Rfam yet. This paper aims to circumvent such problems by presenting a simple step-by-step approach to look at a genome and evaluate the presence of an ncRNA of interest within that unannotated genome. It is targeted towards a general audience with minimal bioinformatics skills, although some basic knowledge of shell command lines would be useful.

1.3 Search for Variants of a Known Motif in All Bacterial Genomes

Occasionally, researchers that are very knowledgeable about a specific riboswitch have reasons to hypothesize that more instances exist. For example, divergence from the structure consensus of a riboswitch class could prevent Infernal from finding such a riboswitch's sub-family. In such cases, it could be desirable to perform a new Infernal search with less stringent criteria to reveal these "hidden" hits [35]. While Rfam lists most of the instances of riboswitches and ribozymes that can easily be found with a relatively high confidence (low E-values), it can occasionally ignore cases that diverge from the consensus. This has been previously illustrated several times, notably for the *glmS* and the hammerhead ribozymes [21, 35–40]. In the case of the *glmS* ribozyme, Infernal was used with a very high E-value tolerance, as high as 5,000 on all microbial genomes from NCBI's Refseq38 sequence dataset. The resulting hits therefore included a vast majority of spurious hits, but also a number of previously unannotated *glmS* ribozyme instances [35]. Homology searches with very relaxed parameters should not be performed on a routine basis, but rather if there are good hints that additional riboswitches or ribozymes could be found in this manner. Indications on how to manage such searches and the resulting hits will be provided in Subheading 3.3.

2 Materials

Infernal requires the Linux/UNIX system to run the program and is accessible on Janelia'server [41]. Another alternative is the use of Mac OS X, which is a certified UNIX platform (*see* **Note 1**).

3 Methods

3.1 Browsing Rfam: The RNA Families Database

The simplest way to verify the presence of a specific riboswitch in target bacteria is to look in Rfam. As long as the bacterial genomes have been sequenced and annotated by Rfam, browsing the genomes section [42] would allow anyone to rapidly find which of the known riboswitches are found in a given genome by examining the ncRNAs found in the "chromosomes" tabs. Conversely, browsing the "families" section provides a quick overview of all species that have a specific riboswitch within their genome. This could be especially useful in the context of the development of a new antibiotic to target only a desired group of bacteria and leave most of the natural microbiota intact. Of course, the presence of a riboswitch in a bacterial species does not warrant microbicidal effect of the newly made antibiotic compound. Indeed, studies have already shown potent compounds capable of binding a specific riboswitch to prevent gene regulation via a competition against its native ligand. This competition can affect the growth of some bacteria that have the riboswitch, while leaving others unscathed although the targeted riboswitch is present in both cases. Depending on which genes are regulated by these riboswitches, significant differences of sensitivity can be observed [29].

In the few cases where Rfam would not be useful, any sequence can be screened for the presence of riboswitches with Infernal, which is described in more detail in Subheadings 3.2 and 3.3.

3.2 Search for a Motif in New Genomes with Infernal

Most of what is described herein can be found with additional details in the Infernal user guide [43] and additional papers [34, 44]. The intent here is to provide inexperienced users a rapid start-up guide. For this tutorial, the motif of purine riboswitches is used as an example where Infernal builds a covariance model from an alignment with structural annotation.

The latest version of Infernal is available to download here [41]. At the time of writing, the latest release of Infernal is 1.0.2 (30 Oct 2009) [45]. Once the source file downloaded, expand the "tar file" at a convenient location. For a basic installation, execute the two commands "configure" and "make" from the "infernal-1.0.2" directory (*see* **Note 2**):

```
# ./configure
# make
```

To run the optional testsuite, execute the following command:

```
# make check
```

Once Infernal installed, the ncRNA covariance model is built. The first step is to generate the "purine" seed alignment in Stockholm format from Rfam at this location [46]. Once the alignment is generated, the file should be downloaded and saved under "purine.sto" (*see* **Note 3**) in the "infernal-1.0.2" directory. The Stockholm format describes the secondary structure of an RNA sequence alignment. Base pairs are annotated as "<" (for the opening base of the pair) and ">" (for the closing base). Other, base pair annotations such as (,), [,], {or} are also used sometimes for base pairs of stems enclosing a multistem junction. Single stranded regions are annotated with other characters, typically ".", but sometimes "_" for loops and "," for junctions. A similar notation is used in Infernal's output for a regular "cmsearch." For simplicity, we assume that all the following commands are executed from the infernal directory.

Build the "cm file" (covariance model) using the command "cmbuild":

```
# src/cmbuild purine.cm purine.sto
```

Execute the command "cmcalibrate" which may take more than 1 h:

```
# src/cmcalibrate purine.cm
```

To search for the presence of that purine motif in a new genome, copy the sequence file of the genome of interest (FASTA format) to the "infernal-1.0.2" directory and use:

```
# src/cmsearch purine.cm genome.fa
```

For the purpose of this example, the search will be performed in a known bacterial genome downloaded from NCBI. The genome sequence of *Bacillus subtilis* (in FASTA format) is downloaded from this link [47]. The file is downloaded by pressing the "send" option while selecting "Destination file" and the "FASTA" format. Rename the file as "sequence_Bsubs.fa" and move it to the same directory as "purine.cm".

Execute the "cmsearch" command:

```
# src/cmsearch --ga purine.cm sequence_Bsubs. fa
```

where the --ga option sets the bit-score cutoff value as the one used by Rfam curators according to the "GA cutoff" value in the purine.sto file downloaded from Rfam. When a large number of hits are expected, "cmsearch" has the useful additional option --tabfile to get a tabular representation of the search results (*see* **Note 4**). However, the current example does not need the option since it is a simple search for one riboswitch in a single genome. To create an output file, the command line would be:

```
# src/cmsearch --ga purine.cm sequence_Bsubs.fa>output.txt
```

Below is the output.

```
# command:    src/cmsearch --ga purine.cm sequence_Bsubs.fa
CM: Purine
>gi|223666304|ref|NZ_CM000487.1|

  Plus strand results:

 Query = 1 - 102, Target = 697666 - 697767
 Score = 85.74, E = 1.742e-18, P = 7.494e-25, GC =  42

          :::::::::::::::::((((((((,,,<<<<<<<_______>>>>>>>,,,,,,,,<<<
        1 aaaaaaaaaaaaaaaaaucacuCgUAUAAucccgggAAUAUGGcccgggaGUUUCUACCag 60
           A+ AAA+ AAAA A    :  C:UAUAAU  :GGGAAUAUGGCCC:  AGUUUCUACC:G
   697666 CAUGAAAUCAAAACACGACCUCAUAUAAUCUUGGGAAUAUGGCCCAUAAGUUUCUACCCG 697725

          <<<<_______>>>>>>>,,))))))))::::::::::::::
       61 gcaaCCGUAAAuugccuGACUAcGagugaaauuauuaaaaau 102
          GCAACCGUAAAUUGCC:GACUA:G  : AAA U +U  A+A+
   697726 GCAACCGUAAAUUGCCGGACUAUGCAGGAAAGUGAUCGAUAA 697767

 Query = 1 - 102, Target = 693731 - 693832
 Score = 85.00, E = 2.717e-18, P = 1.169e-24, GC =  38

          :::::::::::::::::((((((((,,,<<<<<<<_______>>>>>>>,,,,,,,,<<<
        1 aaaaaaaaaaaaaaaaaucacuCgUAUAAucccgggAAUAUGGcccgggaGUUUCUACCag 60
          A AAA+ AAA+AA A+   : CGUAUAAU :CG GAAUAUGGC CG: AGU UCUACCA:
   693731 AGAAAUCAAAUAAGAUGAAUUCGUAUAAUCGCGGGAAUAUGGCUCGCAAGUCUCUACCAA 693790

          <<<<_______>>>>>>>,,))))))))::::::::::::::
       61 gcaaCCGUAAAuugccuGACUAcGagugaaauuauuaaaaau 102
          GC ACCGUAAAU GC:UGACUACG :   A+UU UU+  ++U
   693791 GCUACCGUAAAUGGCUUGACUACGUAAACAUUUCUUUCGUUU 693832

 Query = 1 - 102, Target = 4004455 - 4004556
 Score = 73.67, E = 2.539e-15, P = 1.092e-21, GC =  28

          :::::::::::::::::((((((((,,,<<<<<<<_______>>>>>>>,,,,,,,,<<<
        1 aaaaaaaaaaaaaaaaaucacuCgUAUAAucccgggAAUAUGGcccgggaGUUUCUACCag 60
           A+ ++A AAAAA A +::CU:GUAUA  ::C:G AAUAUGG C:G:: GUUUCUACC::
  4004455 CAUCUUAGAAAAAGACAUUCUUGUAUAUGAUCAGUAAUAUGGUCUGAUUGUUUCUACCUA 4004514
```

```
        <<<<_______>>>>>>>,,))))))))::::::::::::::
     61 gcaaCCGUAAAuugccuGACUAcGagugaaauuauuaaaaau 102
        G:  CCGUAAA  :C::GACUAC:AG::A +UU +++AAA+U
4004515 GUAACCGUAAAAAACUAGACUACAAGAAAGUUUGAAUAAAUU 4004556

 Minus strand results:

Query = 1 - 102, Target = 2319369 - 2319270
Score = 82.19, E = 1.481e-17, P = 6.372e-24, GC =  46

         ::::::::::::::::((((((((,,,<<<<<<<_______>>>>>>>,,,,,,,,<<<
       1 aaaaaaaaaaaaaaaaucacuCgUAUAAucccgggAAUAUGGcccgggaGUUUCUACCag 60
          +A AA+A+AA+A  A+CAC:C:UAUAAU :CG:G AUAUGGC:CG: AGUUUCUACC:G
 2319369 UUACAAUAUAAUAGGAACACUCAUAUAAUCGCGUGGAUAUGGCACGCAAGUUUCUACCGG 2319310

         <<<<_______>>>>>>>,,))))))))::::::::::::::
      61 gcaaCCGUAAAuugccuGACUAcGagugaaauuauuaaaaau 102
          CA CCGUAAA UG C:GACUA:G:GUGA  +++  AA
 2319309 GCA-CCGUAAA-UGUCCGACUAUGGGUGAGCAAUGGAACCGC 2319270

Query = 1 - 102, Target = 625950 - 625851
Score = 65.20, E = 4.193e-13, P = 1.804e-19, GC =  30

         ::::::::::::::::((((((((,,,<<<<<<<_______>>>>>>>,,,,,,,,<<<
       1 aaaaaaaaaaaaaaaaucacuCgUAUAAucccgggAAUAUGGcccgggaGUUUCUACCag 60
         AA++AAA+A  +A++AU   U:GUAUAA:C:C:: AAUAUGG ::G:G:GU UCUACCAG
  625950 AAUUAAAUAGCUAUUAUCACUUGUAUAACCUCAAUAAUAUGGUUUGAGGGUGUCUACCAG 625891

         <<<<_______>>>>>>>,,))))))))::::::::::::::
      61 gcaaCCGUAAAuugccuGACUAcGagugaaauuauuaaaaau 102
         G:  CCGUAAA  :CCUGA UAC:A    + UU++U A A+U
  625890 GAA-CCGUAAA-AUCCUGAUUACAAAAUUUGUUUAUGACAUU 625851
```

The header (not shown here) has information on the version of Infernal, the files used and the run time. The next line is the name of the covariance model used (CM), followed by the sequence name in the FASTA file on another line. For files with multiple sequences in FASTA, the name of the sequence is displayed for each group of corresponding hits in Infernal's output. Also displayed is the strand polarity in which the RNA was found. The corresponding positions of the hits are indicated for the query (which is often the entire query, but can also be only a portion of it, especially in "local searches") and for the target sequence. An evaluation of the validity of the hit is shown on the next line, the "score" reflects how well the hit matches the model, while the "E-value" corresponds to the expected number of hits with that "score" (or better) in a random sequence of the same size as the one you are looking at, i.e., the number of expected false positives. Similarly, a hit with a low P-value has a low probability of being a false positive. Finally, "GC" corresponds to the GC-content (in percentage). The following lines correspond to pairwise alignments where the first line describes the secondary structure, the second line is the "query" sequence, and the third line highlights homologous regions with the fourth line, which is the sequence of the hit. In the case shown above, all the hits have very good E-values (approximately ranging from 10^{-13} to 10^{-18}). However, in the case of ambiguous hits, with E-values closer to 1, manual inspection can provide the additional clues needed to confirm the presence of a riboswitch at that position. For example, a loop–loop base-pairing interaction forms between the two loops of the purine riboswitch (here, the loops are annotated with "_"). This feature is not evaluated by Infernal and therefore does not contribute to the E-value. Observing this interaction in a relatively poor hit, with an E-value of 0.2 for instance, would mean this hit is more likely to be a true riboswitch than suggested strictly by the E-value. For purine riboswitches, detailed knowledge of the riboswitch is also useful to discriminate between adenine and guanine riboswitches. Because these two differ by a single base, Infernal finds both types of riboswitches during the same search. In the output shown above, the first four hits are guanine riboswitches and the last one is an adenine riboswitch. The former have a "C" at the last base of the junction, while the adenine riboswitch has a "U" at that position (shown in bold in the alignments above and annotated with ",").

3.3 Search for Variants of a Known Motif in All Bacterial Genomes

The Infernal suite can be used to find atypical riboswitches or ribozymes, but if many genomes are evaluated for poor E-values, thousands of hits will be generated and will require a lot of CPU time (*see* **Note 1**). Afterwards, knowing the structure of the RNA in detail can help to sort through the haystack of hits that would ensue a search accepting E-values as high as 5,000. Evaluating the presence of pseudoknots or the relevance of the downstream gene being regulated by this RNA are examples of how one can judge whether hits are likely real ncRNAs. This entire process takes much more time than what is described in Subheading 3.2 and is not recommended for all homology searches. This approach is more feasible in a case where only a few genomes are to be scrutinized, although the E-value should be set closer to 1 since it corresponds to the number of false positives expected to be found with that score (or better) in a database of this size. Therefore, a "cmsearch" allowing a maximum E-value of 5 (1 is default) could be performed and the hits could be manually screened one by one with the criteria mentioned hereafter when specifically interested in the genome of one bacteria. However, before performing such a search on all microbial genomes (with a maximal E-value of 1,000 for instance), one should have a strong basis to believe more riboswitches or ribozymes can be found since it would be CPU-time intensive and would generate a lot of false positives requiring a lot of time to sort through.

The steps described in Subheading 3.2 are also valid for searches in all available genomes. However, to get hits with E-values as high as 100 (for example), the "-E" option with "cmsearch" is used (*see* **Note 4** for more information on "--tabfile").

```
# cmsearch -E 100 --tabfile results.tab purine.cm
sequence.fa
```

Where “sequence.fa” could be a large file with all microbial genomes (available from NCBI [48], note that this compressed file is approximately 2 Gb in size). Some adjustments can be useful to determine the best value for “-E” (*see* **Note 5**). Different softwares can help visualizing a large number of hits. In that regard, the “RALEE” major mode in “Emacs” (a text editor program) is very useful [49]. It can color Stockholm alignments according to conservation, stems, or covariation. Both “RALEE” [50] and Emacs [51] are available for any Operating System platform.

When sorting the hits, as many criteria as possible should ideally be used to distinguish potentially good hits from spurious ones. Here are a few noticeable features, that have already proven useful in other works [4, 5, 35]:

1. Pseudoknots: Infernal does not take the base pairs of pseudoknot in consideration. Thus, it cannot account for pseudoknots in its E-value, which means that manually confirming the presence of a known pseudoknot in the ncRNA greatly improves this hit’s likelihood of being real.
2. Essential bases: when the structure has been studied enough to determine which bases are absolutely crucial for the RNA’s function, the hits that do not have these bases can be considered as spurious. However, one must be careful with such criteria since an apparent deleterious mutation at a specific position could be compensated by different bases at other positions, as in the case of the core-conserved C3G8 base-pair within the hammerhead ribozyme core, which was sometime found to be U3A8 [38, 52].
3. Intergenic versus coding sequence (CDS): even though riboswitches could theoretically be found in coding sequences, to our knowledge there is no natural riboswitch found yet that is completely embedded in the coding sequence. Therefore, if the hit is in an intergenic region, it should be regarded as more likely to be real, and, conversely, as spurious if it is in a CDS.
4. Functional relevance: when the riboswitch’s ligand is known, the connection with the genes is often obvious. For example, in the above list of purine riboswitches, a hypoxanthine/guanine permease can be found downstream of a hit, as well as other genes involved in purine synthesis for other hits. The absence of a clear connection between the candidate riboswitch and the function of the downstream gene does not automatically means a false positive, but an obvious connection does help for its validation.
5. Expression platforms: to exert their effect on expression, the aptamer portion of riboswitches is usually close (or even

overlapping) to a terminator (*see* **Note 6**) or ribosome binding site in 5′ untranslated regions (UTR) [53]. Furthermore, if the downstream gene is not in the same orientation as the putative RNA, it is unlikely to be a true riboswitch.

After discarding most false positives, the resulting alignment (or the complete alignment from Rfam) is likely to have subgroups that have some of the diverging positions in common. This subgroup can be used as a secondary alignment for a more specialized "cmsearch." A remarkable example of such a subgroup was noticed in the purine alignment and led to discovery of a novel deoxyguanosine riboswitch, although in that case the secondary alignment did not provide any additional hits [54].

In some cases where ample data is available on the structure requirements of a RNA, such as the hammerhead ribozyme, artificial alignments can be constructed to find a conformation expected to be functional but not known in nature (*see* **Note 7**). For example, the following Stockholm alignment combines two structural types of hammerhead ribozyme sequences to form a hypothetical type, for which artificial constructs were known to be active but not known in nature until recently [21, 37–40]. The types of the hammerhead ribozymes are defined by the identity of the closing stem, the two others being simple hairpins. Here, alignments of type 1 and 3 were combined to search for type 2 hammerhead ribozymes (i.e., where stem 1 and 3 are simple hairpins and stem 2 is the closing stem).

```
# STOCKHOLM 1.0
#=GF ID Hammerhead_2_synthetic_alignment
J1           UGUCC..GAAACGCU.....GCGA.......AGCGUCUAGGC...GUUAU......GCCUACUGAUGA...GGAC
J2           AGGAU..GAAACCAUAC...CAUAGU...GUAUGGUCGGAUAA..UAUUUUU...UUAUCCCUGAUGA...AUCC
J3           UGUCC..GAAA.........CUCG...........UCUGCCCC..UGAUGCC...GGGGCACUGACGAU..GGAC
N1           CGUCAG.GAAACC.......ACUA.........GGUCUGCCCC..UGAUGCC...GGGGCACUGACGAU.CUGAC
N2           .GCUGG.GAAACGUC.....AACA.......GACGUCGUGAUC..UGAAACUC..GAUCACCUGAUGA..CCGGC
N3           UGAC...GAAACACCAACAGUGGGGGCUGUUGGUGUCUGAGCG..UGAUACC...CGCUCACUGAAGAU...GUC
N4           UGAC...GAAACAUCAACAGUGGGGGCUGUUGGUGUUUCGAG...CCA........CACGGCUGAUGAA...GUC
N5           UGCCAG.GAAACC.......CAAUU........GGUUUCGAGC..ACACCG....GCUCUGCUGAUGAA.CUGGC
N6           UGUUC..GAAA.........CUAG...........UCCAGGG...UUA........CCCUGCUGACGA...GAAC
N7           .AUCGG.GAAACGC......UUGG........GCGUCUGGACC..UGAUGCC...GGUCCACUGAUGA..CCGAU
N8           AGCCGG.GAAACCU......UAAC........AGGUCUCCAGAU.GUGU.....GUCUGGACUGAUGA..CCGGC
N9           .GCUGGUGAAACGUC.....AAUC.......GACGUUAAAUG...UCAUACA....CAUUUCUGAGGA.ACCGGC
N10          CGACUAUGAAACCUCU....UUUAU.....AGAGGUUUCGAGC..ACACCG....GCUCUGCUGAUGAAAUAGUC
N11          NGUCC..GAAAC........GUGC..........GUCCAGGG...UUA........CCCUGCUGACGA...GGAC
N12          GGUAU..GAAAAA.......AAAU.........UUUCCGGGAG..UAAUGCU...CUCCCGCUGAUGAU..AUAC
N13          CGA....GAAACAG......GGC.........CUGUUAAAUG...UCAUACA....CAUUUCUGAGGA.....UC
N14          CAUUU..GAAAUAGC.....CAA........GCUGUCGGAAA...GUGUGCGC...UUUCCCUGAUGA...AAAU
N15          AGCUC..GAAACC.......UGUUG........GGUUUCGAG...CCA........CACGGCUGAUGAA..GAGC
N16          CGUCC..GAAAC........CCCA..........GUAUCGGGAU.UGUG.....GUCCCGGCUGAUGA...GGGC
#=GC SS_cons .<<<<<<...<<<<<<<<<<......>>>>>>>>>>.<<<<<<<<........>>>>>>>>........>>>>>>
//
```

After building and calibrating this model, a "cmsearch" can be performed (as seen in Subheading 3.2). Here is a portion of the results of a search against the chromosome 1 of Agrobacterium tumefaciens (NC_003062) [55].

```
CM: Hammerhead_2_synthetic_alignment
>gi|159184118|ref|NC_003062.2|

  Plus strand results:

 Query = 1 - 51, Target = 1183818 - 1183877
 Score = 28.65, E = 0.0002488, P = 2.328e-10, GC =  65

           :((((,,,<<<<......____..>>>>,<<<<<<._______.>>>>>>,,,,,,,.))
         1 uGuCcGAAACau......uaaa..auGUCuggggc.UgAUaCu.gccccaCUGAUGA.gG 49
           U::C GAAAC::      UA    ::GUC: GGG: U+    U :CCC :CUGAUGA  G
   1183818 UACC-GAAACCGgcucccUAGGguCGGUCGUGGGcUUGGCAUgCCCCGCCUGAUGAu-G 1183875

           ))
        50 aC 51
           ::
   1183876 GU 1183877


 Query = 1 - 51, Target = 1956021 - 1956068
 Score = 19.83, E = 0.09822, P = 9.192e-08, GC =  52

           :((((,,,<<<<____.....>>>>,<<<<<<_______>>>>>>,,,,,.,,))))
         1 uGuCcGAAACauuaaa.....auGUCuggggcUgAUaCugccccaCUGAU.GAgGaC 51
           +GUCCGAAA : + A      : U :GG:::       :::CC:CUGAU GAGGAC
   1956021 AGUCCGAAAUC-AUAGaucag-GGUACGGUCA-------UGACCGCUGAUcGAGGAC 1956068

...
```

The first hit is a confirmed ribozyme [38, 40], while the second is more likely to be a false positive because the stem 3 is weak (due to a stem of only 3 base pairs) and there is a "C" insertion in the core (lower case "c" in "CUGAUcGA"). Note that even if this approach works relatively well, the type 2 hammerheads have been discovered by other methods [21]. Also, this approach can only be used for a few RNAs for which the structure is very well understood, which is necessary to manipulate sequence alignments in order to carefully simulate a new structural version of a known ncRNA.

4 Notes

1. While searching a single genome for a ribozyme or riboswitch is typically done within a few hours on a desktop computer (3 GHz for example). It would take several weeks to look at all available genomes. Running the Infernal software on a computer cluster is preferable to perform multiple "cmsearches" on all the sequenced microbes.
2. Using a Mac, the GNU C compiler gcc is not installed by default. Installation of Xcode will resolve this.
3. Alternatively, pre-built and pre-calibrated models are available from Rfam here [56]. The "cm" corresponding to the RNA of

interest can easily be copied and pasted as a text file for which the extension has to be changed to ".cm" before using it.

4. The --tabfile option allows the use of the tab file as an input to the "Easel" miniapp "esl-sfetch" (found in a subdirectory of Infernal). The miniapp "esl-sfetch" extracts the sequences of all hits found from the genome sequence file to a new FASTA file. This file is useful to get a new alignment with the CM file of the motif ("purine.cm" in the example) using the command "cmalign." To get a tabular version of the search results, the command line is:

```
# cmsearch --ga --tabfile results.tab purine.cm sequence_Bsubs.fa
```

Now, to use the tabfile "results.tab" as an input to fetch the hits sequences:

```
# easel/miniapps/esl-sfetch -C -f --tabfile sequence_Bsubs.fa results.tab
```

An error of this type: "`Failed to open SSI index`" may occur. In this case the "sequence_Bsubs.fa" (or the file containing the new genome) has to be indexed. This is done with this step:

```
# easel/miniapps/esl-sfetch --index sequence_Bsubs.fa
```

Now that the file containing the genome sequence is indexed, the "sfetch" command can be re-executed as above. The hits sequences are displayed in FASTA format. To get the output in a new FASTA file, the command line would be:

```
# easel/miniapps/esl-sfetch -C -f --tabfile sequence_Bsubs.fa results.
  tab>hitsequences.fa
```

The tabular version has the format shown below.

```
# model name  target name       start       stop  start   stop    bit sc   E-value  GC%
# ----------  -------------  ---------  ----------  -----  -----  --------  --------  ---
  Purine      gi|223666304|     697666      697767      1    102     85.74  1.74e-18   42
  Purine      gi|223666304|     693731      693832      1    102     85.00  2.72e-18   38
  Purine      gi|223666304|    4004455     4004556      1    102     73.67  2.54e-15   28
  Purine      gi|223666304|    2319369     2319270      1    102     82.19  1.48e-17   46
  Purine      gi|223666304|     625950      625851      1    102     65.20  4.19e-13   30
```

Shown below is the beginning of the file "hitsequences.fa" containing the sequences of the hits found in the *Bacillus subtilis* genome:

```
>gi|223666304|ref|NZ_CM000487.1|/697666-697767/Purine/B85.74/E1.7e-1
8/GC42 Bacillus subtilis subsp. subtilis str. 168 chromosome, whole g
enome shotgun sequence
CATGAAATCAAAACACGACCTCATATAATCTTGGGAATATGGCCCATAAGTTTCTACCCG
GCAACCGTAAATTGCCGGACTATGCAGGAAAGTGATCGATAA
>gi|223666304|ref|NZ_CM000487.1|/693731-693832/Purine/B85.00/E2.7e-
18/GC38 Bacillus subtilis subsp. subtilis str. 168 chromosome, whole g
enome shotgun sequence
AGAAATCAAATAAGATGAATTCGTATAATCGCGGGAATATGGCTCGCAAGTCTCTACCAA
GCTACCGTAAATGGCTTGACTACGTAAACATTTCTTTCGTTT
...
```

These sequences are aligned using the "purine.cm" as a seed to obtain a new motif that can be used in further searches:

```
# src/cmalign purine.cm hitsequences.fa
```

This should give the following (the output was slightly modified to fit on the page):

```
# STOCKHOLM 1.0
#=GF AU Infernal 1.0.2

gi|223666304|...     CAUGAAAUCAAAACACGACCUCAUAUAAUCUUGGGAAUAUGGCCCAUAAG
gi|223666304|...     AGAAAUCAAAUAAGAUGAAUUCGUAUAAUCGCGGGAAUAUGGCUCGCAAG
gi|223666304|...     CAUCUUAGAAAAAGACAUUCUUGUAUAUGAUCAGUAAUAUGGUCUGAUUG
gi|223666304|...     UUACAAUAUAAUAGGAACACUCAUAUAAUCGCGUGGAUAUGGCACGCAAG
gi|223666304|...     AAUUAAAUAGCUAUUAUCACUUGUAUAACCUCAAUAAUAUGGUUUGAGGG
#=GC SS_cons         ::::::::::::::::((((((((,,,<<<<<<<_______>>>>>>>,
#=GC RF              aaaaaaaaaaaaaaaucacuCgUAUAAucccgggAAUAUGGcccgggaG

gi|223666304|...     UUUCUACCCGGCAACCGUAAAUUGCCGGACUAUGCAGGAAAGUGAUCGAU
gi|223666304|...     UCUCUACCAAGCUACCGUAAAUGGCUUGACUACGUAAACAUUUCUUUCGU
gi|223666304|...     UUUCUACCUAGUAACCGUAAAAAACUAGACUACAAGAAAGUUUGAAUAAA
gi|223666304|...     UUUCUACCGGGCA-CCGUAAA-UGUCCGACUAUGGGUGAGCAAUGGAACC
gi|223666304|...     UGUCUACCAGGAA-CCGUAAA-AUCCUGAUUACAAAAUUUGUUUAUGACA
#=GC SS_cons         ,,,,,,,<<<<<<<_______>>>>>>>,,))))))))::::::::::::
#=GC RF              UUUCUACCaggcaaCCGUAAAuugccuGACUAcGagugaaauuauuaaaa

gi|223666304|...     AA
gi|223666304|...     UU
gi|223666304|...     UU
gi|223666304|...     GC
gi|223666304|...     UU
#=GC SS_cons         ::
#=GC RF              au
```

5. To compromise between the number of spurious hits and the potential of getting new valid hits, it can be useful to test a few different E-values. Ideally, an instance that diverges from the consensus and that is known (or strongly suspected) to be a true riboswitch should be used. In that case, you can always expect to find more hits by increasing the E-value until this hit is also detected.

6. Some tools have been developed to detect terminators, "RNIE" being the most up-to-date and is available here [57, 58]. "RNIE" works with the Infernal suite, it can be used to find all transcription terminators in a genome with the following command:

```
# src/cmsearch -T 16 -g --fil-no-qdb --fil-T-hmm 2 --cyk --beta 0.05 CM
  sequence.fa
```

7. Alignments can be modified in many ways. In the case presented above, the first half of stem 2, the GAAA and stem 3 were taken from a type 1 alignment, while stem 1 and the CUGANGA were taken from a stem 3 alignment. A simpler modification that could be made to an alignment could be to change the bases of a conserved pseudoknot with the hopes that the CM built from that new "synthetic alignment" would

find instances that have been missed because their pseudoknot's sequence diverges from the current model.

Acknowledgements

This work was supported by the Natural Sciences and Engineering Research Council of Canada (NSERC) discovery grant (RGPIN 418240-2012) and by a grant from The Banting Research Foundation to JP.

References

1. Wan Y, Kertesz M, Spitale RC, Segal E, Chang HY (2011) Understanding the transcriptome through RNA structure. Nat Rev Genet 12:641–655
2. Benson DA, Karsch-Mizrachi I, Clark K, Lipman DJ, Ostell J, Sayers EW (2012) GenBank. Nucleic Acids Res 40:D48–D53
3. Sun S, Chen J, Li W, Altintas I, Lin A, Peltier S, Stocks K, Allen EE, Ellisman M, Grethe J, Wooley J (2011) Community cyberinfrastructure for Advanced Microbial Ecology Research and Analysis: the CAMERA resource. Nucleic Acids Res 39:D546–D551
4. Weinberg Z, Wang JX, Bogue J, Yang J, Corbino K, Moy RH, Breaker RR (2010) Comparative genomics reveals 104 candidate structured RNAs from bacteria, archaea, and their metagenomes. Genome Biol 11:R31
5. Weinberg Z, Perreault J, Meyer MM, Breaker RR (2009) Exceptional structured noncoding RNAs revealed by bacterial metagenome analysis. Nature 462:656–659
6. Shi Y, Tyson GW, DeLong EF (2009) Metatranscriptomics reveals unique microbial small RNAs in the ocean's water column. Nature 459:266–269
7. Livny J, Waldor MK (2007) Identification of small RNAs in diverse bacterial species. Curr Opin Microbiol 10:96–101
8. Pedersen JS, Bejerano G, Siepel A, Rosenbloom K, Lindblad-Toh K, Lander ES, Kent J, Miller W, Haussler D (2006) Identification and classification of conserved RNA secondary structures in the human genome. PLoS Comput Biol 2:e33
9. Rivas E, Eddy SR (2001) Noncoding RNA gene detection using comparative sequence analysis. BMC Bioinformatics 2:8
10. Washietl S (2007) Prediction of structural noncoding RNAs with RNAz. Methods Mol Biol 395:503–526
11. Gruber AR, Neubock R, Hofacker IL, Washietl S (2007) The RNAz web server: prediction of thermodynamically stable and evolutionarily conserved RNA structures. Nucleic Acids Res 35:W335–W338
12. Washietl S, Hofacker IL, Stadler PF (2005) Fast and reliable prediction of noncoding RNAs. Proc Natl Acad Sci U S A 102:2454–2459
13. Yao Z, Weinberg Z, Ruzzo WL (2006) CMfinder–a covariance model based RNA motif finding algorithm. Bioinformatics 22:445–452
14. Harmanci AO, Sharma G, Mathews DH (2007) Efficient pairwise RNA structure prediction using probabilistic alignment constraints in Dynalign. BMC Bioinformatics 8:130
15. Will S, Reiche K, Hofacker IL, Stadler PF, Backofen R (2007) Inferring noncoding RNA families and classes by means of genome-scale structure-based clustering. PLoS Comput Biol 3:e65
16. Knudsen B, Hein J (2003) Pfold: RNA secondary structure prediction using stochastic context-free grammars. Nucleic Acids Res 31:3423–3428
17. Lorenz R, Bernhart SH, Honer Zu Siederdissen C, Tafer H, Flamm C, Stadler PF, Hofacker IL (2011) ViennaRNA Package 2.0. Algorithms Mol Biol 6:26
18. http://en.wikipedia.org/wiki/List_of_RNA_structure_prediction_software
19. Storz G, Vogel J, Wassarman KM (2011) Regulation by small RNAs in bacteria: expanding frontiers. Mol Cell 43:880–891
20. Cochrane JC, Strobel SA (2008) Catalytic strategies of self-cleaving ribozymes. Acc Chem Res 41:1027–1035
21. Hammann C, Luptak A, Perreault J, de la Pena M (2012) The ubiquitous hammerhead ribozyme. RNA 18:871–885
22. Roth A, Breaker RR (2009) The structural and functional diversity of metabolite-binding riboswitches. Annu Rev Biochem 78:305–334
23. Lee ER, Blount KF, Breaker RR (2009) Roseoflavin is a natural antibacterial compound

that binds to FMN riboswitches and regulates gene expression. RNA Biol 6:187–194

24. Blount KF, Wang JX, Lim J, Sudarsan N, Breaker RR (2007) Antibacterial lysine analogs that target lysine riboswitches. Nat Chem Biol 3:44–49
25. Blount KF, Breaker RR (2006) Riboswitches as antibacterial drug targets. Nat Biotechnol 24:1558–1564
26. Sudarsan N, Cohen-Chalamish S, Nakamura S, Emilsson GM, Breaker RR (2005) Thiamine pyrophosphate riboswitches are targets for the antimicrobial compound pyrithiamine. Chem Biol 12:1325–1335
27. Ott E, Stolz J, Lehmann M, Mack M (2009) The RFN riboswitch of Bacillus subtilis is a target for the antibiotic roseoflavin produced by Streptomyces davawensis. RNA Biol 6:276–280
28. Kim JN, Blount KF, Puskarz I, Lim J, Link KH, Breaker RR (2009) Design and antimicrobial action of purine analogues that bind Guanine riboswitches. ACS Chem Biol 4:915–927
29. Mulhbacher J, Brouillette E, Allard M, Fortier LC, Malouin F, Lafontaine DA (2010) Novel riboswitch ligand analogs as selective inhibitors of guanine-related metabolic pathways. PLoS Pathog 6:e1000865
30. Lunse CE, Schmidt MS, Wittmann V, Mayer G (2011) Carba-sugars activate the glmS-riboswitch of Staphylococcus aureus. ACS Chem Biol 6:675–678
31. http://rfam.sanger.ac.uk
32. Nawrocki EP, Kolbe DL, Eddy SR (2009) Infernal 1.0: inference of RNA alignments. Bioinformatics 25:1335–1337
33. Hoeppner MP, Barquist L, Gardner PP. An introduction to RNA databases. Methods in Molecular Biology (In press)
34. Barquist L, Burge SW, Gardner PP. Building non-coding RNA families. Methods in Molecular Biology (In press)
35. McCown PJ, Roth A, Breaker RR (2011) An expanded collection and refined consensus model of glmS ribozymes. RNA 17:728–736
36. de la Pena M, Garcia-Robles I (2010) Intronic hammerhead ribozymes are ultraconserved in the human genome. EMBO Rep 11:711–716
37. Jimenez RM, Delwart E, Luptak A (2011) Structure-based search reveals hammerhead ribozymes in the human microbiome. J Biol Chem 286:7737–7743
38. Perreault J, Weinberg Z, Roth A, Popescu O, Chartrand P, Ferbeyre G, Breaker RR (2011) Identification of hammerhead ribozymes in all domains of life reveals novel structural variations. PLoS Comput Biol 7:e1002031
39. Seehafer C, Kalweit A, Steger G, Graf S, Hammann C (2011) From alpaca to zebrafish: hammerhead ribozymes wherever you look. RNA 17:21–26
40. de la Pena M, Garcia-Robles I (2010) Ubiquitous presence of the hammerhead ribozyme motif along the tree of life. RNA 16:1943–1950
41. http://infernal.janelia.org/
42. http://rfam.sanger.ac.uk/genome/browse#A
43. Nawrocki EP, Kolbe DL, Eddy SD (2009) Infernal user guide. ftp://selab.janelia.org/pub/software/infernal/Userguide.pdf
44. Nawrocki EP. Annotating functional RNAs in genomes using Infernal. Methods in Molecular Biology (In press)
45. ftp://selab.janelia.org/pub/software/infernal/infernal-1.0.2.tar.gz
46. http://rfam.sanger.ac.uk/family/RF00167#tabview=tab2
47. http://www.ncbi.nlm.nih.gov/nuccore/223666304
48. ftp://ftp.ncbi.nih.gov/genomes/Bacteria/all.fna.tar.gz
49. Griffiths-Jones S (2005) RALEE–RNA ALignment editor in Emacs. Bioinformatics 21:257–259
50. http://personalpages.manchester.ac.uk/staff/sam.griffiths-jones/software/ralee/
51. http://www.gnu.org/software/emacs
52. Przybilski R, Hammann C (2007) The tolerance to exchanges of the Watson Crick base pair in the hammerhead ribozyme core is determined by surrounding elements. RNA 13:1625–1630
53. Barrick JE, Breaker RR (2007) The distributions, mechanisms, and structures of metabolite-binding riboswitches. Genome Biol 8:R239
54. Kim JN, Roth A, Breaker RR (2007) Guanine riboswitch variants from Mesoplasma florum selectively recognize 2'-deoxyguanosine. Proc Natl Acad Sci U S A 104:16092–16097
55. http://www.ncbi.nlm.nih.gov/nuccore/159184118?report=fasta
56. ftp://ftp.sanger.ac.uk/pub/databases/Rfam/CURRENT/Rfam.cm.gz
57. http://github.com/ppgardne/RNIE
58. Gardner PP, Barquist L, Bateman A, Nawrocki EP, Weinberg Z (2011) RNIE: genome-wide prediction of bacterial intrinsic terminators. Nucleic Acids Res 39:5845–5852

Chapter 10

Structure-Based Virtual Screening for the Identification of RNA-Binding Ligands

Peter Daldrop and Ruth Brenk

Abstract

Structure-based virtual screening exploits the 3D structure of the target as a template for the discovery of new ligands. It is a key method for hit discovery and was originally developed for protein targets. Recently, this method has also been applied to RNA targets. This chapter gives an overview of this method and its application in the context of ligand discovery for RNA. In addition, it describes in detail how to conduct virtual screening for RNA targets, making use of software that is free for noncommercial use. Some advice on how to avoid common pitfalls in virtual screening is also given.

Key words Structure-based ligand design, Molecular docking, Virtual screening, RNA–ligand docking

1 Introduction

Virtual screening has become a key method for ligand discovery [1–3]. The starting point for the screening exercise can either be a known ligand or the 3D structure of the target. If the latter is used, the method is often referred to as structure-based virtual screening or molecular docking. Molecular docking predicts the three-dimensional binding mode of given compounds to the receptor binding site and estimates their binding affinity (Fig. 1) [3–5]. Each entry of a small molecule database is sequentially placed in multiple orientations and conformations in the cavity and scored for steric and chemical complementarity. The result is a predicted binding mode for each database entry together with a score. When the database is sorted according to these scores, ligands are enriched among the top ranking molecules.

A plethora of docking algorithms and scoring functions have been developed and excellent reviews on the topic are available [3–5]. While docking was originally only applied for protein targets, its scope was recently extended to ligand discovery for RNA targets [6, 7]. In this field, two alternative approaches are followed:

Daniel Lafontaine and Audrey Dubé (eds.), *Therapeutic Applications of Ribozymes and Riboswitches: Methods and Protocols*, vol. 1103, DOI 10.1007/978-1-62703-730-3_10, © Springer Science+Business Media New York 2014

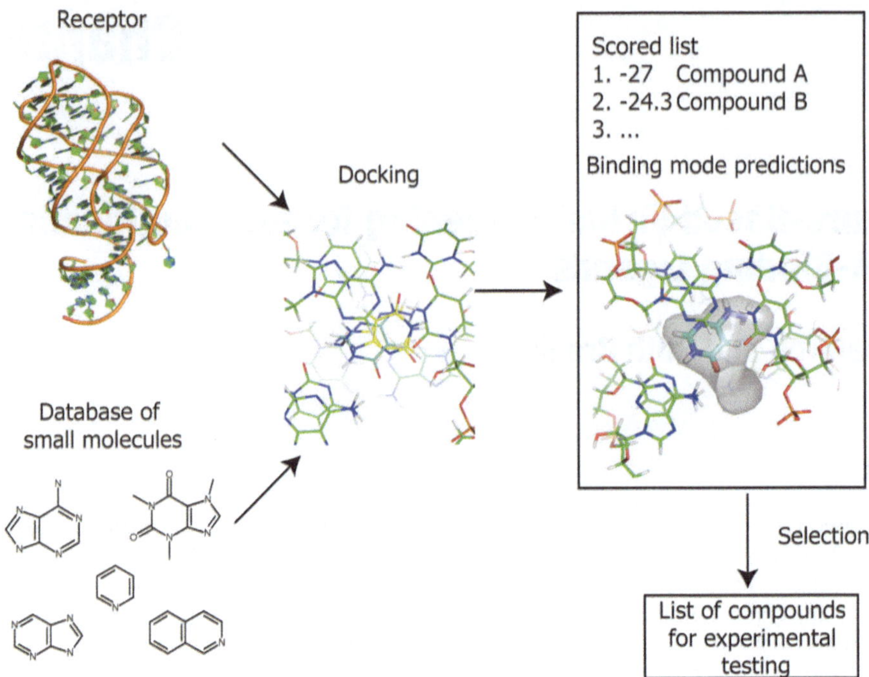

Fig. 1 Outline of molecular docking. Each entry of a small molecule database is sequentially placed in multiple orientations and conformations in the cavity and scored for steric and chemical complementarity. The result is a predicted binding mode for each database entry together with a score. High scoring ligands are visually inspected and selected for experimental testing

1. Adaptation of methods and scoring functions originally developed for protein–ligand docking [8–15].
2. Development of new methods tailored to RNA targets [16–20].

Most of the methods were only evaluated retrospectively using already published data [8–11, 13, 16, 19, 20]. These studies demonstrated that it is possible to correctly predict binding modes of RNA-binding ligands and that correlation between experimental and predicted binding affinities can be achieved.

More stringent tests for docking methods and scoring functions are prospective tests in which new ligands for a given target are predicted and subsequently confirmed experimentally. In a number of such studies, different RNA-tailored versions of DOCK [21] were used, partially in combination with other docking programs. For instance, Chen et al. used this approach for the discovery of ligands binding to the RNA double helix [22]. Yan et al. applied DOCK followed by AutoDOCK [23] for the discovery of RNA tetraloop binding compounds [24]. A similar strategy was adopted by Warui and Baranger to identify ligands for the stem loop 3 ribonucleic acid of the packaging signal Psi of human immunodeficiency virus-1 [25].

By using AutoDOCK in combination with similarity searches the same research group also derived ligands binding to the 3′-dangling end of RNA1 [26]. Lind et al. retreated to RNA-tailored versions of DOCK and ICM [27] for the discovery of ligands that target the 5′ bulge of the transactivation response (TAR) element which is necessary for HIV-1 replication [12]. Finally, we reported recently on the adoption of DOCK 3.5.54 [28, 29] for structure-based virtual screening of a purine riboswitch [15]. In this study, binding of novel ligands was not only confirmed using a binding assay, but also the crystal structures of three new ligands in complex with the riboswitch were determined. Two of the observed binding modes matched closely the predicted ones. Park et al. used another typical protein–ligand docking program, FlexX [30], for the discovery of ligands binding to a pseudoknot inducing ribosomal frame shifting [14]. Morely and Afshar developed RiboDOCK specifically for RNA–ligand docking [18]. The main feature of this software is an adopted scoring function which includes terms for typical RNA–ligand interactions such as guanidinium–RNA interactions and π–π stacking of aromatic rings. RiboDOCK was subsequently used for the discovery of ligands that target the bacterial ribosomal A-site [31]. Another program specifically designed for RNA–ligand docking is MORDOR [17]. MORDOR uses molecular simulation techniques to account for both nucleic acid and ligand flexibility. This software was used for the identification of ligands for human telomerase RNA [32].

Taken together, the retrospective and prospective studies demonstrated that RNA–ligand docking is a valuable tool for the discovery of RNA-binding ligands when the structure of the target is known. As the number of RNA–ligand complexes deposited in the PDB [33] has increased steadily over the past 10 years new opportunities for structure-based virtual screening for RNA targets are opening up [7]. Of particular interest in this context are riboswitches.

Riboswitches are *cis*-acting gene regulatory elements that are mostly found in bacteria [34]. They are located in the 5′ untranslated region (UTR) of mRNAs and consist of an aptamer domain that binds the ligand, and an expression platform that controls the expression of the downstream gene. The RNA can adopt one of several alternative conformations, the relative stability of which is determined by the binding of the ligand to the aptamer domain. Binding of the ligand directs folding of downstream elements in the expression platform that influence expression. Thus, regulation of gene expression is controlled by the concentration of the small molecule ligand via the structure of the RNA. As such, riboswitches constitute potential drug targets [34]. Crystal structures of over ten riboswitches have been determined, making this class of RNA amenable for structure-based drug design [35–41].

Here, we describe how to conduct structure-based virtual screening for RNA targets, making use of software for which the licenses are free for noncommercial use.

2 Materials

1. A PC with a current version of Linux installed.
2. A 2.X version of Python (http://www.python.org/getit/) together with the NumPy module.
3. A version of OpenEye's OEChem TK that is compatible with the installed Python and Linux versions (http://www.eyesopen.com/oechem-tk).
4. The docking program DOCK 3.6 (http://dock.compbio.ucsf.edu/DOCK3.6/).
5. The Molecular Design Software Suite Moloc (www.moloc.ch).
6. A version of PyMOL (0.99 or later).
7. A bundle of Python scripts to assist receptor preparation and for file conversions (http://www.pharmazie.uni-mainz.de/AK-Brenk/usefull_stuff.php).
8. Files with parameters needed to calculate van der Waals and electrostatic energies for RNA–ligand complexes, to setup the directories and calculate the grids needed for docking, and the input parameters for docking (http://www.pharmazie.uni-mainz.de/AK-Brenk/usefull_stuff.php).
9. A PyMOL add on to handle spreadsheets (http://www.pharmazie.uni-mainz.de/AK-Brenk/usefull_stuff.php).
10. A file in PDB format containing the coordinates for an RNA receptor.

3 Methods

3.1 Overview

Structure-based virtual screening consists of five major steps (Fig. 1): First, a database of compounds which shall be screened against a target is needed. This database has to contain the 3D structures of the ligands together with relevant protonation states, tautomers and stereoisomers and has to be stored in a format suitable for docking. Second, the receptor structure is prepared. A decision has to be made if any structural water molecules or metal ions are relevant for bridging ligand-RNA contacts. In this case, they are kept as part of the receptor. All other atoms which are not part of the RNA are deleted. In addition, parameters relevant for docking are calculated. The third step is the central step in which the selected ligands are placed in the binding pocket in multiple conformations and orientations and scored for their fit with the receptor. Fourth, the docking poses of high ranking compounds are visually inspected to eliminate potential false positive predictions and to short list compounds for experimental testing. Finally, the predicted hits are tested for binding. The assay method depends on the target. Therefore, this step is not covered in this chapter.

3.2 Database of Compounds in Dockable Format

There are many ways on how to compile a database suitable for docking. A straightforward option is to download compounds from the ZINC Web site (http://zinc.docking.org), which is a free database of commercially available compounds provided in a format suitable for virtual screening [42]. Relevant protonation states, tautomers, and stereoisomers are also provided (*see* **Note 1**). Tutorials on how to navigate the ZINC database can be found in a recent publication [42]. We recommend starting with the fragment- or lead-like subset to limit the number of compounds for molecular docking and to speed-up the process. It is also advisable to restrict the number of compounds by filtering the subsets for a certain range of physicochemical properties like net charge, and minimum and maximum number of hydrogen-bond donors and acceptors. This will avoid waisting time with compounds which fall clearly outside the desired property range. Appropriate limits can be derived by analyzing the receptor structure and the properties of any known ligands (*see* **Note 4**). Finally, the search can be restricted to affordable compounds with a relative short delivery time by only considering compounds that are classified in ZINC as "in stock only." To proceed with docking using DOCK 3.6 as suggested below, the compiled compound set has to be saved in the "Flexibase" format.

3.3 Preparation of the Receptor for Docking

The output of this step is a receptor file containing hydrogen atoms in the orientation required for ligand binding, and bridging metal ions or water molecules if required. Additionally, files needed to define the location and size of the binding site, to place the ligand into the binding site and to calculate the docking score are generated.

1. Generate a working directory and save the receptor coordinates in the PDB format into this folder.
2. Analyze your receptor structure using a viewer (e.g., PyMOL) and identify water molecules and/or ions that are required to bridge RNA–ligand contacts (*see* **Note 2**). Delete all remaining components except of the RNA molecule and the ligand and all lines that do not start with "ATOM" or "HETATM." This can be done using a text editor.
3. Add hydrogen atoms and minimize their position using Moloc. To follow the steps outlined below it is important that the

software is started from the same directory in which the receptor coordinates are stored.

(a) Start Moloc.

(b) (optional) Right click on the "???" menu at the bottom of the screen to activate menu explanations.

(c) Right click on "..." then "g", "p" and right click into the dialog box to obtain a list of all files in the directory.

(d) Right click on the "n" next to the desired file containing your modified receptor (*see* **step 2** above) and then on the "x" to confirm the selection. Then right click on the "0" next to whole list to load all components of the PDB file and right click "x" to confirm.

(e) Right click to select a color and right click again to finally load the coordinates.

(f) Save your file to change the internal format which is required to correctly assign the protonation states of the atoms in the RNA molecule. Right click on "x" to access the parent menu, then "s" to save. Right click "p" to select the PDB format, then right click the "n" next to "whole list" and right click "x" twice. When asked for a file name, select a file name of your choice and right click into the dialog box. Right click yes and twice more into the dialog boxes.

(g) Delete the active structure from Moloc, by clicking "x" to access the parent menu, "+" for extra options and "d" to delete. Select the "n" next to "whole list" with a right click and right click "x" twice to confirm.

(h) Load the saved PDB file into Moloc by right clicking "x", "g", "p" and the dialog box. Select the correct file with a right click and right click "x" to confirm, then right click the "0" next to "whole list," then "x" to confirm. Right click on a color option and into the dialog box.

(i) Right click on "x" to access the parent menu, then on "a" to change the activity status of the loaded molecules (called entries in Moloc). Right click in the box next to the loaded entry until marked "A" for active and right click "x" to confirm. Right click on "x" again to return to the main menu.

(j) Add hydrogen atoms to the receptor. First, define all atoms present already (the non-hydrogen atoms) as stationary by clicking "opt" "s", "e", hold ctrl key and left click on any atom of the receptor. Right click "x" and again "x" to return to the main menu. Then, add the hydrogen atom by right clicking "dTp", "k", the "n" next to "whole list", "x", "x", and "accept." Right click "x" to return to the main menu.

(k) Minimize the positions of the hydrogen atoms. Right click on "opt" to access the force field menu, "x" to keep the stationary set and then "o" to start the optimization. Hydrogen bonds will appear as colored lines and movements of the hydrogen atoms will be observed. Wait until optimization is finished. Depending on the receptor size this can take a few minutes.

(l) Visually inspect the hydrogen bonding pattern in the binding site making use of the options in the "energy examination" submenu to ensure a correct result. If necessary, manually adjust rotatable hydrogen bonds to obtain a better starting structure free of false local minima using the options in the "forge: coordinate changes" menu accessible from the main menu and rerun the minimisation as described above.

(m) Once satisfied with the result save the minimized coordinates as PDB file. Starting from the main menu, right click "…", then "s", "p" select "all visible entries", "x", choose a file name, press enter and press enter twice more in the dialog boxes. Check that the file has been properly saved before exiting Moloc.

4. If any water molecules or ions are to be kept for docking, the corresponding coordinate lines in the PDB file need to start with "ATOM" as all other lines will be deleted subsequently. Also the atom and residue names in the pdb file need to be edited manually, so they match the water entry in the lookup files (files "amb.crg.oxt" and "prot.table.ambcrg.ambH"). The residue name should be "TIP" and the atom names "OH2", "H1", and "H2". These changes can be carried out using a text editor.

5. Run "python<path to python scripts>/recprep2.py<input file>rec.amb" to reformat the receptor file with <path to python scripts> being the actual path to the directory in which you stored the scripts (*see* Subheading 2) and <input file> the modified PDB file. Create a symbolic link from "rec.amb" to "rec.pdb" (ln –s rec.amb rec.pdb). The "rec.amb" file will now only contain the receptor RNA atoms and, if necessary, ions and/or water molecules which bridge RNA–ligand contacts. Next, the terminal hydrogen atoms on the O3″ and the 5″ phosphate O have to be named "CAP" and "Hfi", respectively, using a text editor. If any terminal phosphate groups are present, the oxygen atom bearing the hydrogen must be renamed as "OP1".

6. The location and size of the binding pocket is defined by a set of dummy atoms (called the sphere set). If you are working

with a structure containing a ligand that fills the entire binding pocket cut-out the ligand of the original PDB file using a text editor, change all "HETATM" entries to "ATOM" (note the two blanks at the end), remove the chain ID if present, and save the file as "xtal-lig.pdb". If the structure you are working with contains no ligand or if the ligand does not fill the binding site, fill the binding site manually with dummy atoms. There are many ways to do this. One option is to use Moloc. Check the program manual for instructions on how to place new atoms in a desired receptor region. Save your file with the name "xtal-lig.pdb".

7. Setup all directories and calculate the grids needed for docking and scoring.
   (a) Make sure the environment variable "$DOCK_BASE" is pointing to the "trunk" directory of your DOCK 3.6 installation.
   (b) Replace the files "amb.crg.oxt" and "prot.table.ambcrg.ambH" in "$DOCK_BASE/scripts/grids"/with the ones containing the parameters for RNA–ligand docking (*see* Subheading 2).
   (c) Download the "Makefile" (*see* Subheading 2) and place it in the directory containing the "rec.pdb", "rec.amb", and "xtal-lig.pdb" files.
   (d) Setup the directory structure and calculate the grids by executing the following command in a shell in the directory containing the PDB files: "make auto –f Makefile".
   (e) Make sure that the receptor atoms were recognized correctly by checking the output below the line "Checking for WARNINGS in OUTPARM." for warnings and errors. If any warnings are displayed the atom types in the receptor were not recognized correctly. Make sure that the atom and residue names match the ones in files "amb.crg.oxt" and "prot.table.ambcrg.ambH". Delete the directories "sph", "grids", and "testing" and run the Makefile again. (If the warnings arise from terminal residues only they can be safely ignored if the residues are not part of the binding site.)
8. The sphere set can be optimized for docking. This set is later need as anchor points for the placement of compounds. During receptor preparation described above a sphere set was generated in which each atom in "xtal-lig.pdb" was converted into a sphere. To generate a less biased sphere set follow the steps outlined below.
   (a) In the directory in which rec.pdb is stored run "python <path to python scripts>/ligwrap.py .xtal-lig.pdb sph/boxed.sph 1 1.5 1". This will generate a sphere set around the atoms in xtal-lig.pdb which are oriented on a regular cubic lattice.

(b) Now run: "python <path to python scripts>/boxnibbler.py sph/boxed.sph rec.pdb sph/sphset.sph 1.7". This will remove all spheres from the generated lattice that are within 1.7 Å of any receptor atom and thus deemed too close for any ligand atoms to be placed here. The parameter can be adjusted as needed.

It is worth checking the sphere set in a molecular viewer to ensure that the desired binding site region is filled. Care should be taken that the number of spheres does not grow excessively, since this will slow down docking significantly (not more than 50 is usually a good starting point). When above scripts are run without any arguments, some information will be displayed about the use of these scripts. The arguments can be adjusted until a satisfactory set of spheres has been generated. Additionally, the sphere set can always be manually edited using a text editor. This allows addition and deletion of spheres as desired. For the sphere set used as matching points for ligand placement, it is not necessary that the entire binding site is filled. It is enough if spheres are placed in the region which has been identified has being key for RNA–ligand interactions and should be targeted by all potential ligands.

3.4 Docking Calculations

In this step, each ligand in the database is placed in multiple orientations and conformations into the binding site and scored for its fit with the receptor.

1. Download the "INDOCK" file (*see* Subheading 2) and save it into the "testing" directory.
2. Make the following adjustment to the INDOCK file using a text editor:
   (a) "ligand_atom_file": add path to the database file to be docked.
   (b) "output_file_prefix": can be freely chosen.
3. Docking is started by executing the command "$DOCK_BASE/bin/Linux/dock.csh" in the "testing" directory. Depending on the size and nature of the database file and the size of the sphere set this step can take anything between seconds and hours. Generally docking will be time-consuming for large ligands and large sphere sets (*see* **Notes 3** and **4**). Progress can be monitored by following the "OUTDOCK" file, e.g., "tail –f OUTDOCK".

3.5 Analysis of Docking Results

After successful completion of the docking calculation the results are visualized and analyzed. For this step, files containing the coordinates of the computed docking poses for each molecule and a table containing the score for each molecule are needed.

1. First, generate the coordinate files:
   (a) Decompress the output file generated by docking by executing the compound "gunzip <prefix>.eel1.gz" where <prefix> is the "output_file_prefix" entered in the "INDOCK" file.
   (b) In the "testing" directory create a subfolder. You can choose the name of this directory freely.
   (c) Change into the just created folder and run "python <path to python scripts>/dock2mol2_high_scoring.py ../<prefix>.eel1", where <prefix> is the name you chose as "output_file_prefix" in the "INDOCK" file. This will generate a mol2 file for each successfully docked and saved compound.
   (d) Change into the parent directory and run "python <path to python scripts>/outdock_score.py OUTDOCK scores.txt" to generate a text file containing the scores table for all docked compounds.
2. Open the scores.txt file in a text editor and insert the absolute path of the directory in which the docked molecules in mol2 format are stored into the first line, e.g., "/home/user/docking/testing/results/" and delete the line with the column headings.
3. Open PyMOL with the installed MSS add on, and press the load button. Open the "scores.txt" file. The docking results should now be visible in the MSS window as a spreadsheet. Selecting a line in the spreadsheet and pressing "show" will display the selected ligand in the PyMOL viewing window.
4. Load the receptor file used for docking ("rec.pdb").
5. Docking results can be sorted by score, molecule name, rank, etc. and visualized in the PyMOL viewing window.
6. Visually inspect high scoring compounds for appropriate hydrogen-bonding networks with the receptor and steric fit. Short-list promising compounds for purchase (*see* **Note 5**).

4 Notes

There are many different software programs and tools that can be used for RNA–ligand docking. Outlined above is just one option that has proven to be successful for us. However, independent of which methods you use, there are some points you should consider:

1. Make sure relevant tautomers, protonation states and stereoisomer of the ligands are either present in your docking database or generated by the docking program.

2. Analyze your receptor structure carefully to make sure you include important water molecules and ions if necessary. Also, if explicit hydrogen atoms are used for docking, make sure they are in an orientation required for ligand binding. It might make sense to use several different receptor setups taking into account the presence or absence of water molecules and ions and different hydrogen atom orientations.
3. Before embarking on docking a large number of ligands, dock a small test, preferably containing known ligands, to ensure that the receptor was prepared properly and sensible binding modes and scores are generated. If the output is not satisfying but there are no errors in the input files, adjust the docking parameters. What exactly you have to do is dependent on your docking program. For DOCK 3.6, you can increase sampling by increasing the ligand and receptor bin sizes (see http://wiki.bkslab.org/index.php/INDOCK_for_DOCK_3.6). You can also favor interactions with certain atoms by changing their partial charges. To do this; you have to give these atoms unique atom names in the "rec.amb" file. Add these atom types to the files "amb.crg.oxt" and "prot.table.ambcrg.ambH" and adjust the partial charges in these files. Make sure that the total charge of the residues is not altered by also adjusting the partial charges of neighboring atoms in the same residues that are not in direct contact with a ligand. Rerun the "Makefile" to recalculate the grids.
4. Carefully select the potential binders to be docked. Compounds for testing are usually short listed by visual inspection. There is no need to include any compounds into docking that you would later reject based on their chemical structure anyway. In addition, filtering the library for physicochemical properties is advantageous to reduce the numbers of compounds to be docked. Typical filters are net charge, number of non-hydrogen atoms, number of rotatable bonds, number of hydrogen-bond acceptor and donor groups and logP.
5. To select potential binders from the docking results we advise to sort by score and look at the best scoring compounds. Then select based upon a critical visual inspection of the binding modes. Ensure that the compounds have good shape complementarity with the binding sites and there are no unsatisfied hydrogen-bond donors or acceptors in either the ligand or the receptor. If the score sorted list seems overly biased by compounds size, sorting by score/MW or score/SQRT(MW) has been beneficial for the authors.

Acknowledgments

Research in the BCDD is supported by the Wellcome Trust (WT083481). P.D. was supported with a Wellcome Trust studentship (083930/Z/07/).

References

1. Ripphausen P, Nisius B, Peltason L et al (2010) Quo vadis, virtual screening? A comprehensive survey of prospective applications. J Med Chem 53:8461–8467
2. Schneider G (2010) Virtual screening: an endless staircase? Nat Rev Drug Discov 9: 273–276
3. Mcinnes C (2007) Virtual screening strategies in drug discovery. Curr Opin Chem Biol 11: 494–502
4. Kolb P, Ferreira RS, Irwin JJ et al (2009) Docking and chemoinformatic screens for new ligands and targets. Curr Opin Biotechnol 20:429–436
5. Klebe G (2006) Virtual ligand screening: strategies, perspectives and limitations. Drug Discov Today 11:580–594
6. Fulle S, Gohlke H (2010) Molecular recognition of RNA: challenges for modelling interactions and plasticity. J Mol Recognit 23:220–231
7. Tuccinardi T (2011) Binding-interaction prediction of RNA-binding ligands. Fut Med Chem 3:723–733
8. Detering C, Varani G (2004) Validation of automated docking programs for docking and database screening against RNA drug targets. J Med Chem 47:4188–4201
9. Kang X, Shafer RH, Kuntz ID (2004) Calculation of ligand-nucleic acid binding free energies with the generalized-born model in DOCK. Biopolymers 73:192–204
10. Lang PT, Brozell SR, Mukherjee S et al (2009) DOCK 6: combining techniques to model RNA-small molecule complexes. RNA 15: 1219–1230
11. Li Y, Shen J, Sun X et al (2010) Accuracy assessment of protein-based docking programs against RNA targets. J Chem Inf Model 50: 1134–1146
12. Lind KE, Du Z, Fujinaga K et al (2002) Structure-based computational database screening, in vitro assay, and NMR assessment of compounds that target TAR RNA. Chem Biol 9:185–193
13. Moitessier N, Westhof E, Hanessian S (2006) Docking of aminoglycosides to hydrated and flexible RNA. J Med Chem 49:1023–1033
14. Park SJ, Jung YH, Kim YG et al (2008) Identification of novel ligands for the RNA pseudoknot that regulate -1 ribosomal frameshifting. Bioorg Med Chem 16:4676–4684
15. Daldrop P, Reyes FE, Robinson DA et al (2011) Novel ligands for a purine riboswitch discovered by RNA-ligand docking. Chem Biol 18:324–335
16. Barbault F, Zhang LR, Zhang LH et al (2006) Parametrization of a specific free energy function for automated docking against RNA targets using neural networks. Chemometr Intell Lab Syst 82:269–275
17. Guilbert C, James TL (2008) Docking to RNA via root-mean-square-deviation-driven energy minimization with flexible ligands and flexible targets. J Chem Inf Model 48:1257–1268
18. Morley SD, Afshar M (2004) Validation of an empirical RNA-ligand scoring function for fast flexible docking using Ribodock. J Comput Aided Mol Des 18:189–208
19. Pfeffer P, Gohlke H (2007) DrugScoreRNA–knowledge-based scoring function to predict RNA-ligand interactions. J Chem Inf Model 47:1868–1876
20. Zhao X, Liu X, Wang Y et al (2008) An improved PMF scoring function for universally predicting the interactions of a ligand with protein, DNA, and RNA. J Chem Inf Model 48:1438–1447
21. Ewing TJA, Kuntz ID (1997) Critical evaluation of search algorithms for automated molecular docking and database screening. J Comput Chem 18:1175–1189
22. Chen Q, Shafer RH, Kuntz ID (1997) Structure-based discovery of ligands targeted to the RNA double helix. Biochemistry 36:11402–11407
23. Morris GM, Goodsell DS, Halliday RS et al (1998) Automated docking using a Lamarckian genetic algorithm and an empirical binding free energy function. J Comput Chem 19:1639–1662
24. Yan Z, Sikri S, Beveridge DL et al (2007) Identification of an aminoacridine derivative that binds to RNA tetraloops. J Med Chem 50:4096–4104
25. Warui DM, Baranger AM (2009) Identification of specific small molecule ligands for stem loop 3 ribonucleic acid of the packaging signal Psi of human immunodeficiency virus-1. J Med Chem 52:5462–5473

26. Ramisetty SR, Baranger AM (2010) Cooperative binding of a quinoline derivative to an RNA stem loop containing a dangling end. Bioorg Med Chem Lett 20:3134–3137
27. Abagyan R, Totrov M, Kuznetsov D (1994) Icm - a new method for protein modeling and design - applications to docking and structure prediction from the distorted native conformation. J Comp Chem 15:488–506
28. Lorber DM, Shoichet BK (1998) Flexible ligand docking using conformational ensembles. Protein Sci 7:938–950
29. Wei BQ, Baase WA, Weaver LH et al (2002) A model binding site for testing scoring functions in molecular docking. J Mol Biol 322: 339–355
30. Rarey M, Kramer B, Lengauer T et al (1996) A fast flexible docking method using an incremental construction algorithm. J Mol Biol 261:470–489
31. Foloppe N, Chen IJ, Davis B et al (2004) A structure-based strategy to identify new molecular scaffolds targeting the bacterial ribosomal A-site. Bioorg Med Chem 12:935–947
32. Pinto IG, Guilbert C, Ulyanov NB et al (2008) Discovery of ligands for a novel target, the human telomerase RNA, based on flexible-target virtual screening and NMR. J Med Chem 51:7205–7215
33. Berman HM, Westbrook J, Feng Z et al (2000) The protein data bank. Nucleic Acids Res 28: 235–242
34. Deigan KE, Ferre-D'amare AR (2011) Riboswitches: discovery of drugs that target bacterial gene-regulatory RNAs. Acc Chem Res 44:1329–1338
35. Schwalbe H, Buck J, Furtig B et al (2007) Structures of RNA switches: insight into molecular recognition and tertiary structure. Angew Chem Int Ed Engl 46:1212–1219
36. Serganov A (2010) Determination of riboswitch structures: light at the end of the tunnel? RNA Biol 7:98–103
37. Huang L, Serganov A, Patel DJ (2010) Structural insights into ligand recognition by a sensing domain of the cooperative glycine riboswitch. Mol Cell 40:774–786
38. Smith KD, Shanahan CA, Moore EL et al (2011) Structural basis of differential ligand recognition by two classes of bis-(3'-5')-cyclic dimeric guanosine monophosphate-binding riboswitches. Proc Natl Acad Sci U S A 108: 7757–7762
39. Smith KD, Lipchock SV, Livingston AL et al (2010) Structural and biochemical determinants of ligand binding by the c-di-GMP riboswitch. Biochemistry 49:7351–7359
40. Trausch JJ, Ceres P, Reyes FE et al (2011) The structure of a tetrahydrofolate-sensing riboswitch reveals two ligand binding sites in a single aptamer. Structure 19:1413–1423
41. Pikovskaya O, Polonskaia A, Patel DJ et al (2011) Structural principles of nucleoside selectivity in a 2'-deoxyguanosine riboswitch. Nat Chem Biol 7:748–755
42. Irwin JJ, Sterling T, Mysinger MM et al (2012) ZINC - A Free Tool to Discover Chemistry for Biology. J Chem Inf Model 52:1757–1768

Chapter 11

Probing Riboswitch Binding Sites with Molecular Docking, Focused Libraries, and In-line Probing Assays

Francesco Colizzi, Anne-Marie Lamontagne, Daniel A. Lafontaine, and Giovanni Bussi

Abstract

Molecular docking calculations combined with chemically focused libraries can bring insight in the exploration of the structure–activity relationships for a series of related compounds against an RNA target. Yet, the in silico engine must be fueled by experimental observations to drive the research into a more effective ligand-discovery path. Here we show how molecular docking predictions can be coupled with in-line probing assays to explore the available chemical and configurational space in a riboswitch binding pocket.

Key words Structure-based, Ligand design, Virtual screening, Structure–activity relationship (SAR), Docking, Focused library, In-line probing

1 Introduction

In this volume, Daldrop and Brenk have described how to prioritize a virtual library to discover novel compounds modulating the activity of the RNA target of interest. Here we continue discussing on the possibility of exploring the structure–activity relationships of any putative hit compound both from a structure-based and biochemical perspective. In particular, we describe how a standard molecular docking algorithm can be usefully combined with chemically focused ligand libraries and in-line probing assays, and bring to the identification of novel modulators. We will address important aspects that are worth considering in the elaboration of a virtual screen protocol.

The discovery of cognate-ligand analogs can be performed through biochemical assays where an analog binding the desired target and showing favorable biological features can be considered for further optimization [1, 2]. After the identification of such a compound, the investigation of the structure–activity relationships is a

Daniel Lafontaine and Audrey Dubé (eds.), *Therapeutic Applications of Ribozymes and Riboswitches: Methods and Protocols*, vol. 1103, DOI 10.1007/978-1-62703-730-3_11, © Springer Science+Business Media New York 2014

desirable process for the characterization of the scaffold bioactivity [1]. The prediction of the relative affinity for a series of derivatives bearing different chemical decorations can be pursued using a variety of structure-based approaches [2–4]. In this context, molecular docking procedures can offer a straightforward support for driving the synthetic effort of new derivatives as well as for the prioritization of a library of compounds to be tested for biological activity [4]. As far as the library screening is concerned, the use of "brute-force" strategies, in which the largest possible number of compounds is collected and screened against the desired target, usually does not fulfill expectations. Empirical experience has shown that focused library can save time and money by reducing the number of compounds to be experimentally tested, also improving the ligand discovery success rate by identifying more potent and specific binders [5]. As a result, published reports in recent years have increasingly described methods for designing, selecting or synthesizing variously focused or biased libraries [6–8]. In this respect and related to the herein discussion, chemically focused libraries are those in which the selection process is mostly based on the identification of prototype-chemical scaffolds, or active ligands for which structure–activity relationships or bioisosteric investigations are desired. To make the binding hypothesis quickly assessable, a desirable feature of the library would be the one of containing commercially/synthetically available compounds that investigators can promptly acquire and test for biological activity.

2 Materials

2.1 Molecular Docking

1. Zinc database. A free database of commercially available compounds for virtual screening (http://zinc.docking.org).
2. GOLD *suite* for docking. GOLD is a program for computing the docking modes of small molecules in target binding sites. Hermes is used to prepare inputs and visualize docking outcomes (http://www.ccdc.cam.ac.uk).
3. Marvin Calculator plugin. The Web server is freely accessible and allows a variety of physicochemical properties (e.g., pKa, relative population of tautomeric forms) of small molecules to be computed (http://www.chemaxon.com).

3 Methods

As an example of a virtual screening process, here we discuss the exploration of structure–activity relationships using the recently discovered pyrimidine compound (PC1) that can modulate the biosynthesis of purines in nosocomial bacterial pathogens, which was shown to have therapeutically relevant applications [9].

1. Generate a chemically focused subset of putative ligands structurally related to the PC1 compound. Such a subset of commercially available compounds can be easily assembled using the ZINC database [10, 11]. ZINC provides an easy-to-use and intuitive interface that gradually brings the user into the collection of the focused subset. The subset will contain compounds in a ready-to-dock format, and for each compound a direct connection with vendors for quotation request will also be supplied (*see* **Note 1**). It is usually recommended to use a small number of "calibrating" compounds during the first steps of the screening process (i.e., ~5,000 compounds).

3.1 Molecular Docking: Assessing Performances

1. Molecular docking is performed using GOLD v5.0.1 [12, 13]. GOLD utilizes a genetic algorithm to generate putative ligand–target complexes and it has been recently shown that a standard version of the software can fulfill general docking tasks against RNA, and can provide useful insight in RNA-based ligand discovery [14]. Here, the ChemScore is used to drive and rank the genetic algorithm search, and the native GoldScore is then also employed to re-score the generated binding poses for each ligand (*see* **Note 2**).

   GOLD consists of three main parts: (1) a *scoring function* to rank different binding modes (during the years several scoring functions have been implemented and included in GOLD). (2) a mechanism for *placing* the ligand in the binding site. GOLD's method is based on fitting points: it adds fitting points to hydrogen-bonding groups on target and ligand, and maps acceptor and donor points on both counterparts. Additionally, hydrophobic fitting points are generated in the target cavity and ligand aliphatic groups are mapped into those points. (3) A *search algorithm* to explore possible binding modes. GOLD uses a genetic algorithm to optimize ligand rotatable bonds, some dihedrals of the target, and the mappings of the fitting points (i.e., the position of the ligand in the binding site).
2. For this virtual screening, the crystal structure of the *xpt* guanine riboswitch of *Bacillus subtilis* bound to hypoxanthine is used as the reference target structure (PDB: 1U8D) [15]. The docking algorithm is tested for *pose-fidelity*, which is the ability to reproduce experimentally observed poses within some tolerance limit. The algorithm is also evaluated for the *enrichment*, which is the ability to enrich active compounds from among a database of decoys, where a decoy is a member of the database that does not bind to the target [16, 17]. Property-matched decoys have similar physical properties but different topologies that one would not expect to be recognized by the

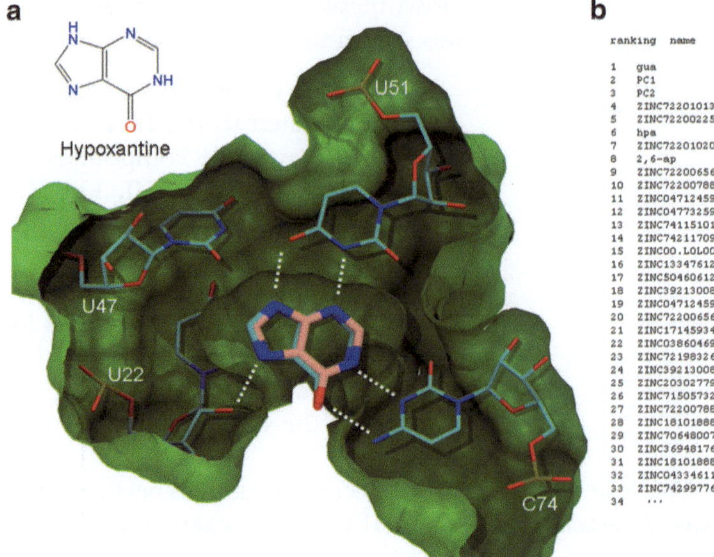

Fig. 1 Assessing docking performances. (**a**) Comparison between the binding mode of hypoxanthine in the docked pose (carbon atoms in *pink*) and the one solved by X-ray crystallography (carbon atoms in *cyan*, pdb code: 1u8d). The *white dotted lines* show H-bond interactions established with the aptamer. (**b**) The known active ligands are top-ranked when docked among a set of property matched decoys. Compounds abbreviation as follows: guanine (gua), hypoxanthine (hpa), 2,6-diaminopurine (2,6-ap), PC1 and PC2 are pyrimidine compounds discovered by Mulhbacher et al. 2010, decoys are labeled with the "ZINC" prefix. See main text for further information

target (*see* **Note 3**). As shown in Fig. 1, the docked binding mode of hypoxanthine as predicted by GOLD is closely resembling the one experimentally observed (root mean square deviation between crystal structure and docking pose <1 Å), and the known active ligands (guanine, PC1, PC2, hypoxanthine, and 2,6-diaminopurine) obtain the highest ranking when docked among a set of property-matched decoys related to PC1.

The assessment of the docking protocol is a critical preliminary step before embarking on the screening of larger dataset of compounds. If the outcome is not satisfactory there are several possible actions: (1) check that atom types have been correctly assigned both at the ligand and target level (*see* **Note 4**). (2) Use a different scoring function that more accurately takes into account the properties of the binding site and those of the native ligands, and the interactions involved between the counterparts. (3) Use ligand-dependent Genetic Algorithm parameter settings and increase the "search efficiency" (it can be done using the front end). This will reduce the speed of docking, increase the sampling and likely augment the reliability of the results. (4) Consider adding customized hydrophobic fitting-points to those generated by default. Such an approach would ensure the sampling of all regions of interest in the cavity, providing sufficient alternatives for placement of hydrophobic ligand atoms within the cavity.

3.2 Virtual Library Screening and Its Critical Interpretation

Virtual screening can be used to gather additional insights about RNA–ligand interactions such as in the case of the guanine riboswitch, which was shown to be potentially interesting for the development of novel antibiotics [9]. A small subset (here ~5,000) of compounds structurally related to the hit compound (PC1) is large enough to perform a first screen.

Despite that the docking algorithm prioritizes the database of ligands according to their complementarity with the binding site, the experience of the user and the critical judgment of the docking outcome will play a central role in selecting compounds for biological testing. As an empirical rule of thumb, when analyzing the outcome of docking screens and judging the quality of a ligand in the context of the predicted binding mode, one should consider both enthalpic and entropic contributions to binding, compound solubility, possible protonation states, possible tautomeric forms, and not least the flexibility of the ligand and the likeliness of the docked conformers (*see* **Note 5**). Furthermore, particular attention should be paid in the identification and evaluation of missed interacting opportunities for polar groups. For instance, chemical groups that are typically well solvated (e.g., hydroxyl group) will need to find good interacting counterparts within the binding site (e.g., hydrogen-bonding network). If their interactions are poor in the binding site, the desolvation cost for those moieties will be too high and their introduction in the chemical scaffold would likely negatively affect the binding affinity.

3.3 Structure–Activity Aspects in Virtual Screening Outcomes

Before commenting on some structure–activity aspects of the in silico screening, it might be useful to explicitly focus on the interaction opportunities between the target and its putative ligands. To this aim, here we briefly describe the binding mode predicted for PC1 against the *xpt* aptamer. PC1 is a structural simplification of guanine and conserves the main interaction pattern of the cognate ligand with the riboswitch aptamer [18].

As shown in Fig. 2, PC1 is embedded into a dense network of H-bond interactions and can establish canonical Watson-Crick pairing with the conserved C74 as well as interact with U51. Moreover, the exocyclic nitrogen in position 5 may also fall within H-bond distances from the 2-hydroxyl group of the U22 ribose moiety. Despite this apparent wealth of H-bond interactions with the aptamer, the binding affinity of PC1 is about three orders of magnitude lower than that of guanine for the *xpt* aptamer [9]. One might speculate that a possible reason for this reduced activity may

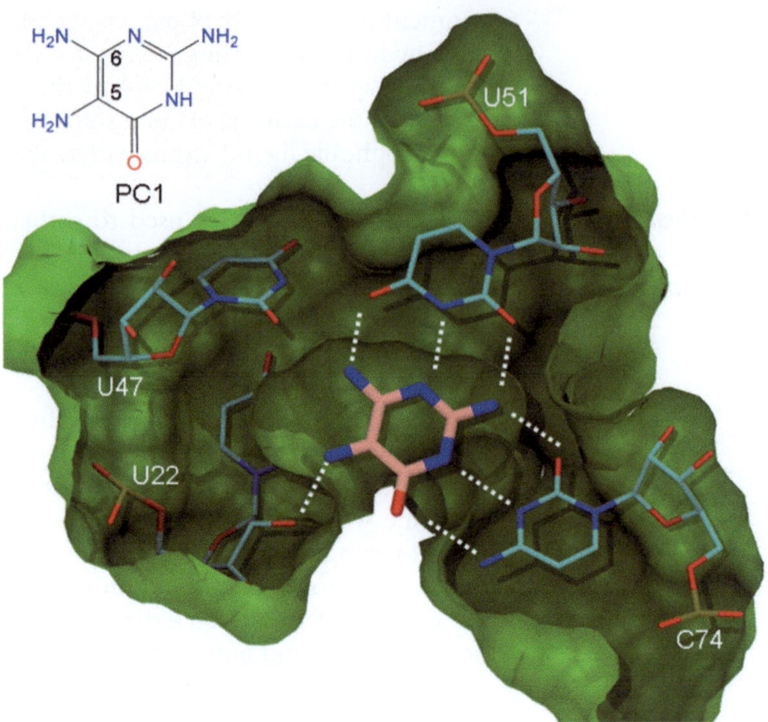

Fig. 2 Exploring the structure–activity relationships of PC1. Binding mode predicted by molecular docking for PC1 (carbon atoms in *pink*). The *white dotted lines* show H-bond interactions established with the *xpt* aptamer (carbon atoms in *cyan*)

include desolvation-related issues. The exocyclic amino groups at positions 5 and 6 of the pyrimidine ring may not find proper interacting counterparts in the binding site and therefore the desolvation cost is higher than the gain in interactions with the binding site. In other words, those polar moieties will have little propensity to be buried into the binding site thus penalizing the binding of the entire molecule. In this view the outcome of docked PC1 analogs may provide interesting suggestions to probe the structure–activity relationships related to the pyrimidine scaffold. Alternatively, the visual inspection of the top-ranked hundreds ligands in the database could also suggest complementary strategies (Fig. 3). For example, several guanine derivatives were present among the screened library and the introduction of an amino group in position 8 appeared to be compatible with binding establishing additional H-bond interactions with O2 of U47 (Fig. 3a). Interestingly, position 8 has been referred to as sterically hindered on the guanine scaffold as it has been probed with a bulky methyl substituent leading to a drastic drop in the binding affinity of the *xpt* aptamer [19]. Moreover, the introduction of an easy-to-solvate group may lead to undesired results if the available interactions within the

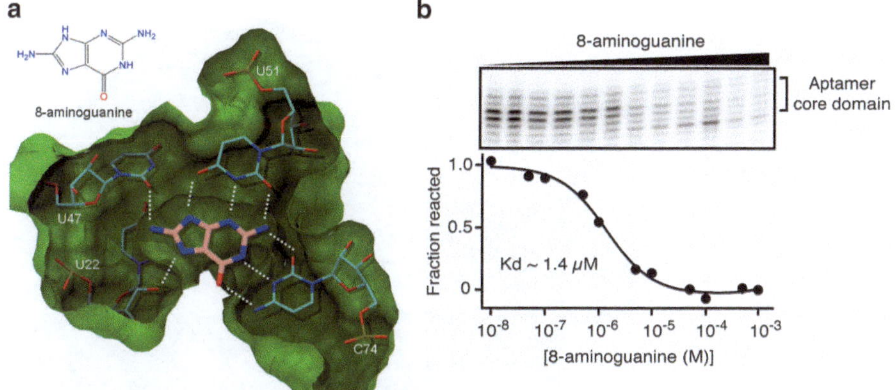

Fig. 3 Computational and experimental characterization of the binding properties for a guanine derivative. (**a**) Binding mode predicted by molecular docking for 8-aminoguanine (carbon atoms in *pink*). The *white dotted lines* show H-bond interactions established with the *xpt* aptamer (carbon atoms in *cyan*). B) In-line probing analysis of the guanine aptamer guaA variant. The concentrations used are 0, 10, 50, 100, and 500 nM, 1, 5, 10, 50, 100, and 500 μM, and 1 mM. A plot showing the normalized fraction of reacted RNA against the ligand concentration reveals an apparent K_d value of ~1.4 μM, which is ~280-fold higher than reported for guanine [19]

binding pocket are not satisfied. This additional solvation penalty, together with a certain extent of steric hindrance, may be responsible for the reduced binding affinity measured for 8-aminoguanine (K_d ~ 1.4 μM, Fig. 3b) when compared to guanine (K_d ~ 5 nM against the *xpt* aptamer). From a ligand-design standpoint, the substitution of the 8-amino group with moieties (e.g., halogens) having a lower desolvation penalty, but still able to interact with polarized groups of the target (such as e.g., C=O and C–H), might also be evaluated [20–22]. In this respect, the introduction of halogen atoms at position 8 of the guanine scaffold might help exploring the electronic and steric features of the aptamer pocket lined by U22, U47, and U51, and may result in a useful strategy to specifically target the polar C=O group of U47 (O2) at a little desolvation cost. Interestingly, halogenated compounds have been described in the field of purine riboswitch ligand discovery [23, 24]. For example, 2-fluoroadenine has been reported as a modulator of the activity of an adenine-responsive riboswitch, and X-ray crystallography experiments have shown the 2-fluoro atom at interacting distance from the O2 atoms of U74 and U51 [23]. However, halogens differ from both electronic and steric properties and, in a separate study, 2-chloroadenine did not bind to the riboswitch [24], suggesting electronic and steric hindrances for this substitution.

3.4 Relevance of Ligand Protonation States

Recent studies continue to demonstrate the importance of using the relevant protonated [25] and tautomeric [26, 27] molecular form for docking, and accordingly, in the ZINC database several pH-dependent representations of a ligand are available.

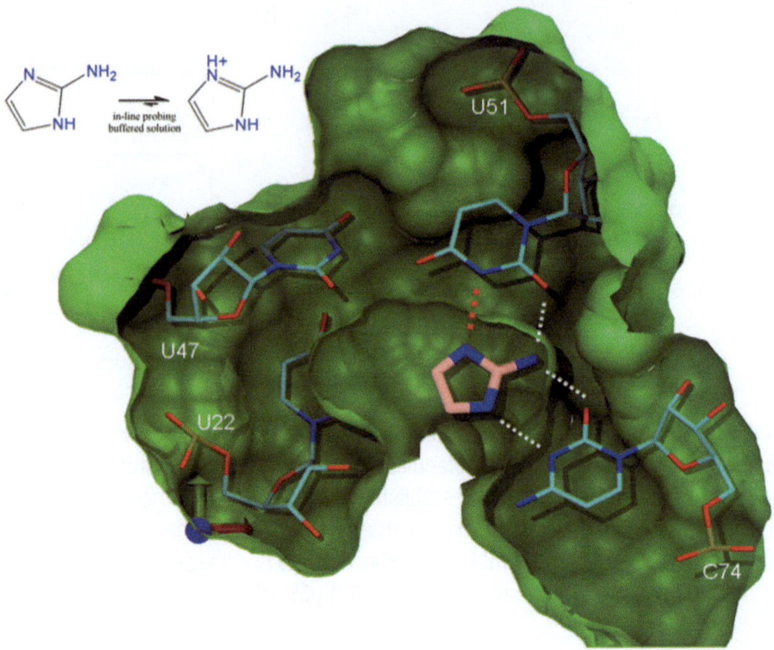

Fig. 4 Effects of ligand pK_a on binding capabilities. (**a**) Putative binding mode for the neutral form of 2-aminoimidazole (carbon atoms in *pink*). The *white dotted lines* show H-bond interactions established with the *xpt* aptamer (carbon atoms in *cyan*). The N3 in U51 has predominant H-bond donor features and the *red dotted line* shows the H-bond interaction which is compromised in the binding of the protonated state of the ligand. Structural models were visualized and rendered with VMD [28]

Whereas this is essential to include biologically relevant representations for docking, sometimes it requires the user to critically judge the protonation state or the tautomeric form for the docked ligands (as well as of the target). Here we report the case of a screened ligand, 2-aminoimidazole, for which a highly ranked binding mode was predicted for a protonation state that is likely poorly represented at pH 8.5 at which the in-line probing experiments are performed. Indeed, at the buffered pH the 2-aminoimidazole is mostly protonated (~90 % of the species) while the neutral form is poorly (~10 %) represented (for prediction of ligands pK_a and tautomeric forms population *see* Subheading 2). The docking predictions for the two species, protonated and neutral, suggest that only the neutral form can establish proper H-bond interactions with the aptamer while the positively charged form disrupts such a network (Fig. 4). Accounting for proper protonation states is a common pitfall and should be carefully considered when docking ligands with acid or basic centers.

As a final remark for possible ligand design strategies, it should be reminded how basicity can be finely tuned by heteroatom substitution (e.g., by replacing a hydrogen with fluorine) which may also lead to the introduction of desired pharmacophoric features [29, 30].

In conclusion, in this paragraph we have shown how molecular-docking calculations combined with chemically focused libraries and in-line probing assays can bring a structure-based insight in the exploration of the structure–activity relationships for a series of related compounds against an RNA target.

4 Notes

1. The practical procedure on how to obtain a focused library is carefully and didactically explained in the "Chemistry4Biology" channel of the video-sharing Web site YouTube (Fig. 5).
2. The GoldScore, ChemScore and other fitness functions that have been implemented in GOLD are dimensionless, and the scale of the score gives a guide as to how good the pose is; the higher the score, the better the docking pose has been evaluated. The GoldScore fitness function is the original scoring function provided with GOLD, and takes into particular account putative H-bonding energy. Using a different parameterization approach, several energetic terms have been incorporated in the ChemScore. In small ligands with few rotatable bonds, the docking with ChemScore can be several times faster than with GoldScore without significantly affecting the accuracy of the computation.

   A combination of scoring functions usually provides better results than one alone. This is due to the fact that erroneously highly ranked configurations generated by one function will likely receive a poor score when evaluated by functions that have been parameterized differently. In other words, although an incorrect ligand–target configuration might be generated and highly scored by one function, the other function can be used to filter out these false positives.
3. Property-matched decoys can be generated using the DUD-E Web-server from http://dude.docking.org.
4. For further information about setting up target and/or ligand(s) consult the GOLD online documentation (http://www.ccdc.cam.ac.uk/products/life_sciences/gold/).
5. For further suggestions, see http://wiki.uoft.bkslab.org/index.php/Hit_picking_party.

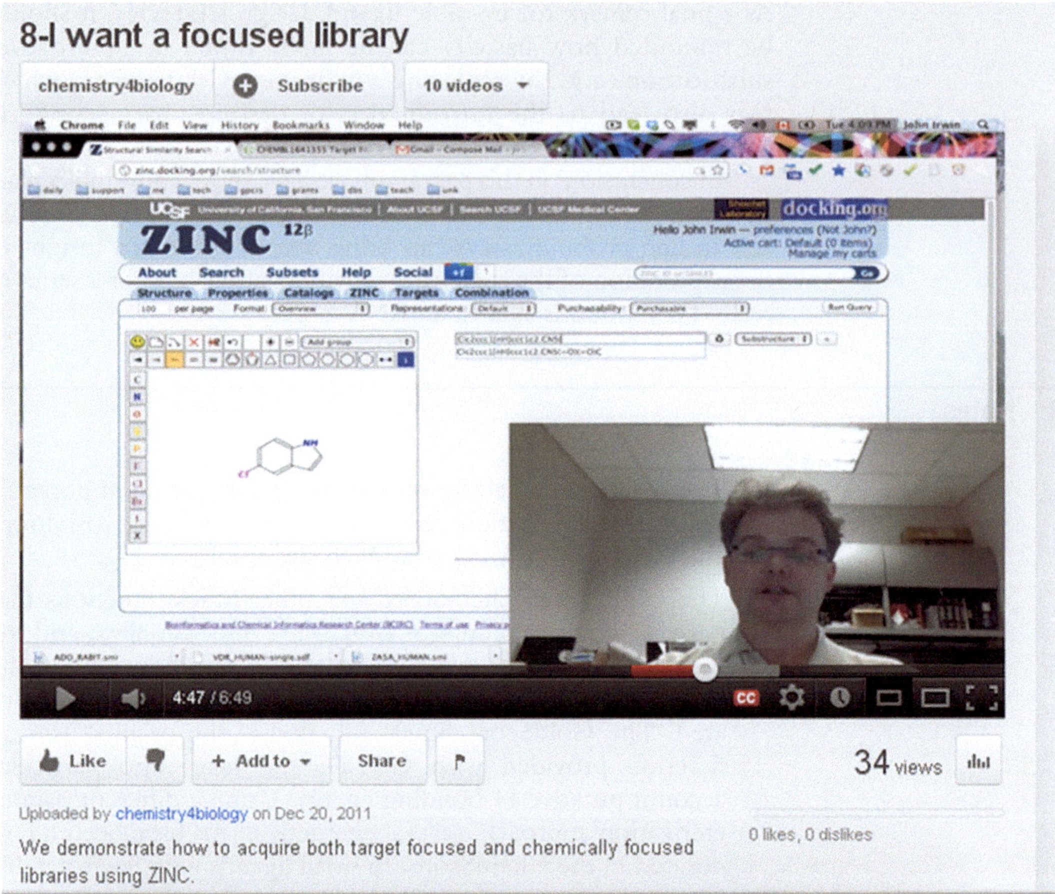

Fig. 5 ZINC is a free public resource for ligand discovery. The database contains millions of commercially available molecules in biologically relevant representations that may be downloaded in popular ready-to-dock formats and subsets. A full range of technical support is provided

Acknowledgements

This work was supported by the National Sciences and Engineering Research Council of Canada (NSERC). D.A.L. is a Canadian Institutes of Health Research (CIHR) New Investigator Scholar. F.C. was supported by the "Young SISSA Scientist" grant (FISB.647 2011) for independent research development. F.C. and G.B. acknowledge the European Research Council for funding through the Starting Grant S-RNA-S (no. 306662).

References

1. Guha R (2013) On exploring structure-activity relationships. Methods Mol Biol 993:81–94. doi: 10.1007/978-1-62703-342-8_6
2. Bleicher KH, Bohm HJ, Muller K, Alanine AI (2003) Hit and lead generation: beyond high-throughput screening. Nat Rev Drug Discov 2:369–378
3. Jorgensen WL (2004) The many roles of computation in drug discovery. Science 303: 1813–1818
4. Shoichet BK (2004) Virtual screening of chemical libraries. Nature 432:862–865
5. Orry AJ, Abagyan RA, Cavasotto CN (2006) Structure-based development of target-specific compound libraries. Drug Discov Today 11: 261–266
6. Leach AR, Hann MM (2000) The in silico world of virtual libraries. Drug Discov Today 5: 326–336
7. Valler MJ, Green D (2000) Diversity screening versus focussed screening in drug discovery. Drug Discov Today 5:286–293
8. Villar HO, Koehler RT (2000) Comments on the design of chemical libraries for screening. Mol Divers 5:13–24
9. Mulhbacher J, Brouillette E, Allard M, Fortier LC, Malouin F, Lafontaine DA (2010) Novel riboswitch ligand analogs as selective inhibitors of guanine-related metabolic pathways. PLoS Pathog 6:e1000865
10. Irwin JJ, Shoichet BK (2005) ZINC – a free database of commercially available compounds for virtual screening. J Chem Inf Model 45: 177–182
11. Irwin JJ, Sterling T, Mysinger MM, Bolstad ES Coleman RG (2012) ZINC: a free tool to discover chemistry for biology. J Chem Inf Model (Epub ahead of print)
12. Jones G, Willett P, Glen RC, Leach AR, Taylor R (1997) Development and validation of a genetic algorithm for flexible docking. J Mol Biol 267:727–748
13. Verdonk ML, Cole JC, Hartshorn MJ, Murray CW, Taylor RD (2003) Improved protein-ligand docking using GOLD. Proteins 52: 609–23
14. Li Y, Shen J, Sun X, Li W, Liu G, Tang Y (2010) Accuracy assessment of protein-based docking programs against RNA targets. J Chem Inf Model 50:1134–1146
15. Batey RT, Gilbert SD, Montange RK (2004) Structure of a natural guanine-responsive riboswitch complexed with the metabolite hypoxanthine. Nature 432:411–415
16. Irwin JJ, Shoichet BK, Mysinger MM, Huang N, Colizzi F, Wassam P, Cao Y (2009) Automated docking screens: a feasibility study. J Med Chem 52:5712–5720
17. Mysinger MM, Carchia M, Irwin JJ, Shoichet BK (2012) Directory of useful decoys, enhanced (DUD-E): better ligands and decoys for better benchmarking. J Med Chem 55: 6582–6594
18. Serganov A, Yuan YR, Pikovskaya O, Polonskaia A, Malinina L, Phan AT, Hobartner C, Micura R, Breaker RR, Patel DJ (2004) Structural basis for discriminative regulation of gene expression by adenine- and guanine-sensing mRNAs. Chem Biol 11:1729–1741
19. Mandal M, Boese B, Barrick JE, Winkler WC, Breaker RR (2003) Riboswitches control fundamental biochemical pathways in Bacillus subtilis and other bacteria. Cell 113:577–586
20. Bissantz C, Kuhn B, Stahl M (2010) A medicinal chemist's guide to molecular interactions. J Med Chem 53:5061–5084
21. Muller K, Faeh C, Diederich F (2007) Fluorine in pharmaceuticals: looking beyond intuition. Science 317:1881–1886
22. Voth AR, Khuu P, Oishi K, Ho PS (2009) Halogen bonds as orthogonal molecular interactions to hydrogen bonds. Nat Chem 1:74–79
23. Gilbert SD, Reyes FE, Edwards AL, Batey RT (2009) Adaptive ligand binding by the purine riboswitch in the recognition of guanine and adenine analogs. Structure 17:857–868
24. Mandal M, Breaker RR (2004) Adenine riboswitches and gene activation by disruption of a transcription terminator. Nat Struct Mol Biol 11:29–35
25. Irwin JJ, Raushel FM, Shoichet BK (2005) Virtual screening against metalloenzymes for inhibitors and substrates. Biochemistry 44: 12316–12328
26. Milletti F, Vulpetti A (2010) Tautomer preference in PDB complexes and its impact on structure-based drug discovery. J Chem Inf Model 50:1062–1074
27. Martin YC (2009) Let's not forget tautomers. J Comput Aided Mol Des 23:693–704
28. Humphrey W, Dalke A, Schulten K (1996) VMD: visual molecular dynamics. J Mol Graph 14(33–8):27–28
29. Bohm HJ, Banner D, Bendels S, Kansy M, Kuhn B, Muller K, Obst-Sander U, Stahl M (2004) Fluorine in medicinal chemistry. Chembiochem 5:637–643
30. Wuitschik G, Carreira EM, Wagner B, Fischer H, Parrilla I, Schuler F, Rogers-Evans M, Muller K (2010) Oxetanes in drug discovery: structural and synthetic insights. J Med Chem 53:3227–3246

Chapter 12

Discovery of Small Molecule Modifiers of microRNAs for the Treatment of HCV Infection

Valerie T. Tripp and Douglas D. Young

Abstract

While RNA has traditionally been viewed as a mechanism to transfer genetic information, its importance and functionality has only truly been demonstrated in the past 20 years. One prime example of this significance can be found within microRNAs (miRNAs), which are involved in the regulation of a substantial number of human genes. Consequently, these miRNAs represent a novel target for therapeutic agents towards the treatment of a variety of diseases and disorders. Specifically, misregulation of miR-122 has been demonstrated to be relevant in the cellular propagation of Hepatitis C (HCV), and thus modulators of miR-122 can have a substantial effect on viral loads. This protocol describes the development of an assay for the discovery of small molecule regulators of miR-122, and ultimately HCV therapeutics. Due to the excellent pharmacokinetic properties of small molecule libraries, these regulators represent a substantial advantage over traditional antisense agents.

Key words miRNA, High-throughput screens, Small molecule libraries, Luciferase reporter assay, HCV

1 Introduction

MicroRNAs (miRNAs) are single-stranded, noncoding RNAs that are found in a variety of species and have recently been implicated in gene regulation [1, 2]. Typically miRNAs regulate mRNA expression through binding at the 3′ untranslated region (UTR), resulting in either inhibition of mRNA translation or degradation of the mRNA transcript [3]. The various human miRNAs control 30 % of all genes, playing key roles in development and cellular differentiation, at least in human pathologies such as viral infection and cancers [4, 5].

The biogenesis of miRNAs begins with genomic transcription to form an 80-nucleotide RNA stem loop, known as primary miRNA (pri-miRNA) [6]. There are several key players that interact with and modify the pri-miRNAs. The RNase III enzyme, Drosha, Exportin 5, and Dicer generate a shorter RNA factor from

Daniel Lafontaine and Audrey Dubé (eds.), *Therapeutic Applications of Ribozymes and Riboswitches: Methods and Protocols*, vol. 1103, DOI 10.1007/978-1-62703-730-3_12, © Springer Science+Business Media New York 2014

the nucleus and further process the miRNA hairpin to yield a 21–23-nucleotide miRNA duplex [6]. Overall, each pri-miRNA precursor generates a single miRNA duplex. The miRNA duplex enters the RNA-induced silencing complex (RISC), where it is unwound by an associated helicase. One strand of miRNA remains in the activated RISC and targets protein-encoding RNA through complementary base pairing [6].

Though some miRNAs are ubiquitously expressed, others are limited to specific tissues. The expression of microRNA miR-122 is confined to the liver, where it constitutes 70 % of the total miRNA population [3]. Within the liver, miR-122 has been implicated in cholesterol and lipid metabolism, and was identified as a regulator for systematic iron homeostasis [7, 8]. Moreover, miR-122 has also been demonstrated to be necessary for the replication and infectious production of hepatitis C virus (HCV). Binding of miR-122 to the 5′ noncoding region of the HCV genome upregulates expression, causing accumulation of viral RNA in liver cells [9]. HCV infection is one of the major causes of liver disease worldwide, including cirrhosis and hepatocellular carcinoma [9]. The essential interaction between miR-122 and HCV suggests that miR-122 could be an excellent therapeutic target for the treatment of HCV infections. Moreover, recent mouse and primate models demonstrate a decrease in HCV RNA levels after knockdown of miR-122 expression, without any toxic side effects [8, 9].

The discovery of small molecule inhibitors of miR-122 function demonstrates a novel approach to inhibit HCV replication in liver cells [10]. These compounds display more efficient pharmacokinetic properties than comparable antisense therapies. Furthermore, the advance of small molecule based therapeutics provides a promising means to target the critical miR-122-HCV mRNA interaction, without the development of viral resistance. This can be done in a specific or nonspecific fashion, by targeting an individual miRNA or by interfering with the miRNA biogenesis pathway, respectively [11]. A specific approach presents more advantages as a therapy as it would not disrupt levels of other important miRNAs within the cell. Previously discovered miR-122 specific compounds reduce both mature-miR-122 and pri-miR-122 levels, but do not appear to affect the levels of other miRNAs within the cell. These small molecules offer unique tools for exploring the biogenesis of miR-122 while also providing structures for the development of future therapeutics [10].

This protocol describes the development of a functional cell-based assay for the elucidation of biologically active small molecules on the miRNA pathway. The assay can be employed to discover both specific and nonspecific inhibitors, and is based on a dual luciferase reporter system (Fig. 1). Upon transfection of a reporter plasmid containing a miRNA binding site in the 3′ UTR

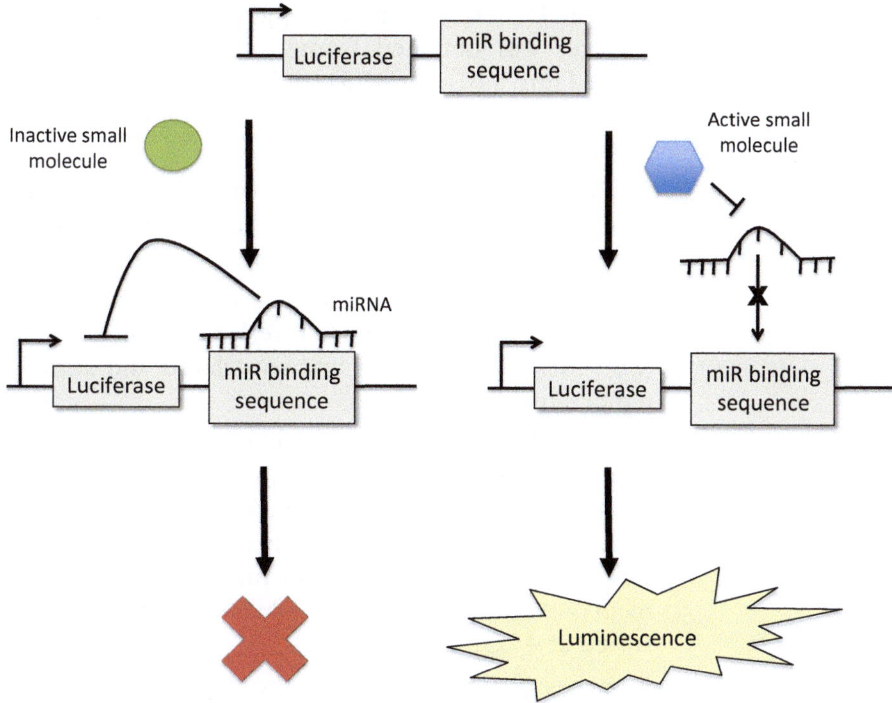

Fig. 1 Schematic of the assay system employed for the screening of small molecule regulators of miRNAs. The presence of an active small molecule results in effective translation of the luciferase gene, yielding functional protein that can be easily employed as a reporter system

of one of the luciferase genes, the presence of a small molecule inhibitor of the miRNA of interest (or the miRNA pathway in general) inhibits the ability of the miRNA to bind the mRNA transcript, resulting in the expression of functional luciferase. If the small molecule screened has no effect on miRNA levels, then normal inhibition of the gene occurs, leading to minimal levels of fluorescence. Due to the "knock-in" nature of this assay and the presence of an internal luciferase control signal, the results are easily interpretable and the system is optimized for high-throughput screening of large libraries.

2 Materials

All materials were prepared using ultrapure water (18 MΩ) sterilized in an autoclave. Most reagents were prepared, filter sterilized (2 μm pore size, CN membrane) and stored at 4 °C unless otherwise noted. Analytical grade reagents were employed in the preparation of most buffers and media. Both bacterial and mammalian cells were sterilized after use and disposed of accordingly.

2.1 Plasmid Preparation

1. The psiCHECK-2 plasmid (20 μg) harboring the dual luciferase reporter was obtained from Promega (C8021) and transformed into a BL-21(DE3) strain of *Escherichia coli* chemically competent cells (New England Biolabs) for plasmid propagation.
2. Ampicillin stock solutions: 0.5 g of ampicillin was dissolved in 10 mL (50 mg/mL), sterile filtered, and dispensed into 500 μL aliquots stored at −20 °C and used as 1,000× stocks [12].
3. 2×YT Broth: Pre-mixed powder (Fisher Scientific) was dissolved according to manufacturer's recommendation in water (500 mL). The solution was autoclaved and stored at room temperature.
4. LB agar plates: Agar (Fisher Scientific) was dissolved according to manufacturer's recommendation in water (500 mL). The solution was sterilized in an autoclave and cooled in a 50 °C water bath. Ampicillin was added (500 μL) and the medium was used to pour plates (100 mm O.D. × 15 mm; ~20 mL media/plate). Plates were cooled to room temperature and stored at 4 °C.
5. Enzymes: *Sgf*I endonuclease (Promega) and *Pme*I endonuclease and T4 DNA ligase (New England Biolabs) were used along with associated buffers.
6. DNA: Oligonucleotides (Integrated DNA technologies) were designed to contain the forward and reverse miRNA binding site sequences for miR-122 (harboring the P*meI* and *Sgf*I digestion sites). Forward sequence: 5′ *CGC*AGTAG AGCTCTAGTA<u>CAAACACCATTGTCACACTCCA</u>*GTTT* 3′; Reverse sequence: 5′ *AAAC*<u>TGGAGTGTGACAATGGTG TTTG</u>TACTAGAGCTCTACT*GCGAT* 3′ (*see* **Notes 1** and **2**). Stocks were made (100 μM) with based on synthesis yield provided by the manufacturer and stored at −20 °C.
7. TBE Buffer (for agarose gels): Tris base (54 g) and boric acid (27.5 g) were dissolved in water (800 mL). EDTA (0.5 M; pH 8.0; 20 mL) was added. The pH of the buffer was adjusted to 8.3 and brought to a volume of 1 L following complete dissolution of the salts. The buffer was used as a 5× stock and stored at room temperature [12].
8. Ethidium Bromide: Ethidium bromide (1 g) was added to water (100 mL) with vigorous stirring. The solution was transferred to a vial, wrapped in foil to prevent exposure to light, and stored at room temperature. Due to the potentially toxic nature, all solutions/gels containing ethidium bromide were disposed of properly [12].

2.2 Cellular Assay/ Compound Screening

1. Huh7 cells were obtained from the ATTC (PTA-8561) and stored in a liquid nitrogen freezer.
2. Screening compounds were either synthesized independently, or obtained from the National Cancer Institute (NCI Diversity Set II) and stored as 1 mM stock solutions in DMSO.
3. Dulbecco's Modified Eagle Medium (DMEM; Hyclone) was prepared from the commercially available powder with water to 400 mL. Sodium bicarbonate (1.2 g; Fisher Scientific) and Fetal Bovine Serum (FBS; 50 mL; Hyclone) were added followed by 200× Penicillin/Streptomycin solution (10 mL; MP Biomedicals). The pH was adjusted to 7.2 and brought to 500 mL with water. The media was then sterile filtered and stored at 4 °C.
4. Stock DMSO was prepared by sterile filtration of commercially available DMSO (Aldrich).
5. Transfections were performed with commercially available X-tremGENE (Roche) in Opti-MEM media (Invitrogen).
6. The Dual Luciferase Assay Kit (Promega) was purchased and components were stored according to manufacturer's directions.
7. Assays were conducted in 96-well PS Cellstar white TC plates (Greiner, Bio-One; USA Scientific).
8. Assays were conducted on a Wallac VICTOR^{3}V luminometer using a pre-set software program with a measurement time of 1 s and a delay time of 2 s.
9. Quantitative Real Time PCR analysis was conducted using the commercially available TaqMan microRNA Assay for miR122 (Applied Biosystems). RNAs were isolated using the mirPremier microRNA isolation kit (Aldrich) according to manufacturer's protocols and all assays were conducted on a Bio-Rad MyiQ RT-PCR thermocycler.
10. PCR primers for the detection of pri-mir-122 were obtained from Integrated DNA Technologies. Forward primer: 5′ GCTCTTCCCATTGCTCAAGATG 3′; Reverse primer: 5′ GTATGTAACAACAGCATGTG 3′. Stocks were made (100 μM) with based on synthesis yield provided by the manufacturer and stored at −20 °C. The PCR was performed using commercially available iQ SYBR Green Supermix (Bio-Rad) which contains the SYBR Green, the polymerase, dNTPs, and buffer as a 2× stock.

2.3 HCV Assay

1. The pHtat2Neo/QR/KR/SI plasmid for the generation of HCV RNA was provided by Dr. Stanley Lemon (UTMB).
2. Huh7 cells were obtained from the ATTC (PTA-8561) and stored in a liquid nitrogen freezer.

3. Enzymes for plasmid digestion (XbaI) and transcription (T7 RNA Polymerase) were obtained from New England Biolabs with corresponding buffers.
4. Cell media was prepared as previously described, and transfections were performed with commercially available X-tremGENE (Roche) in Opti-MEM media (Invitrogen).
5. PCR primers (Integrated DNA Technologies) were designed for the detection of HCV. Forward primer: 5′ CGGGAGAGCCATAGTGGTCTGCG 3′; Reverse primer: 5′ CTCGCAAGCACCCTATCAGGCAGTA 3′. GADPH control primers were also obtained to serve as a control for RNA levels. Forward primer: 5′ TGCACCACCAACTGCTTAGC 3′; Reverse primer: 5′ GGCATGGACTGTGGTACTGAG 3′. Stocks were made (100 μM) with based on synthesis yield provided by the manufacturer and stored at −20 °C.
6. Quantitative Real Time PCR analysis for HCV transcript was conducted using the commercially available iQ SYBR Green Supermix (Bio-Rad) which contains the SYBR Green, the polymerase, dNTPs, and buffer as a 2× stock RNAs were isolated using the mirPremier microRNA isolation kit (Aldrich) according to manufacturer's protocols and all assays were conducted on a Bio-Rad MyiQ RT-PCR thermocycler.

3 Methods

3.1 Reporter Plasmid Construction

1. The psiCHECK-2 plasmid (1 μg) was added to sterile water (43 μL), enzyme buffer (5 μL) and initially digested with *Sgf*I (10 U, 1 μL) for 2 h at 37 °C. The enzyme was then inactivated at 70 °C for 20 min, and the reaction was purified using a Qiagen PCR Purification kit according to manufacturer's directions.
2. The singly digested plasmid (~1 μg, 30 μL) was then brought to volume with sterile water (14 μL), enzyme buffer (5 μL), BSA (0.5 μL) and digested with *Pme*I (10 U, 1 μL) for 2 h at 37 °C (*see* **Note 3**). The enzyme was then inactivated at 70 °C for 20 min.
3. A 1 % agarose gel was poured and the entire digest reaction was loaded (with 5 μL 10× loading dye) and run at 120 V for 30 min in 1× TBE buffer (supplemented with ethidium bromide). The linearized plasmid was then visualized and excised on a transilluminator. Agarose slices containing the desired plasmid DNA were then purified using a Qiagen Gel Extraction Kit according to manufacturer's protocols (Fig. 2).
4. The 2 strands of synthetic DNA encoding the miRNA binding sequence (10 μL each) were combined and heated to 90 °C

Fig. 2 *PmeI*/*SgfI* digested psiCHECK-2 plasmid purification on a 1 % agarose gel. Following imaging, the DNA band was extracted and isolated with a Qiagen Gel Extraction Kit

then cooled to 4 °C over a 5 min period. This heat/cool cycle was repeated five times to ensure effective hybridization, and then allowed to incubate at 4 °C for 1 h (*see* **Note 4**).

5. The hybridized insert (from step 4) was mixed with digested plasmid vector in a 5:1 insert to vector ratio (2 μL total volume). The mixture was then diluted with sterile water (6 μL), followed by the addition of T4 DNA Ligase buffer (1 μL) and T4 DNA Ligase (40 units, 1 μL). The ligation was incubated at 4 °C for 16 h then transformed (2 μL) via a heat shock into chemically competent cells based on the manufacturer's protocols (*see* **Note 5**).
6. After the appropriate recovery time, an aliquot (50 μL) of the transformation was plated on LB Agar/Ampicillin plates and incubated at 37 °C for 16 h.
7. Several colonies were selected and grown in liquid culture LB/Ampicillin (4 mL) overnight at 37 °C. The cultures were then microcentrifuged (1,600 × *g*, 8 min) and plasmids were purified using a Qiagen Mini-Prep Kit according to manufacturer's protocols.
8. Purified plasmids were verified for successful cloning via DNA sequencing using the sequencing primer. Upon the confirmation of miRNA-122 binding sequence, the plasmid was utilized for miRNA small molecule regulation assays.

3.2 Small Molecule Screening Assay

1. A culture of Huh7 cells was started by the slow addition of a frozen aliquot of cells into DMEM (10 mL) in a tissue culture flask, followed by incubation at 37 °C in a CO_2 incubator. The adherent cells were monitored over the course of several days with the medium changed every 48 h until the culture reached 90–100 % confluency. The cells were then passaged via medium removal, trypsinization (1 mL), dilution (10 mL), and transfer into a fresh tissue culture flask (0.5 mL culture to 9.5 mL fresh media). The cells were passaged two additional times over a 2-week period.
2. Cells were then passaged into each well of a 96-well PS Cellstar white TC plate (20 μL of diluted cells/180 μL DMEM) and allowed to grow to approximately 60 % confluency. The DMEM was removed and replaced with Opti-MEM medium (180 μL/well).
3. Transfection reagent was prepared by mixing X-tremGENE (100 μL) with Opti-MEM (1.4 mL) in a sterile eppendorf tube. In a separate tube, psiCHECK-miR122 plasmid (from Subheading 3.1; 500 ng/transfection) was mixed with Optim-MEM (to a final volume of 1.5 mL). Both tubes were incubated at room temperature for 5 min, then combined and mixed, followed by a 20-min room temperature incubation.
4. The transfection mixture (20 μL/well) was transferred to the cells and incubated for 4 h at 37 °C in a CO_2 incubator in order to effectively introduce the reporter plasmid into the Huh7 cells. The transfection media was removed and replaced with DMEM (198 μL/well) followed by the compound of interest from the NCI Diversity Set II (2 μL/well; 10 μM final concentration; 1 % DMSO). Cells were then incubated for 48 h at 37 °C.
5. The DMEM was then removed and cells were lysed according to manufacturer's protocols using the Promega Dual Luciferase Assay Kit, followed by luciferase measurements on a Wallac VICTOR^{3}V luminometer for both Firefly and *Renilla* luciferase activity.
6. The raw luminescence data from each well was used to calculate a ratio of *Renilla* luciferase to Firefly luciferase in each well. Wells demonstrating a substantially higher signal (relative to a DMSO no compound control) were recorded as potential miRNA pathway inhibitors (Fig. 3).
7. The potential hit compounds were then re-subjected to the previously described assay (Subheading 3.2, **steps 1–6**) in triplicate to ascertain their reproducibility and statistical relevance as pathway inhibitors.
8. Based on the results of the previous screen, the effect of the small molecule modifiers on the miRNA levels was assessed via

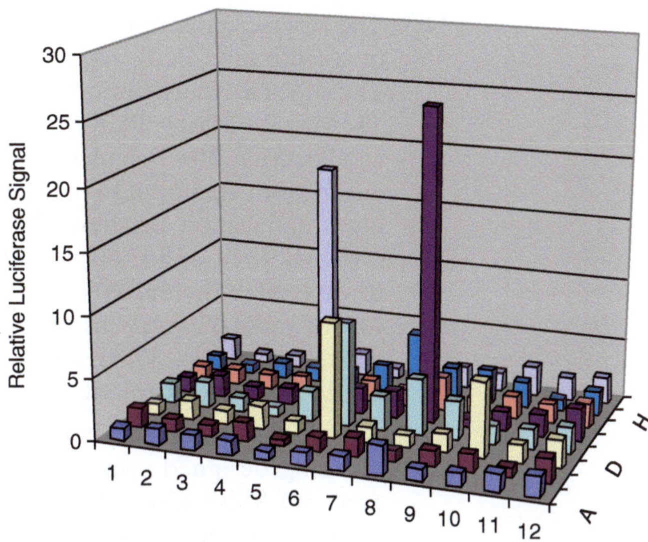

Fig. 3 Sample normalized 96-well-plate output for a small molecule assay based on Firefly and *Renilla* luciferase readings

quantitative real-time PCR analysis. Huh 7 cells were passaged into a 96-well plate (200 μL cells in 3 mL DMEM/well) and grown to approximately 60 % confluency.

9. The medium was removed and replaced with fresh DMEM (1.98 mL) and either compound (20 μL; 10 μM; 1 % DMSO final concentration) or diluted DMSO (1 % final concentration). Each compound was tested in triplicate to ensure statistical relevance. The cells were then incubated at 37 °C in a CO_2 incubator for 48 h.

10. The media was removed and the cells were washed with PBS buffer (2×2 mL, pH 7.4). The RNA was then isolated using a mirPremier microRNA Isolation Kit (Aldrich) according to manufacturer's protocols.

11. Isolated RNA was quantitated using a NanoDrop ND-1000 spectrophotometer in order to normalize total RNA concentrations for reverse transcription.

12. The isolated RNA was then reverse transcribed using a TaqMan microRNA Reverse Transcription Kit with the provided primers. To 10 ng of isolated RNA per reaction, dNTPs (100 mM stock; 0.15 μL), reverse transcription buffer (10×; 1.5 μL), RNAsin (0.18 μL), TaqMan Primer (for either a miRNA or a housekeeping gene; 5× stock; 3 μL), Reverse Transcriptase (1.0 μL) and brought to a final volume of 15 μL. The reverse transcription was cooled to 16 °C for 30 min, followed by 42 °C for 30 min, and then the enzyme was denatured at 85 °C for 5 min.

13. The reverse transcription reaction was then directly employed in quantitative PCR according to manufacturer's protocols. The reverse transcription product (1.3 μL) was added to TaqMan 2× Master PCR Mix (10 μL), distilled water (7.7 μL) and TaqMan 20× Polymerase (1 μL). The reaction was placed inside a Bio-Rad MyiQ RT-PCR thermocycler with the following amplification program: 95 °C, 10 min; followed by 40 cycles of 95 °C, 15 s; 60 °C, 60 s. Threshold cycles were used to determine the miRNA copy number allowing the comparison of miR122 levels to the levels of miR21 controls: copy number = $10[(C_t - 37.4)/-3.3]$. The data was then normalized to the DMSO control samples.
14. A similar protocol was then repeated to assess pri-miR122 levels using designed primers, and instead of the TaqMan PCR assay, iQ SYBR Green Supermix was employed: reverse transcription product (1.3 μL) iQ SYBR Green Supermix (10 μL), primers (2 μL each), sterile water (3.7 μL), with an amplification program of 95 °C, 30 min followed by 40 cycles of 95 °C, 15 s; 60 °C, 60 s. Data was analyzed as previously discussed.

3.3 Assessment of Small Molecules on HCV Replication

1. The pHtat2Neo/QR/KR/FV/SI plasmid (10 ng) was linearized using *Xba*I (10 U, 1 μL) in NEB Buffer 4 (5 μL) and BSA (0.5 μL) brought to 50 μL with sterile water. The digestion was incubated at 37 °C for 2 h followed by enzyme deactivation at 70 °C for 10 min.
2. The restriction digest was purified using a Qiagen PCR Purification Kit according to manufacturer's protocols. The linearized plasmid was then subjected to a transcription reaction using plasmid (20 μL), T7 Transcription Buffer (5 μL), NTPs (8 μL, 10 μM each final concentration), T7 RNA Polymerase (1 μL), and sterile water (16 μL) for 6 h at 37 °C. The transcript was then purified on a Microcon 10 centrifugation column (5 × 100 μL PBS washes) and quantitated on a NanoDrop ND-1000 spectrophotometer.
3. Huh7 cells were passaged into a 6 well tissue culture plate (200 μL cells in 3 mL DMEM/well) and grown to approximately 60 % confluency. The DMEM was removed and replaced with Opti-MEM (900 μL). In separate tubes 1 μg of HCV RNA was brought to a volume of 50 μL with Opti-MEM per reaction, and 5 μL X-tremGENE was diluted in 45 μL of Opti-MEM, with each tube being incubated for 5 min at room temperature. The tubes were then combined and incubated an additional 20 min at room temperature before being added to the cells, which were then incubated for an additional 4 h at 37 °C.
4. The transfection media was then removed and replaced with DMEM (1.98 mL) supplemented with either hit compound (20 μL, 10 μM, 1 % DMSO final concentration) or DMSO.

The cells were then incubated at 37 °C for 48 h in a CO_2 incubator.

5. HCV levels in the cells were then assessed by qRT-PCR according to the previously described protocol (Subheading 3.2, **steps 10–14**). PCR analysis was performed using HCV primers (Subheading 2.3, **item 5**) and GADPH housekeeping gene primers as an RNA control.

4 Notes

1. By simply altering the sequence of DNA ordered (with appropriate *Sgf*I and *Pme*I cloning sequences) to that of a different miRNA binding sequence, this assay can rapidly be adapted to screen for small molecule inhibitors of a multitude of known miRNAs.
2. Italicized sequence of the synthetic DNA represents *Sgf*I and *Pme*I restriction sites. Underlined sequences denote the specific binding sequence of miR122.
3. Double-digestion of the plasmid led to ineffective transformations (most likely to incompatible digestion conditions), therefore the issues were remedied via a sequential digest.
4. The DNA was designed to harbor the cut restriction sites, requiring no enzymatic digest.
5. A control transformation was also performed using the same concentration of digested plasmid in the absence of insert to monitor for incomplete plasmid digestion.

References

1. Tomari Y, Zamore PD (2005) Perspective: machines for RNAi. Genes Dev 19:517
2. Deiters A (2010) Small molecule modifiers of the microRNA and RNA interference pathway. AAPS J 12:51
3. Jopling CL, Yi M, Lancaster AM, Lemon SM, Sarnow P (2005) Modulation of hepatitis C virus RNA abundance by a liver-specific MicroRNA. Science 309:1577
4. Esquela-Kerscher A, Slack FJ (2006) Oncomirs – microRNAs with a role in cancer. Nat Rev Cancer 6:259
5. Lu J et al (2005) MicroRNA expression profiles classify human cancers. Nature 435: 834
6. Chang J et al (2004) miR-122, a mammalian liver-specific microRNA, is processed from hcr mRNA and may downregulate the high affinity cationic amino acid transporter CAT-1. RNA Biol 1(2):106–13
7. Castoldi M et al (2011) The liver-specific microRNA miR-122 controls systemic iron homeostasis in mice. J Clin Invest 121:1386
8. Elmén J et al (2008) LNA-mediated microRNA silencing in non-human primates. Nature 452:896
9. Lanford RE et al (2010) Therapeutic silencing of microRNA-122 in primates with chronic hepatitis C virus infection. Science 327:198
10. Young DD, Connelly C, Grohmann C, Deiters A (2010) Small molecule modifiers of MicroRNA miR-122 function for the treatment of hepatitis C virus infection and hepatocellular carcinoma. J Am Chem Soc 132:7976
11. Georgianna WE, Young DD (2011) Development and utilization of non-coding RNA-small molecule interactions. Org Biomol Chem 9:7969
12. Sambrook J, Russell DW (2011) Molecular cloning: a laboratory manual, 3rd edn. Cold Spring Harbor Laboratory Press, Cold Spring Harbor, NY

Chapter 13

Bacterial Flavin Mononucleotide Riboswitches as Targets for Flavin Analogs

Danielle Biscaro Pedrolli and Matthias Mack

Abstract

Roseoflavin is a toxic riboflavin (vitamin B_2) analog and naturally is produced by *Streptomyces davawensis*. Roseoflavin is converted to roseoflavin mononucleotide (RoFMN) by promiscuous flavokinases (EC 2.7.1.26). Flavin mononucleotide (FMN) riboswitches control the expression of genes involved in riboflavin biosynthesis and/or transport. RoFMN triggers FMN riboswitches and negatively (or positively) affects expression of the downstream genes. RoFMN binding to the aptamer portion of FMN riboswitch RNAs occurs in the course of transcription by cellular RNA polymerases. We developed an in vitro test system to functionally characterize the interaction between riboflavin/FMN analogs such as roseoflavin/RoFMN and FMN riboswitches in the context of an actively transcribing RNA polymerase.

Key words FMN riboswitches, Riboflavin analogs, Roseoflavin, *Streptomyces davawensis*, In vitro transcription/translation assays, Fluorescence quenching

1 Introduction

The antibiotic roseoflavin (RoF) is synthesized by the gram-positive bacterium *Streptomyces davawensis* [1]. RoF is toxic to gram-positive but also to gram-negative bacteria if RoF is able to enter the cell [1–4]. Notably, RoF is also active against the human bacterial pathogen *Listeria monocytogenes* [5]. RoF has been shown to target flavin mononucleotide (FMN) riboswitches and to affect gene expression [5–8]. Additional support for flavins (or flavin analogs) targeting FMN riboswitches came from structural models for the FMN riboswitch of the gram-negative bacterium *Fusobacterium nucleatum* bound to riboflavin (RF), FMN and RoF [9]. RoF and its precursor 8-amino-8-demethyl-riboflavin (AF) are produced by *S. davawensis* from RF, and are so far the only known natural RF analogs with antibiotic activity [1, 10, 11] (Fig. 1). Flavin analogs are activated to the corresponding cofactor analogs by cellular flavokinases and FAD synthetases [12, 13]. Thus, RoFMN and AFMN rather than the non-phosphorylated

Daniel Lafontaine and Audrey Dubé (eds.), *Therapeutic Applications of Ribozymes and Riboswitches: Methods and Protocols*, vol. 1103, DOI 10.1007/978-1-62703-730-3_13,

riboflavin roseoflavin 8-amino-riboflavin

Fig. 1 The structures of 7,8-dimethyl-10-(D-1′-ribityl)isoalloxazine (riboflavin, RF; vitamin B_2), 8-dimethylamino-8-demethyl-riboflavin (roseoflavin, RoF), and 8-amino-8-demethyl-riboflavin (AF)

flavins are the active (toxic) compounds in the cytoplasm of target cells.

Four steps in riboswitch formation (Fig. 2) are critical for gene expression modulation: Correct aptamer folding, speed of the transcription process, association rate between cognate metabolite and riboswitch aptamer and folding of the expression platform. The speed of transcription depends on intrinsic cell characteristics like RNA polymerase speed, the presence of transcriptional pause sites in the DNA template but also on the physiological state of the cell, ribonucleotide availability and/or energy charge [14, 15]. The metabolite binding rate depends on the ligand structure itself and also on the extension of rearrangements in the riboswitch aptamer necessary to accommodate the cognate metabolite. Aptamer and expression platform folding probably depends on the presence of cations (especially Mg^{2+} and K^{+}) [9] and on the mRNA primary structure, since alternative folding structures can compete with the aptamer or the terminator/sequestor for the same nucleotides [15]. In the latter case, it might be important that the aptamer has adopted its final (metabolite binding) structure before a long stretch of downstream RNA has been synthesized. The same in principle is true for the formation of the transcription terminator/ribosomal binding site sequestor. Moreover, for an effective control, metabolite binding must occur during the short time frame between aptamer formation and synthesis of the terminator/sequestor sequence by RNA polymerase [14, 15]. Consequently, it is important to study riboswitch function in the context of an active transcription process. In order to be able to quantify the response of FMN riboswitches from different bacteria upon treatment with different flavins and flavin cofactor analogs we developed an in vitro transcription/translation assay based on the plasmid pT7luc (Promega, Mannheim, Germany) and the corresponding kit.

RoFMN, AFMN, RoFAD, and other cofactor analogs are not commercially available. Enzymatic synthesis of phosphorylated and adenylylated flavins is simple, fast, and specific enough to yield only one isomeric form of the final product. The human enzymes

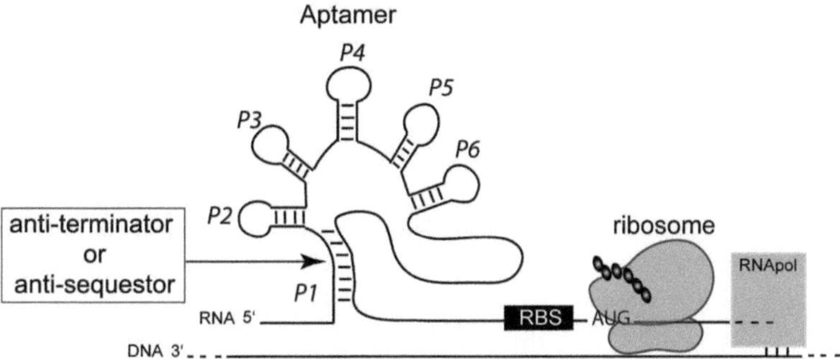

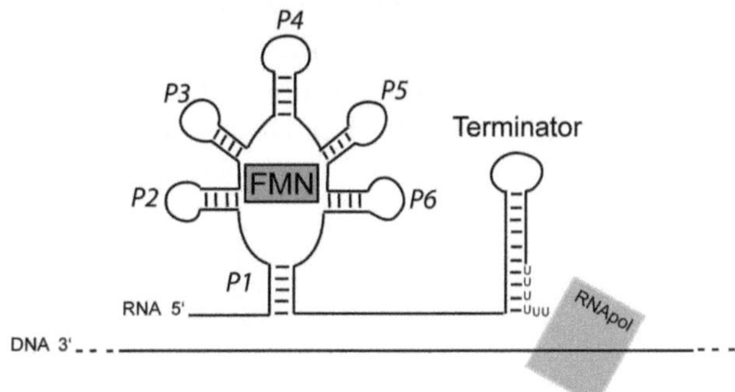

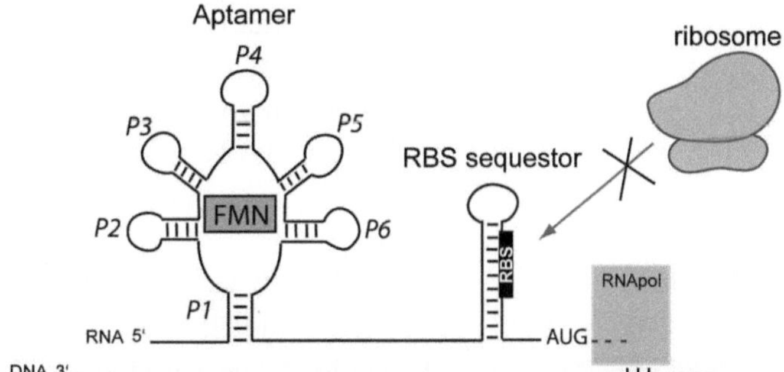

Fig. 2 Mechanisms of FMN riboswitch-mediated gene regulation (according to [15] and [9]). (**a**) At low FMN concentrations the aptamer portion of the riboswitch RNA does not bind the regulating metabolite and adopts the anti-termination or anti-sequestor structure. Transcription is able to proceed, translation may initiate and protein synthesis occurs; (**b**) at high FMN concentrations the aptamer binds the metabolite which leads to a specific structure allowing the terminator stem to form causing transcription termination and dissociation of RNA-polymerase from the DNA-template; (**c**) alternatively at high FMN concentrations the aptamer binds the metabolite and folds into a specific structure allowing the RBS sequestor stem to form. Although transcription still proceeds, protein synthesis does not occur since the ribosome cannot access the ribosomal binding site (RBS)

flavokinase (huRFK) and FAD synthetase (hFADS2) are especially useful because they are monofunctional enzymes, and, under the applied conditions, reach almost 100 % of substrate conversion [13].

2 Materials

All solutions should be prepared with high-quality sterile water. The reagents are stored at room temperature unless otherwise specified. In general, chemicals with the highest possible quality should be used. The Subheadings 2.2, 2.3, 3.2 and 3.3 involve RNA synthesis and general precautions should be taken to avoid contamination with ubiquitous ribonucleases. Disposable gloves should be worn at all times. Distilled water and aqueous solutions should be autoclaved twice. Stock solutions to be used, pipette tips, and others materials should be kept separately from other laboratory equipment and should be used exclusively for RNA experiments. The latter materials should not be handled without gloves even before autoclaving.

2.1 Preparation of Phosphorylated and Adenylylated Flavins

1. 20 mM ATP, prepared freshly or kept frozen in small aliquots in order to avoid multiple freeze–thaw cycles.
2. 240 mM sodium dithionite, prepared freshly.
3. 20 mM RoF (Chemos, Regenstauf, Germany) solution in dimethyl sulfoxide (DMSO) (*see* **Notes 1** and **2**).
4. Recombinant human riboflavin kinase (huRFK) 1 mg/mL [13]. Store at –20 °C.
5. Recombinant human FAD synthetase isoform 2 (hFADS2) 5 mg/mL [13]. Store at –20 °C.
6. HPLC connected to a diode array detector (DAD).

2.2 In vitro Transcription/ Translation Assays (TK/TL)

1. pT7luc vector (Promega, Mannheim, Germany) (*see* map in Figs. 3 and also 4).
2. pT7luc^{mod} vector (Fig. 4). In modified pT7luc the start codon ATG of the luciferase reporter gene was changed to GCA (using QuikChange II XL Kit, Invitrogen, Darmstadt, Germany) in order to allow the construction of translational fusions of FMN riboswitches.
3. *E. coli* Top10 (Invitrogen, Darmstadt, Germany) competent cells.
4. Lysogeny Broth (LB-agar) *(8)* containing 100 μg/mL ampicillin.
5. Plasmid DNA purification kit (GeneJET™ Plasmid Miniprep Kit, Fermentas, St Leon Rot, Germany).
6. Nuclease-free water.

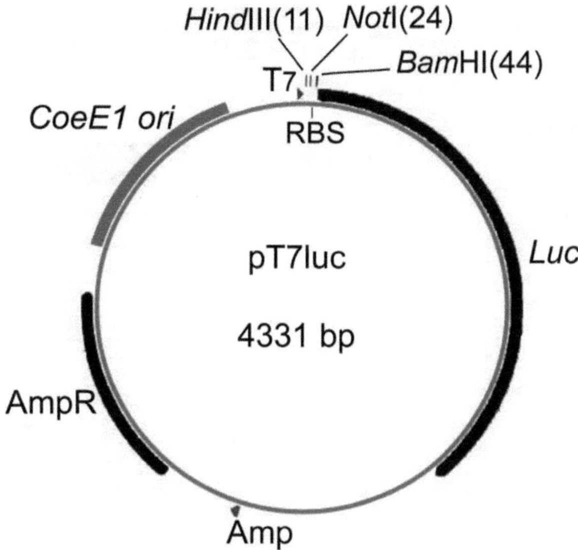

Fig. 3 Map of pT7luc used for the in vitro transcription/translation assays. Ligation of the riboswitch sequence to *Hind*III/*Not*I digested plasmid maintains the intrinsic ribosomal binding site (RBS) and allows transcriptional fusions with the gene for luciferase (*luc*). In contrast, ligation to the *Hind*III/*Bam*HI digested plasmid eliminates the RBS and allows translational fusion with *luc* (note that in pT7luc^{mod} the start codon (ATG) of *luc* was changed to GCA). *Arrowheads* represent promoters

7. *Escherichia coli* T7 S30 Extract System for Circular DNA Kit (Promega, Mannheim, Germany). The kit components are stored in small aliquots at −80 °C. Avoid multiple freeze–thaw cycles.
8. 10× Flavin solution in water in concentrations varying from 100 μM to 2 mM. Use flavin 5′-monophosphate (FMN; Sigma, Munich, Germany) and enzymatically prepared RoFMN and RoFAD prepared in the Subheading 3.1.
9. BSA solution containing: 1 mg/mL bovine serum albumin, 2 mM DTT, 25 mM Tris–phosphate, pH 7.8. Store at −20 °C.
10. Luciferase Assay Reagent (Promega, Mannheim, Germany). Dissolve the substrate as suggested by the supplier and store it in small aliquots at −80 °C.
11. 96-well white microplates (OptiPlate-96, Perkin Elmer, Rodgau, Germany).
12. Microtiter plate reader (Tecan Genios Pro microplate reader, Tecan, Mainz, Germany).

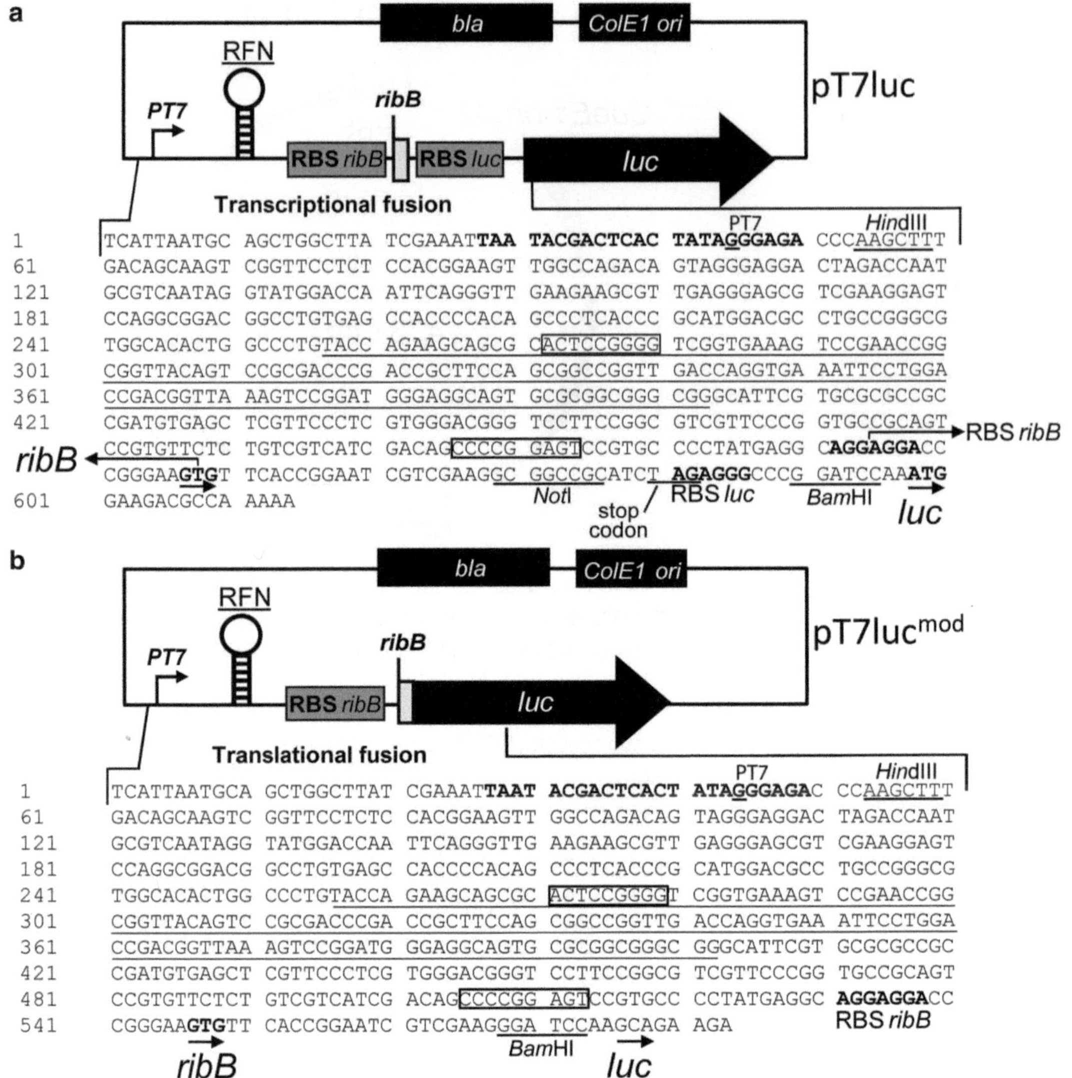

Fig. 4 Scheme of the reporter plasmid used for testing FMN riboswitches using in vitro transcription/translation. (**a**) Plasmid (pT7luc) used for the generation of transcriptional fusions. The DNA sequence exemplarily is shown for the transcriptional fusion construct used for testing the *ribB* FMN riboswitch of *S. davawensis*. (**b**) Plasmid (pT7luc^{mod}) used for the generation of translational fusions. The DNA sequence exemplarily is shown for the translational fusion construct used for testing the *ribB* FMN riboswitch of *S. davawensis*. The sequence in *bold* (*top*) is the promoter from bacteriophage T7, transcription starts with G (*underlined*). The underlined sequence (*middle*) is the FMN riboswitch aptamer; the *boxed sequences* are complementary to each other and form the anti-RBS sequestor. The ribosomal binding sites RBS are in *bold*: RBS *ribB* is part of the FMN riboswitch and RBS *luc* was originally present in the plasmid. The *underlined sequence* between the *NotI* and *BamHI* restriction sites in (**a**) represents a stop codon

2.3 Flavin Fluorescence Quenching Measurements

1. Restriction endonuclease *Eco*RV (Fermentas, St Leon Rot-Germany).
2. Plasmid DNA purification kit (GeneJET™ Plasmid Miniprep Kit, Fermentas, St Leon Rot, Germany).

3. Nuclease-free water.
4. DNaseI (RNase-free) (Fermentas, St Leon Rot, Germany).
5. Transcription Optimized 5× buffer: 40 mM Tris, pH 7.9, 6 mM $MgCl_2$, 10 mM NaCl, 2 mM spermidine (Fermentas, St Leon Rot, Germany).
6. Ribonuclease inhibitor (recombinant RNasin) (Fermentas, St Leon Rot, Germany).
7. 10 μM flavin solutions in water. Use flavin 5′-monophosphate (FMN; Sigma, Munich-Germany) and RoFMN prepared in the Subheading 3.1.
8. rNTP mix: 2.5 mM rATP, 2.5 mM rUTP, 2.5 mM rGTP and 2.5 mM rCTP.
9. T7 RNA polymerase (Fermentas, St Leon Rot, Germany).
10. RNA purifying columns (Fermentas, St Leon Rot, Germany).
11. 50 mM potassium phosphate pH 7.4 containing 2 mM $MgCl_2$ and 2 mM KCl.
12. 96-well black microplates (FluoroNunc, Nunc A/S, Denmark).

3 Methods

3.1 Preparation of Phosphorylated and Adenylylated Flavins

1. To a sterile plastic tube add: 28.25 mL 100 mM potassium phosphate pH 7.4, 5 mL 120 mM NaF, 5 mL 60 mM $MgCl_2$, 5 mL 240 mM sodium dithionite, 5 mL 20 mM ATP, 750 μL 20 mM flavin, and 1 mL enzyme solution. For RoFMN production, add RoF to huRFK. RoFAD can be prepared from RoFMN using hFADS or alternatively from RoF using both enzymes in one reaction.
2. Incubate the reactions at 37 °C for 4 h after addition of hRFK or 24 h after addition of hFADS. If preparing RoFAD from RoF, add huRFK and incubate the reaction mixture for 4 h; subsequently add hFADS and incubate for another 24 h.
3. Stop the reaction with TCA added to a final concentration of 1 % (w/v). Vigorously mix the reaction (*see* **Note 3**).
4. Centrifuge the samples at 10,000 × *g* for 30 min (*see* **Note 4**).
5. Remove the supernatant carefully and save it. Discard the pellet.
6. Purify the flavins by preparative HPLC (or any other suitable method). It is recommended to confirm the identity of the compounds by liquid chromatography/mass spectrometry as described earlier [11].

3.2 In Vitro Transcription/Translation Assays (TK/TL)

3.2.1 Preparation of Plasmids

1. PCR amplify the DNA of interest, which will be tested for FMN riboswitch activity: amplify a DNA sequence corresponding to the FMN riboswitch sequence starting upstream of the aptamer until the 22nd nucleotide from the downstream ORF (which means the six first codons of the gene of interest including the initial codon plus one nucleotide to get in frame with the *luc* gene) (*see* example in Fig. 4).
   (a) To prepare transcriptional fusions introduce a *Hind*III restriction endonuclease site to the forward primer and a *Not*I restriction endonuclease site to the reverse primer (Fig. 4a) (*see* **Note 5**).
   (b) To prepare translational fusions add a *Hind*III restriction endonuclease site to the forward primer and a *Bam*HI restriction endonuclease site to the reverse primer (Fig. 4b) (*see* **Note 5**).
2. Ligate the restriction endonuclease treated DNA fragment for the transcriptional fusion to *Hind*III/*Not*I treated pT7luc and the restriction endonuclease treated DNA fragment for translational fusion to *Hind*III/*Bam*HI digested pT7luc^{mod}.
3. Transform *E. coli* with the ligation reactions. *E. coli* Top10 cells (Invitrogen) are highly recommended.
4. Select transformed cells on LB-plates containing ampicillin (100 μg/mL) and proof-sequence the corresponding plasmids.
5. Isolate plasmids from overnight cultures of the selected *E. coli* strain. Double purify the plasmids using DNA purifying columns and elute plasmid DNA with nuclease-free water. This step is important to avoid salts and other substances that could disturb the subsequent assays (*see* **Note 6**).

3.2.2 Assay (TK/TL)

1. Slowly thaw *E. coli* T7 S30 Extract System for Circular DNA Kit components on ice. Place the DNA plasmid solution (prepared in Subheading 3.2.1), the flavin stock solution, the BSA solution, and the nuclease-free water on ice as well.
2. Prepare the TK/TL mix freshly from thawed aliquots mixing the "T7 extract solution," the "S30 premix," and the "amino acid mix" at a ratio of 3:4:1 (v/v/v).
3. In a test tube placed on ice, set up the in vitro TK/TL reaction by mixing: 8 μL freshly prepared TK/TL mix, 1 μL 10× flavin solution (*see* **Notes 7** and **8**) or 1 μL nuclease-free water (control), and 1 μL plasmid solution (1–100 ng/μL) (*see* **Note 9**). It is important to follow the above described order.
4. Immediately after plasmid addition incubate the reaction mixture at 30 °C for 5 min in a water-bath.

5. After 5 min stop the reaction by adding 90 μL of the BSA solution and place the tube on ice (*see* **Note 10**).
6. Keep the stopped TK/TL reaction on ice and proceed to the next section. Do not freeze the reaction to avoid luciferase inactivation.

3.2.3 Firefly Luciferase Activity Assay

1. Set the plate reader equipment to allow the incubation of the microplate at 26 °C for 10 s with agitation followed by kinetic measurements (4–6 reads with minimal interval) at 26 °C performing a 2-s measurement delay followed by a 10-s measurement read for luciferase activity.
2. In a white microplate mix 20 μL of the TK/TL reaction with 100 μL of the "Luciferase Assay Reagent." The resulting solution has to completely cover the bottom of the microplate wells.
3. Start the incubation-reading process. Take measurements until light emission is constant (*see* **Note 11**).

3.3 Flavin Fluorescence Quenching Measurements Upon FMN Riboswitch Binding

Fluorescence quenching assays are performed on the basis of the intrinsic fluorescence of FMN and RoFMN (*see* **Note 12**), which is quenched upon the specific interaction of the flavins with the riboswitch aptamer RNAs. Fluorescence quenching can be employed in order to study the basis of ligand–riboswitch interactions. Fluorescence quenching measurements can be performed with complete mRNA molecules (Subheading 3.3.2) or during the time course of the transcription process (Subheading 3.3.3). The same plasmids used in the in vitro TK/TL assays (*see* above) can be used.

3.3.1 Preparing DNA Templates

1. Linearize pT7luc constructs by treatment with *Eco*RV.
2. Purify the linearized plasmid twice using the GeneJET™ Plasmid Miniprep Kit (Fermentas) and elute the DNA from the columns with nuclease-free water.

3.3.2 Flavin Fluorescence Quenching Measurements Using Complete RNA Molecules

1. Place all the reaction components on ice (Subheading 2.3).
2. In a test tube, set the reaction by mixing: 10 μL transcriptional 5× buffer, 5 μL 100 mM DTT, 50 U ribonuclease inhibitor (recombinant RNasin), 10 μL rNTP mix, 1 μg linearized DNA (prepared in the Subheading 3.3.1), 20 U T7 RNA polymerase, and nuclease-free water up to 50 μL. Addition of T7 RNA polymerase must be the last step.
3. Immediately after addition of RNA polymerase incubate the reaction mixture at 37 °C for 4 h in a water bath.
4. Digest the DNA plasmid in the reaction mixture with DNaseI (RNase-free enzyme).

5. Purify the mRNA using RNA purifying columns. Determine the RNA concentration. Store RNA solution at 4 °C, do not freeze it.
6. Set the microplate reader for intensity of fluorescence emission measurement at wavelength 535 nm with an excitation at 485 nm at 25 °C. Consult the manual instructions of the equipment for gain (*see* **Note 12**).
7. To a sterile nuclease-free black microplate add 80 μL 50 mM potassium phosphate pH 7.4 containing 2 mM $MgCl_2$ and 2 mM KCl and 10 μL flavin solution (10 μM). Homogenize the mixture by pipetting slowly up and down.
8. Add 10 μL mRNA solution (10 μM) prepared in **steps 1–5**. Start measurements right after addition of mRNA solution.

3.3.3 Flavin Fluorescence Quenching Measurements of Nascent RNA Molecules

1. Place all the reaction components on ice (Subheading 2.3).
2. Set the microplate reader for intensity of fluorescence emission measurement at wavelength 535 nm with an excitation at 485 nm at 30 °C. Consult the manual instructions of the equipment for gain (*see* **Note 12**).
3. In a sterile nuclease-free black microplate, set the reaction by mixing: 10 μL transcriptional 5× buffer, 5 μL 100 mM DTT, 50 U ribonuclease inhibitor (recombinant RNasin), 10 μL rNTP mix, 5 μL flavin solution (10 μM), 1.5 nM linearized DNA (Subheading 3.3.1), 20 U T7 RNA polymerase, and nuclease-free water up to 50 μL. Addition of T7 RNA polymerase must be the last step.
4. Start measurement immediately after RNA polymerase addition. Follow the reaction for 30 min with kinetic measurements each 30 s.

4 Notes

1. RF and RoF are poorly soluble in water (only up to about 250 μM).
2. All flavins must be protected from light at all times and should be stored at −20 °C.
3. Before the reaction is stopped by adding TCA it is recommended to analyze a small sample by HPLC in order to confirm that the reaction is complete.
4. The addition of reducing agents such as sodium dithionite to flavin solutions leads to decolorization of the flavin samples. Vigorous mixing in the presence of air restores the color of the flavins.
5. Transcriptionally controlled FMN riboswitches can be tested using either transcriptional or translational fusions. However,

translationally controlled FMN riboswitches can be tested only by employing translational fusions.

6. High salt concentrations present in the in vitro TK/TL reaction can decrease the reaction efficiency.
7. Flavin concentrations above 200 μM in the in vitro TK/TL reaction are not recommended since this can disturb the reaction.
8. If different flavin concentrations are to be tested, the T_{50} values can be estimated by fitting the plot of the luciferase activity percentage versus the flavin concentration, using for example SigmaPlot 9 software (Systat Software, Erkrath, Germany). The data fit to a first-order exponential decay equation. T_{50} represents the amount of flavin needed for a 50 % reduction of luciferase activity in the transcription/translation assays and is a measure for the apparent ligand affinity of the FMN riboswitch aptamers in the context of an active transcription/translation process.
9. The concentration of the plasmid solution should be adjusted in order to produce a response signal (light emission by luciferase) that is within the linear range of detection of the luminometer equipment. Moreover, similar luciferase activities should be generated for all control reactions.
10. The BSA solution is used in order to dilute the reaction components. A 10-times dilution is sufficient to stop the reaction. Moreover, BSA stabilizes luciferase avoiding rapid inactivation of the enzyme.
11. The light intensity of the reaction is nearly constant for about 1 min. After 1 min a slow decrease of the signals can be observed, with a half-life of approximately 10 min.
12. Notably, RoFMN fluorescence is about 20 times weaker when compared to FMN.

References

1. Otani S et al (1974) Letter: roseoflavin, a new antimicrobial pigment from *Streptomyces*. J Antibiot (Tokyo) 27(1):88–89
2. Grill S et al (2007) Identification and characterization of two *Streptomyces davawensis* riboflavin biosynthesis gene clusters. Arch Microbiol 188(4):377–387
3. Vogl C et al (2007) Characterization of riboflavin (vitamin B2) transport proteins from *Bacillus subtilis* and *Corynebacterium glutamicum*. J Bacteriol 189(20):7367–7375
4. Hemberger S et al (2011) RibM from *Streptomyces davawensis* is a riboflavin/roseoflavin transporter and may be useful for the optimization of riboflavin production strains. BMC Biotechnol 11(1):119
5. Mansjo M, Johansson J (2011) The riboflavin analog roseoflavin targets an FMN-riboswitch and blocks *Listeria monocytogenes* growth, but also stimulates virulence gene-expression and infection. RNA Biol 8(4):674–680
6. Lee ER, Blount KF, Breaker RR (2009) Roseoflavin is a natural antibacterial compound that binds to FMN riboswitches and regulates gene expression. RNA Biol 6(2):187–194
7. Ott E et al (2009) The RFN riboswitch of Bacillus subtilis is a target for the antibiotic roseoflavin produced by *Streptomyces davawensis*. RNA Biol 6(3):276–280
8. Pedrolli DB et al (2012) A highly specialized flavin mononucleotide riboswitch responds

differently to similar ligands and confers roseoflavin resistance to *Streptomyces davawensis*. Nucleic Acids Res 40(17):8662–8673

9. Serganov A, Huang L, Patel DJ (2009) Coenzyme recognition and gene regulation by a flavin mononucleotide riboswitch. Nature 458(7235):233–237
10. Otani S, Kasai S, Matsui K (1980) Isolation, chemical synthesis, and properties of roseoflavin. Methods Enzymol 66:235–241
11. Jankowitsch F et al (2011) A novel N, N-8-amino-8-demethyl-D-riboflavin Dimethyltransferase (RosA) catalyzing the two terminal steps of roseoflavin biosynthesis in *Streptomyces davawensis*. J Biol Chem 286(44):38275–38285
12. Grill S et al (2008) The bifunctional flavokinase/flavin adenine dinucleotide synthetase from *Streptomyces davawensis* produces inactive flavin cofactors and is not involved in resistance to the antibiotic roseoflavin. J Bacteriol 190(5):1546–1553
13. Pedrolli DB et al (2011) The antibiotics roseoflavin and 8-demethyl-8-amino-riboflavin from Streptomyces davawensis are metabolized by human flavokinase and human FAD synthetase. Biochem Pharmacol 82(12):1853–1859
14. Wickiser JK et al (2005) The speed of RNA transcription and metabolite binding kinetics operate an FMN riboswitch. Mol Cell 18(1):49–60
15. Breaker RR (2012) Riboswitches and the RNA world. Cold Spring Harb Perspect Biol 4(2)

Chapter 14

Construction and Application of Riboswitch-Based Sensors That Detect Metabolites Within Bacterial Cells

Casey C. Fowler and Yingfu Li

Abstract

A riboswitch is an RNA element that detects the level of a specific metabolite within the cell and regulates the expression of co-transcribed genes. By fusing a riboswitch to a reporter protein in a carefully designed and tested construct, this ability can be exploited to create an intracellular sensor that detects the level of a particular small molecule within live bacterial cells. There is a great deal of flexibility in the design of such a sensor and factors such as the molecule to be detected and the downstream experiments in which the sensor will be applied should guide the specific blueprint of the final construct. The completed sensor plasmid needs to be rigorously tested with appropriate controls to ensure that its dynamic range, signal strength, sensitivity and specificity are suitable for its intended applications. In this chapter, methods for the design, assessment and use of riboswitch sensors are provided along with those for one example application for which riboswitch sensors are ideally suited.

Key words Riboswitch, Aptamer, RNA, Biosensor, Intracellular sensor, Metabolite, Small molecule detection

1 Introduction

Riboswitches are RNA elements that regulate gene expression in cis through structural changes that occur in response to directly binding specific small molecule ligands [1, 2]. Over the course of the last decade, dozens of naturally occurring riboswitches have been discovered and several methods for creating artificial riboswitches have been explored [3–8]. Due to their inherent ability to both detect a specific metabolite with great sensitivity and specificity and to transduce this information into a detectable signal, riboswitches have great potential as molecular sensors. Genetic constructs that place riboswitches in control of the expression of reporter proteins have begun to be exploited as metabolite sensing systems that detect the level of a target molecule within the tremendously complex context of a live bacterial cell (*see* Fig. 1) [9]. These systems have several advantages over traditional methods for

Daniel Lafontaine and Audrey Dubé (eds.), *Therapeutic Applications of Ribozymes and Riboswitches: Methods and Protocols*, vol. 1103, DOI 10.1007/978-1-62703-730-3_14, © Springer Science+Business Media New York 2014

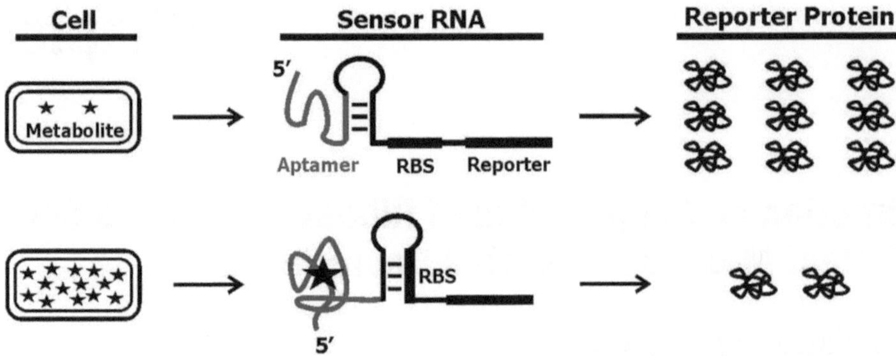

Fig. 1 Schematic of an intracellular sensor made of a riboswitch and a reporter protein. The example depicted is an off-switch. When the level of the metabolite to be detected is low, the aptamer domain (*gray line*) is in the ligand-free state; the expression platform (*black line*) adopts a structure that exposes the ribosomal binding site (RBS) for the translation of the reporter gene, resulting in a high level of the reporter protein and thus a strong signal. When the metabolite concentration is high, the aptamer domain of an increasing proportion of the riboswitches is in the ligand-bound state, which makes the expression platform adopt an alternative structure in which the RBS is masked, preventing the translation of the reporter gene and leading to a weak signal

monitoring metabolite levels: the assays are fast, easy and compatible with high throughput analyses, they do not require sophisticated equipment and highly specialized expertise associated with most metabolite detection processes and, perhaps most importantly, the detection takes place within live cells and not following lysis when metabolite levels can be skewed by degradation, aggregation or association with cellular components.

The design of riboswitch-based sensors is very flexible and a number of creative features and adjustments can be made to enhance their sensing capabilities. Depending upon the downstream applications envisioned, the riboswitch being used, the host organism and the target metabolite to be detected, the specifics of the design and verification required to create an efficient and reliable riboswitch sensor could vary drastically. Additionally, the diverse nature of metabolites ensures that the detection of each will come with its own unique set of challenges. Accordingly, this chapter begins by providing general information on the design of riboswitch sensors for use in bacterial cells. In order to provide detailed protocols for the testing and application of the sensors, later sections apply a scenario employing a validated, naturally occurring riboswitch and a β-galactosidase reporter gene to detect a metabolite within the Gram-negative γ-proteobacterium *Escherichia coli* (*E. coli*). These sections provide detailed protocols for cell growth and reporter assays, and for testing and analyzing the detection properties of riboswitch sensors. The chapter concludes by describing a sample application of the sensors, which involves scaling up to high throughput assays and probing the activity of proteins involved in the uptake of the target metabolite.

2 Materials

2.1 Construction of Riboswitch Sensors

The materials required to produce the riboswitch sensor plasmid will vary depending upon the precise nature of the final construct and the molecular cloning strategies used. Typically, creating the riboswitch sensor plasmid will require the following reagents:

1. High fidelity PCR reagents, including DNA primers, to amplify the elements to be included in the plasmid.
2. DNA templates that encode each of the selected elements including: reporter gene(s), promoter(s), riboswitch(es), and transcriptional terminator(s).
3. Reagents for agarose gel electrophoresis and a kit suitable for the purification of DNA fragments from agarose gels.
4. DNA modifying enzymes, which may include restriction endonucleases, T4 DNA ligase, alkaline phosphatase, polynucleotide kinase, T4 DNA polymerase (or DNA polymerase I, Klenow fragment).
5. A competent strain of *E. coli* engineered for use in molecular cloning (DH5α, for example).
6. L-broth (LB), LB-agar plates and appropriate antibiotics to selectively maintain the plasmids used.

For those unfamiliar with molecular cloning practices, a number of excellent resources are available, including a widely used text that thoroughly describes a range of relevant principles [10]. Another option that is becoming increasingly affordable is to have the entire construct chemically synthesized, a service offered commercially by many companies.

2.2 Growth Media

The precise composition of the growth media required will depend on the nature of the metabolite being detected and the experiment being conducted. This section provides two medium options that we have previously employed with *E. coli* cells, one which contains minimal nutrient content and one which is a richer option. There are limitless options for medium composition that might suit the needs of the particular system being used, but in all cases, chemically undefined media such as LB should be avoided.

2.2.1 Chemically Defined, Low Nutrient Medium

Create the following stocks and sterilize by autoclaving unless otherwise indicated.

1. 5× M9 salts: dissolve 56.4 g of M9 salts (Sigma-Aldrich, M6030) per liter of distilled water.
2. 1 M $MgSO_4$: dissolve 12.0 g $MgSO_4$ (Sigma-Aldrich, M7506) in 100 ml of distilled water.
3. 0.1 M $CaCl_2$: dissolve 1.1 g $CaCl_2$ (Sigma-Aldrich, C4901) in 100 ml of distilled water.

4. 20 % glucose: dissolve 20 g of D-glucose (Sigma-Aldrich, G7528) in 100 ml of distilled water and sterilize by filtration with a 0.2 μm filter.

For 500 ml of medium add 100 ml of 5× M9 salts, 10 ml of 20 % glucose, 1 ml of 1 M $MgSO_4$, and 500 μl of 0.1 M $CaCl_2$ to 388.5 ml of sterile, distilled water.

2.2.2 Rich, Chemically Defined Medium

This medium is based on the low nutrient medium described above, but is supplemented with amino acids, vitamins and nucleic acid precursors. Stock solutions of these supplements can be made individually or in large mixes. Prior to adding these supplements to the medium they should be sterilized by filtration through a 0.2 μm filter. Making up individual stocks of these many compounds is laborious and so making and storing larger batches is an appealing option. Sterilized supplement mixes can be stored for several weeks at 4 °C and for longer periods when frozen.

1. Amino acids should be added to the medium at the following final concentrations: 100 μg/ml DL-alanine (Sigma-Aldrich, A7502), 22 μg/ml L-arginine (Sigma-Aldrich, A5006), 100 μg/ml L-asparagine (Sigma-Aldrich, A0884), 100 μg/ml L-aspartic acid (Sigma-Aldrich, A9256), 22 μg/ml L-cysteine (Sigma-Aldrich, 168149), 100 μg/ml L-glutamic acid (Sigma-Aldrich, 49449), 100 μg/ml L-glutamine (Sigma-Aldrich, 49419), 100 μg/ml glycine (Sigma-Aldrich, 50046), 22 μg/ml L-histidine (Sigma-Aldrich, 53319), 20 μg/ml L-leucine (Sigma-Aldrich, 61819), 88 μg/ml L-lysine (Sigma-Aldrich, L5501), 20 μg/ml L-isoleucine (Sigma-Aldrich 58879), 20 μg/ml L-methionine (Sigma-Aldrich, 64319), 20 μg/ml L-phenylalanine (Sigma-Aldrich, P2126), 30 μg/ml L-proline (Sigma-Aldrich, 81709), 100 μg/ml L-serine (Sigma-Aldrich, 84959), 80 μg/ml L-threonine (Sigma-Aldrich, 89179), 20 μg/ml L-tryptophan (Sigma-Aldrich, 93659), 20 μg/ml L-tyrosine (Sigma-Aldrich, 93829), 40 μg/ml L-valine (Sigma-Aldrich, 94619).
2. The following vitamins should be added to the medium at the indicated concentrations: 0.5 μg/ml biotin (Sigma-Aldrich, B4501), 1 μg/ml calcium pantothenate (Sigma-Aldrich, P2250), 1 μg/ml niacin (Sigma-Aldrich, N4126), 1 μg/ml pyridoxine-HCl (Sigma-Aldrich, P9755), 1 μg/ml thiamine-HCl (Sigma-Aldrich, T4625).
3. The following nucleic acid precursors should be added to the medium at a final concentration of 40 μg/ml: adenine (Sigma-Aldrich, A8626), thymine (Sigma-Aldrich, T0376), thymidine (Sigma-Aldrich, T9250), and uracil (Sigma-Aldrich, 0750).

2.3 β-Galactosidase Reporter Assays

1. Permeabilization solution: 100 mM Na_2HPO_4, 20 mM KCl, 2 mM $MgSO_4$, 0.8 mg/ml hexadecyltrimethyl ammonium bromide (CTAB), 0.4 mg/ml sodium deoxycholate, 5.4 μl/ml β-mercaptoethanol. Used for both low and high throughput β-galactosidase assays. Permeabilization solution can be stored for about 1 week at 4 °C.
2. Substrate solution: 60 mM Na_2HPO_4, 40 mM NaH_2PO_4, 1 mg/ml O-nitrophenyl β-D-galactopyranoside (ONPG), 2.7 μl/ml β–mercaptoethanol. Used for low throughput β-galactosidase assays. Substrate solution can be stored for about 1 week at 4 °C.
3. Quenching solution: 1 M Na_2CO_3. Used for low throughput β-galactosidase assays. Can be stored indefinitely at room temperature.
4. Galacto-Star™ Chemiluminescent Reporter Gene Assay system (Applied Biosystems, T1014).
5. Clear 96-well plates used for bacterial growth, OD_{600} measurement and permeabilization during high-throughput assays (Corning Life Sciences, 3370).
6. Black 96-well plates used for luminescence measurements in high throughput β-galactosidase assays (Corning Life Sciences, 3915).

3 Methods

3.1 The Selection of Riboswitch Sensor Components and the Design of Final Construct

The design of any riboswitch sensor will need to be tailored to the individual riboswitch being utilized. Before beginning such a project, it is imperative to thoroughly review all available information on the structure, mechanism and function of the riboswitches of the class to be employed. It is also essential to have a clear idea in mind of the types of experiments for which it will be employed before beginning the project, as this will strongly affect the design, requirements and optimization of the sensing construct. This section provides a guide on selecting and assembling the elements that comprise a riboswitch sensor.

1. *Selection of a riboswitch that detects your target metabolite of interest.* The list of metabolites for which naturally occurring riboswitches have been identified is large and growing (*see* Table 1). For most of the metabolites on this list, many individual riboswitches have been identified that regulate different genes in different organisms. Testing a panel of riboswitches to identify one that best suits the specific sensing requirements is very strongly recommended. Based on our limited experience, it appears that in some instances riboswitches maintain activity

Table 1
Riboswitch classes discovered to date

Riboswitch class	Metabolite	Apatmer affinity (K_D)	Aptamer size (nt)
Adenine	Adenine	300 nM [25]	70
AdoCbl	Adenosylcobalamin (AdoCbl)	300 nM [26]	200
Cyclic di-GMP-I	Cyclic di-GMP	1 nM [27]	110
Cyclic di-GMP -II	Cyclic di-GMP	200 pM [28]	90
Deoxyguanosine	Deoxyguanosine	80 nM [29]	70
Fluoride	Fluoride	60 μM [30]	80
FMN	Flavin mononucleotide (FMN)	5 nM [23]	120
GlcN6P	Glucosamine-6-phosphate (GlcN6P)	200 μM [31]	170
Glutamine	Glutamine	150 μM [32]	70
Glycine	Glycine	30 μM [22]	110
Guanine	Guanine	5 nM [33]	70
Lysine (L-box)	Lysine	1 μM [24]	175
Mg2+ I (mgtA)	Mg2+	N.D.	100
Mg2+ II (M-Box)	Mg2+	N.D.	70
Moco riboswitch	Molybdenum cofactor (Moco)	N.D.	130
PreQ1-I	Pre-queuosine1 (PreQ1)	50 nM [34]	40
PreQ1-II	PreQ1	100 nM [35]	85
SAH	S-adenosylhomocysteine (SAH)	20 nM [36]	65
SAM-I	S-adenosylmethionine (SAM)	4 nM [37]	100
SAM-II	SAM	1 μM [38]	60
SAM-III	SAM	N.D.	80
SAM-IV	SAM	15 μM [39]	60
THF	Tetrahydrofolate (THF)	70 nM [40]	100
TPP (Thi-box)	Thiamine pyrophosphate (TPP)	100 nM [11]	120

The in vitro binding affinity (KD) and size of the aptamer domain are provided for each class. The binding affinity for each aptamer is approximate and dependent on the specific construct and conditions tested. Similarly, the size of each aptamer can vary considerably from one riboswitch to the next. *N.D.* not determined

when transferred between bacterial species, while in other instances they might not (unpublished observations). There are many possible reasons for this including the direct or indirect influence of other genetic factors within the host organism on riboswitch activity or differences in the organism's native environment that might influence RNA folding, such as temperature or the concentration of metal ions or salt. Additionally, the concentration range over which any riboswitch is sensitive is limited and it is crucial that this dynamic range overlaps with the concentration range that is relevant for the purposes of your study (*see* **Note 1**). Depending upon the gene(s) a particular riboswitch is controlling and the organism in which it resides, the sensitive range could vary greatly for different riboswitches that detect a given metabolite. The binding range measured in vitro might not always match up with a riboswitch's dynamic range within a live cell, but when available this information might be useful in guiding your selection. It is intuitive—and supported by experimental evidence—that the efficiency of regulation will vary between riboswitches depending on the evolutionary pressures applied [11]. While some riboswitches will only subtly tune downstream gene expression, others will act much more like an on–off switch. The ratio of the level of expression between the fully activated and the fully repressed scenarios is a very significant consideration in the choice of riboswitch as this will dictate the signal amplitude of the resulting detection system. Whether a particular riboswitch activates or represses gene expression in response to its ligand and the regulatory mechanism of the riboswitch are also aspects to consider, as these factors will influence the sensing characteristics of the system as well as the fusion of the riboswitch to the reporter gene (*see* below). Finally, several different approaches have been used to isolate synthetic riboswitches that respond to a molecule of choice [4–8]. This is beyond the scope of the present chapter, but is a possible solution to detecting a molecule not recognized by any of the naturally occurring riboswitches identified to date. These protocols could also be applied to modify natural riboswitches to exhibit desired properties such as altered ligand specificity or more stringent regulatory control [12] (*see* **Note 2**).

2. *Selection of the reporter protein.* The choice of reporter protein for a given sensor should be strongly influenced by the nature of the experiment for which the sensor will ultimately be used. An enormous amount of research has gone into the study of reporter proteins and the engineering of modified proteins and substrates that elicit beneficial properties for certain research applications; this subject should be reviewed prior to selecting a reporter. Fluorescent proteins are a major class of reporters that have many benefits. Unlike enzymatic reporters they do not require a substrate to produce a signal, which can lead to easier and less expensive assays and can facilitate scaling up to higher throughput experiments. Also, the wide selection of proteins available with distinct spectra presents the possibility of using a second reporter protein as an internal control to

account for phenomena that affect reporter expression independent of riboswitch activity (*see* **step 6**, this section) [13]. A drawback of fluorescent reporters is that, unlike their enzymatic counterparts, one protein cannot process multiple substrates to produce an amplified signal. This can lead to problems with low signal intensity, which can be particularly troublesome for proteins that emit in the blue–green range where autofluorescence introduces significant background for many bacterial species. Additionally, while mutants have been created that attempt to address these issues, fluorescent proteins generally fold slowly and have a long half-life in the cell [14, 15]. This results in very poor temporal sensitivity and thus sensors that respond very slowly to changing conditions.

Enzymatic reporters come in many variations, most of which work by catalyzing a reaction that produces a colorimetric, fluorescent or luminescent product. Luminescent reporters, such as firefly luciferase or the bacterial *lux* system, are generally very sensitive due to the nature of the signal they emit and are capable of producing a linear output over a broad dynamic range [16, 17]. Eukaryotic luciferases (such as firefly luciferase) require a relatively expensive substrate that must be added exogenously during the assay. The assay produces a time-sensitive signal that requires coordination between substrate addition and signal acquisition. The biosynthetic genes for the *lux* substrate have been identified and the substrate can be produced within the host organism and does not need to be added exogenously. This can be also viewed as a potential problem, however, since any conditions that affect substrate synthesis could interfere with the detection process. *E. coli* β-galactosidase is a very well characterized reporter enzyme whose natural function is the breakdown of lactose. Numerous chemical analogs of natural substrates have been synthesized that produce a measurable signal upon cleavage. For quantitative analysis, colorimetric substrates—most famously O-nitrophenyl β-D-galactopyranoside (ONPG)—have typically been employed to indicate β-galactosidase activity in robust and reliable assays [18]. More expensive luminescent substrates are also available that provide increased sensitivity (*see* Subheading 3.4). β-galactosidase expression can also be qualitatively monitored in bacterial colonies grown on solid medium using the substrate 5-bromo-4-chloro-3-indolyl β-D-galactopyranoside (X-gal), which has been dubbed "blue-white screening". It is important to consider that your host organism might naturally encode a β-galactosidase enzyme and it is important to work within a strain that lacks this activity.

3. *Selection of a plasmid to house the riboswitch sensor*. The plasmid selected needs to have a resistance cassette that is compatible with the experiments in which it will be employed.

The copy number, guided by the origin of replication, can vary dramatically between plasmids. While a high copy number will help increase the strength of the signal, such vectors are more physiologically intrusive. A sufficiently high copy number plasmid with a strong enough promoter, coupled with a scarce target metabolite, could create the problematic scenario in which the molecule being detected is not in excess over your sensor. As a general guide, choose a plasmid with the lowest copy number that still maintains a strong enough signal for your purposes. Additionally, it is important that spurious promoters from the vector do not read through and transcribe the sensing elements, as such promoters might respond to stimuli other than your target metabolite and interfere with the signal (*see* below). It is useful if strong transcriptional terminators precede and follow the sensing elements.

4. *Selection of a promoter to drive the expression of the riboswitch sensor*. Regardless of the regulatory mechanism of the riboswitch, promoter-driven changes in the level of transcriptional initiation will affect the amount of reporter protein that is produced. Since this regulation is not dependent on riboswitch activity it will interfere with the detection process. An ideal promoter would therefore not be subject to any form of regulation. While many promoters are commonly referred to as "constitutive," it is difficult to rule out the possibility of a condition in which a promoter is regulated in some fashion. Indeed, all promoters will be subject to variation in transcriptional initiation under certain conditions, such as extreme starvation. While controls will be required to account for this, it is important to select a promoter that is as stable as possible under the conditions in which the sensor will be employed. Many riboswitches appear to be the sole—or at least primary—form of regulation for simple feedback loops and thus in some cases the riboswitch's native promoter might be a good choice if it is compatible with the bacterial species in which it will be employed [19]. Alternatively, a well characterized "constitutive" promoter would make a good choice.

5. *Design of the sensor plasmid*. Figure 2a shows a schematic of the individual elements discussed above and their assembly into the final sensing construct. RNA folding is the crux of a riboswitch's function and changing the sequence context of a riboswitch can disrupt this folding. The synthetic nature of the sensor plasmid dictates that the riboswitch will be removed from its natural setting, however steps should be taken to minimize the potential for disruption. To be as safe as possible, the entire unaltered 5′ untranslated region (UTR) can be fused to the reporter protein, maintaining its native start (+1) nucleotide, its ribosome binding site (RBS) and start codon.

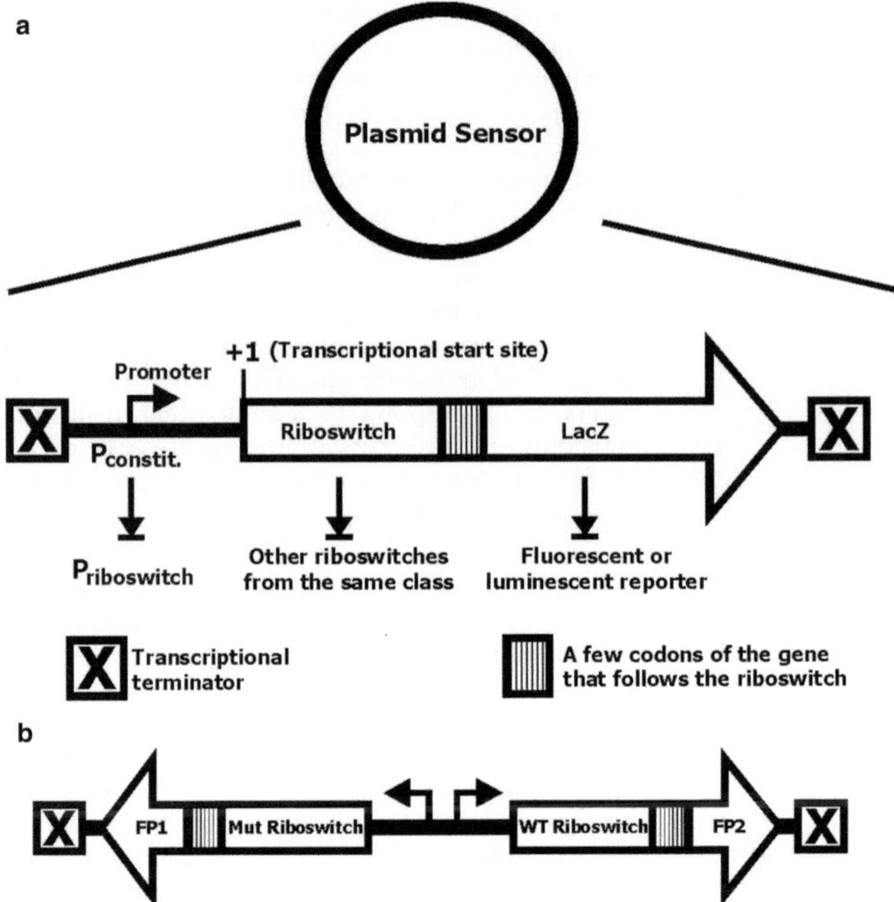

Fig. 2 Design of sensor plasmids. A plasmid sensor that contains a single reporter (**a**) or two fluorescent protein (FP) reporters (**b**). *See* Subheading 3.1 for detailed discussion on the selection of riboswitches, reporter proteins, plasmids, and promoters for the design of an appropriate sensor plasmid

When possible, fusing the first several codons of the protein naturally controlled by the riboswitch to the reporter protein is also wise. Fusion restraints for riboswitches that employ a transcriptional termination or an RNA cleavage mechanism might be more flexible than those that work at the level of translation and could, for example, allow for the introduction of an optimal RBS to boost expression. However, any RNA sequence introduced in the neighborhood of the riboswitch will have the capacity to interfere with folding in an unexpected way and should be avoided when possible. For all sensor constructs, parallel plasmids carrying mutant riboswitches should also be created. The mutant riboswitches should contain a minimal number of mutations that prevent the binding of the target metabolite. Additionally, a third version of the vector should be made that completely lacks the riboswitch element, but maintains an identical promoter and reporter. These two control plasmids are useful controls for future experiments (*see* Subheading 3.3).

6. *Special additions to the base design.* One enticing addition to the basic design of a riboswitch sensor is a second reporter protein that acts as an internal control for changes in reporter activity that are not related to riboswitch activity. For the second reporter to be as accurate as possible, it should be housed on the same plasmid and driven by the same promoter. If a suitable mutant is available, this second reporter could have an identical 5′ UTR as well, with a mutation introduced that disrupts the riboswitch's ligand binding capabilities and thus the metabolite-induced response (*see* Fig. 2b). Using a two-reporter method, the ratio of the two reporters would be used to ascertain the levels of the target molecule in cells, since fluctuations in its concentration should be the only factor that should influence that ratio. For this to be true, the two reporters would need to behave as similarly as possible and, importantly, have similar maturation times and lifetimes. Certain pairs of fluorescent proteins might satisfy this stipulation. Another intriguing possibility is the use of multiple riboswitches that respond to the same metabolite to control a single reporter gene. Most riboswitches bind their metabolites in a simple one to one fashion and the output of riboswitch sensors is dictated by the dynamics of such an interaction [3, 9]. Tandem riboswitches could be used to produce sensors that are more sensitive to subtle changes in ligand concentration and that have a greater signal amplitude [3].

3.2 Growth Conditions and Reporter Assay Protocols

From this point forward, detailed protocols are provided for the use of a riboswitch sensor employing β-galactosidase to detect a target metabolite in *E. coli* cells. Some examples of defined growth media that we have successfully employed are provided in the materials section that might be suitable as a base medium. The individual experiment will, of course, dictate the specific details of growth and factors such as medium content, pH, oxygen level, growth temperature and growth time can be adjusted accordingly.

1. Compose a defined medium in which to conduct your assays. Rich and simple media options are provided in the Materials section (*see* Subheading 2.2).
2. Transform the sensor plasmid into the desired strain(s) and plate onto LB-agar plates containing the appropriate antibiotic(s) to obtain single colonies.
3. Pick three single colonies for each sample and grow with shaking overnight (~12–20 h) at 37 °C in ~2 ml of the medium described in **step 1** and the appropriate antibiotic(s).
4. Inoculate fresh 2–3 ml cultures by adding the overnight culture at a dilution of 1:1,000 into fresh medium. Grow at 37 °C

with shaking until late log phase (~5 h or more depending on the growth conditions).

5. Measure the cell density for each sample by taking absorbance measurements at 600 nm and recording the values.
6. Label one 1.5 ml tube for each sample. Label one extra tube as "blank".
7. Add 80 μl of permeabilization solution to each tube. Add 20 μl of culture for each sample to the permeabilization solution and pipette up and down to mix. For the blank sample, add 20 μl of culture from a strain that lacks a source of β-galactosidase activity. Alternatively, 20 μl of growth medium can be added as the blank. Incubate at room temperature for 30 min.
8. Add 600 μl of the substrate solution to each tube and mix by inverting the tube several times. Start a timer (counting up) as soon as you add the substrate solution to the first tube. The reaction needs to be carefully timed and an accurate start time for each sample is important. When processing a large number of samples, it might be important to stagger the start time for reactions. This is particularly true with samples exhibiting high levels of reporter activity as these reactions will need to be quenched quickly (*see* below).
9. In somewhere between a few seconds and a few hours, your samples should begin to turn yellow. Once a sample has reached an obvious yellow color (OD_{420} ~ 0.1–0.5), quench that reaction by adding 700 μl of quenching solution and inverting the tube several times to mix. Upon quenching a reaction, record the total reaction time for that sample. If you have significant changes in expression from sample to sample, you will need to quench different samples at different time points. If samples turn a bright yellow color in less than a few minutes, reduce the volume of culture being added to the reaction accordingly. If samples are not visibly yellow after ~2–3 h, the culture can be concentrated by pelleting the bacteria (centrifuge for 10 min at 5,000 × *g*) and resuspending in a smaller volume. This should not be taken to extremes, since sufficiently dense cultures might interfere with the reaction. If low signal intensity is a problem, a more sensitive detection method might be required.
10. Once all samples have been quenched, centrifuge at top speed (~18,000 × *g*) in a tabletop microfuge for 5 min.
11. Add 1 ml of each sample to a disposable cuvette. Blank the spectrometer using the blank sample at an absorbance of 420 nm. Record the absorbance of each sample at 420 nm.
12. Calculate β-galactosidase activity as Miller units using the following equation:

$$\text{Miller units} = 1000 \times \left(\text{Abs}_{420}\right) / \left(\text{Abs}_{600} \times \text{culture volume added (ml)} \times \text{reaction time (minutes)}\right).$$

3.3 Testing the Detection Capabilities of Sensors

Before a sensor plasmid can be used it must be carefully tested in order to ensure it is functional and to understand its restrictions and key detection parameters. Because detection takes place within the complex and uncontrolled context of a cell and because the sensors are dependent on the successful translation and folding of the reporter protein, extra caution must be used. An essential component of testing the sensor will be to create conditions wherein the concentration of the metabolite within cells can be varied in a controlled fashion. While some molecules will be at relatively low levels or absent from cells if a suitable growth medium is used, others will require the use of strains with mutations in appropriate metabolic genes to create low target metabolite levels. Similarly, many molecules can be introduced at relatively high levels within cells simply by adding them (or a metabolic precursor) to the growth medium. In rare cases other measures might be required, such as the use of strains featuring mutations in efflux systems. Below is a description of an experimental scheme to test the activity of the riboswitch sensors and to uncover several key parameters. When appropriate, an abbreviated version of this type of analysis should be repeated for new experimental schemes in which the sensors will be employed, including when introducing significant changes to the strain being used (when possible), the growth medium, the growth phase, the growth temperature, etc. At the end of this section, the analysis of these tests is described.

1. Identify conditions in which the target metabolite is absent from cells or at a relatively low concentration (*see* above). If your target metabolite is unstable, not commercially available or not taken up by the strain being used, identify a suitable precursor that can be added to the medium to create relatively high ligand levels within cells.
2. Transform the sensor plasmid into the desired strain and plate onto LB-agar plates containing the appropriate antibiotic(s) to obtain single colonies.
3. Pick three single colonies and grow them with shaking overnight (~12–20 h) at 37 °C in ~2 ml of the medium selected that produces minimal metabolite levels; this medium will henceforth be referred to as "low metabolite medium".
4. Choose a maximal medium concentration for the target metabolite (or precursor). This concentration should be safely in excess of that predicted to be required to saturate target molecule levels in cells.
5. Add the target metabolite (or precursor) to low metabolite medium at the maximal concentration selected in **step 4**. Carry out serial threefold dilutions of this into low metabolite medium producing a total of eight dilutions, each with a volume of ~6–10 ml.

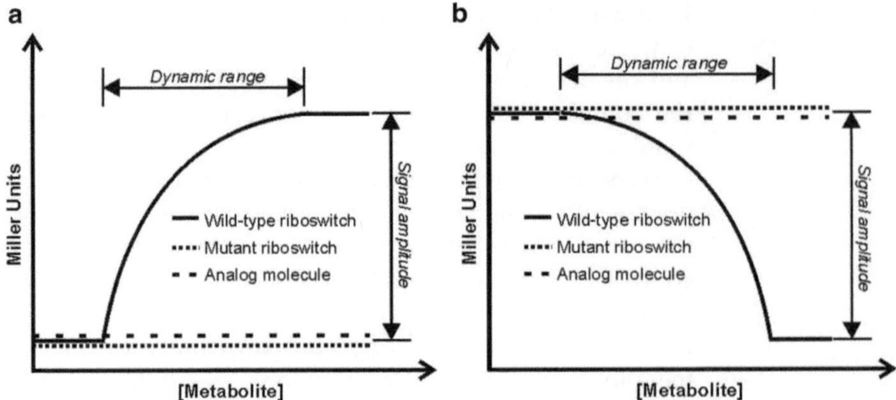

Fig. 3 Expected signal-metabolite concentration profiles of a sensor that has β-galactosidase as the reporter whose expression is controlled by a riboswitch. The riboswitch (**a**) activates or (**b**) represses the gene expression of the downstream reporter in response to metabolite binding. As the concentration of the metabolite increases, the signal (measured by Miller units) increases in (**a**) but decreases in (**b**) due to the production of a higher (**a**) or lower (**b**) amount of β-galactosidase

6. Set up three tubes containing 2-3 ml of medium for each of the eight metabolite concentrations above and an additional set for low metabolite medium (a total of 27 tubes).
7. Continue with these samples as described from **step 4** onward in Subheading 3.2.
8. Plot the Miller units for each set of samples as a function of the concentration of target metabolite (or precursor) provided in the medium.
9. Upon plotting the data it might become apparent that these dilutions have not ideally captured the dynamic range of the sensor. If so, repeat using a modified dilution scheme.
10. Repeat the steps above using the parallel plasmid that contains the mutant version of the riboswitch and the version that lacks the riboswitch element.
11. Repeat the steps above by similarly titrating a close chemical derivative of the target molecule. This is to ensure that the riboswitch is specific and does not respond to chemical analogs that might be present in the cell under relevant experimental conditions. Unless involved in downstream experiments, synthetic chemical analogs not usually encountered by the cell are not relevant; this test should focus on natural small molecules that could potentially interfere with riboswitch activity.
12. Analyze the concentration curves. The expected trend for a functional sensor is shown in Fig. 3. For a riboswitch that represses downstream gene expression in response to metabolite binding, reporter values should decrease as a function of target concentration (Fig. 3a), whereas for a riboswitch that

activates expression upon binding its ligand, reporter values should increase at higher ligand values (Fig. 3b). As shown in Fig. 3a, b reporter activity should be stable over increasing concentrations of the selected analogs and the mutant riboswitch should be non-responsive to the target metabolite. Other important features to note include the signal amplitude and the dynamic range, shown schematically in Fig. 3a. The dynamic range should roughly follow the 1:1 simple binding dynamics of a ligand and a receptor and should therefore span ~2 logs of metabolite concentrations as the riboswitch population progresses from the essentially completely unbound to the essentially completely bound state [3]. It is important to note, however, that the media concentration of the metabolite will not always coincide with its free, intracellular concentration (*see* **Note 3**). The signal amplitude is the ratio of the reporter activities at either end of the sensor's dynamic range, which will be a function of the efficiency of the riboswitch's regulation. A large signal amplitude is a vital characteristic for a sensor of this nature (*see* **Note 4**). For a discussion of the scenarios in which the mutant sensor is responsive to the target metabolite or the wild-type sensor is sensitive to the analog molecule(s), *see* **Notes 5** and **6**.

3.4 Sample Riboswitch Sensor Application: High Throughput Assays to Probe the Activity of Transport Mutants

There are countless possible experiments that could exploit the ability of riboswitch sensors to monitor metabolite levels in living bacterial cells. Recently, we used a riboswitch sensor that detects an active form of vitamin B_{12} to probe the activity of a vitamin B_{12} uptake protein. We created libraries of mutants to this protein that targeted specific residues of special interest for understanding the mechanistic details of this intricate transport system. The biological activity of these mutants was assessed using high throughput assays that used riboswitch sensors to determine how much vitamin B_{12} was taken up over a given period of cell growth. The ability to monitor small molecule transport in live cells is a significant strength of riboswitch sensors that holds a great deal of promise to aid in future research on biological transport. In the final section of this chapter, we provide a detailed protocol for the application of riboswitch sensors to monitor the activity of a library of mutant transporters. These assays are conducted in 96-well plate format using a luminescent β-galactosidase substrate, and so this section also provides methodology for these different assay formats.

Prior to beginning the protocol below, the wild-type version of the transport gene to be examined needs to be deleted from the chromosome and a plasmid-based system that complements this deletion must be established. The mutant library must then be created and cloned into this vector along with wild-type and empty vector control plasmids. Be sure that the origins of replication and the antibiotic resistances of the riboswitch sensor plasmid and the transporter-containing plasmid are compatible. For larger experiments to be practical, multi-channel pipettes or automated liquid transfer instruments are required. The protocol described suits a metabolite that is absent from cells (or at a concentration below the dynamic range of the sensor) when not provided in the growth medium. Modifications to obtain relatively low metabolite levels like those described in the preceding section will be required to monitor the transport of some molecules.

1. Prepare competent cells for the selected transporter deletion strain. Transform the riboswitch sensor plasmid into this strain to obtain single colonies.
2. Prepare competent cells for this sensor-containing transporter deletion strain. Throughout this process, be sure to grow cells in medium containing the required antibiotic to maintain the sensor plasmid. Be sure to create enough aliquots of these cells for each of the mutants and controls that will be tested.
3. Transform the plasmids containing the mutant transporters, along with the wild-type transporter plasmid and the empty vector control. Obtain single colonies of each.
4. Pick single colonies to inoculate single wells of a sterile 96-well plate containing 250 μl of LB containing antibiotics to maintain both plasmids. Pick four individual colonies of each mutant and eight colonies for the wild-type and empty vector controls. This setup allows you to assay 20 mutants per plate. Ideally, the replicates for each sample should be dispersed randomly throughout the plate in order to minimize possible "edge effects" that can cause the growth conditions to vary at different locations of the plate. Grow overnight (~16 h) at 37 °C with shaking.
5. Seed a 96-well plate that will house frozen stocks with 100 μl/well of LB + 30 % glycerol. Add 100 μl of the overnight cultures to the corresponding wells of this plate and pipette up and down to mix. This represents the stock plate for your library test plate. Store at −80 °C.
6. Pin the frozen stock plate to inoculate a new plate containing 250 μl of rich defined medium that contains both required antibiotics in each well. This can be done using a pinning tool or using a multi-channel pipette with sterile tips. Pin twice and make two identical plates that will represent technical replicates. Grow cultures overnight (~16 h).
7. For each replicate plate, inoculate fresh 250 μl cultures of defined rich medium with a minimal volume (~2 μl) of saturated overnight culture. In parallel, similarly inoculate a second set of plates that contain rich defined medium supplemented with the target molecule in the same manner.

This leaves a total of four plates, two technical replicates for each medium condition (with and without the target molecule). Grow all plates at 37 °C with shaking for several hours until late log phase or early stationary phase.

8. Remove the plates from the incubator and measure the absorbance at 600 nm.
9. In four fresh plates, add 80 μl of permeabilization solution per well. Add 20 μl of culture per well and mix by pipetting up and down. Incubate at room temperature for ~30 min.
10. Mix the concentrated Galacto-Star™ substrate and dilution buffer as indicated in the instructions provided with the kit. Add 100 μl of 1× substrate per well to four black assay plates.
11. Add 10 μl of permeabilized sample to the assay plate and mix by pipetting up and down. Incubate covered from light for 1 h at room temperature.
12. Measure the total luminescence.
13. For each sample (well), determine the luminescence/OD_{600}. This provides a normalized reporter value for each sample.
14. For each replicate, divide the luminescence/OD_{600} value for each well in the "– target" plate by the value for the corresponding "+ target" value. If your sensor produces greater reporter values as target concentrations increase, invert this equation. This provides a controlled measure of the target uptake for each sample; larger values represent greater substrate uptake and a value of one represents no uptake. The wild-type control should have a relatively high value and the empty vector control will have a value of ~1 if the protein is essential for target uptake. This value might be greater than 1, which indicates that the target can make its way into cells at some level in the absence of the protein under investigation. In either scenario the empty vector control represents a complete lack of activity for the protein under investigation.
15. This analysis provides four biological replicates and two technical replicates for each mutant. Using the uptake values calculated above, the activity of each mutant can now be analyzed statistically to determine its effect on the transport of the target metabolite.

4 Notes

1. Riboswitch sensors are best suited to determine the relative levels of a small molecule within cells, which for many experiments will be the pertinent metric. However, in some instances it could be important to determine the absolute concentration of the target molecule within cells. One way to go about getting absolute

information with riboswitch sensors would be to conduct an analysis like the one described in Subheading 3.3, using an in vitro method such as mass spectrometry to determine absolute metabolite concentrations in the same samples in parallel. The reporter values from the riboswitch sensors could then be plotted against the in vitro concentration data to obtain a standard curve that relates reporter values to concentrations. Since the reporter activity per cell data provided by the riboswitch sensors is subject to fluctuation under different conditions, this analysis would only be useful if conditions were held relatively constant.

2. Because gene expression is required for the activity of riboswitch sensors, they might not be suitable for use with cells that are dormant or under extreme starvation conditions. In the example provided in section below in **Note 5**, amino acids might not be detectable using a riboswitch sensor when they drop to very low levels. While this might affect the analysis of the sensor's activity, creating sensors to detect such metabolites at levels above those that cause starvation responses should still be possible. However the detection of any molecule that is tightly tied to gene expression or plasmid replication will require careful controls for fluctuations in these processes, such as the two reporter sensor described in Subheading 1.6.

3. The concentration of a small molecule provided in the growth medium will not necessarily correspond to its free, intracellular concentration. Many molecules are actively transported into and/or out of cells, in some cases by multiple different protein complexes. This can create a complex relationship between the medium concentration and the intracellular concentration and thus skew the observed response of the sensor in the reporter activity vs. target concentration analysis described in Subheading 3.3. As long as the target's cellular concentration can be manipulated over the sensitive range of the riboswitch in order to assess its activity and this range overlaps with the target concentration that is relevant for downstream experiments, this phenomenon should not pose any serious issues.

4. The sensor's ability to dependably detect metabolite levels will rely upon the ratio between the signal amplitude and the experimental error. While there is no set number required for the signal amplitude, sensors experiencing less than eightfold change in signal from minimum to maximum metabolite levels would require very carefully controlled experiments with minimal error in order to be useful and are unlikely to be capable of distinguishing between intermediate target levels in many experimental schemes. Sensors with a low signal amplitude may be the result of a riboswitch that naturally finely tunes expression rather than acting as an "on/off switch". Mechanistically, this inefficiency may be due to a weak sequestration of the RBS or an inefficient transcriptional terminator. To overcome this

problem, different riboswitches that respond to your target metabolite should be tested within your sensing construct. Alternatively, it might be possible to modify the riboswitch to be more efficient using engineering approaches such as those used to isolate synthetic riboswitches.

5. If the mutant riboswitch is responding to increasing target concentrations but the control vector lacking a riboswitch is not, it is likely that the mutations introduced did not fully disrupt metabolite binding. In this case, a different mutant, or perhaps a double mutant, might provide a more suitable control. In the event that both control plasmids exhibit a response to changing target concentrations it is likely that introducing the metabolite affects cell growth, plasmid replication, or alters gene expression in a non-specific manner. An example of a metabolite that could trigger such effects is an essential, core metabolite such as an amino acid. Amino acid starvation will shut down gene expression and when the levels are restored this restraint is relieved and expression will go up (*see* **Note 2**) [20]. In a scenario such as this, a sensor with a second reporter (Subheading 1.6 and Fig. 2b) might be required. Since the relevant sensor output will be the ratio of the two reporter signals, generic effects on cell physiology should be cancelled out.

6. When using a naturally occurring riboswitch it is unlikely that it will respond significantly to analogous metabolites, particularly when testing native analogs at physiologically relevant levels. It is well documented that naturally occurring riboswitches have evolved exquisite specificity against even very close chemical derivatives of their target metabolite [11, 21–24]. This is essential to their function in the cell. In some cases, however, experiments might involve using molecules not normally encountered by the organism from which the riboswitch was taken or introducing analogous molecules at concentrations that exceed those that drove the evolution of the riboswitch. Specificity is an essential feature of any sensor and if any observed crosstalk could be relevant in your experimental scheme, the sensor will need to be redesigned. A different riboswitch within the same class might be more specific. Alternatively, this might require reengineering the riboswitch by selecting for specificity against the problematic analogs.

References

1. Barrick JE, Breaker RR (2007) The distributions, mechanisms, and structures of metabolite-binding riboswitches. Genome Biol 8:R239
2. Winkler WC, Breaker RR (2003) Genetic control by metabolite-binding riboswitches. ChemBioChem 4:1024–1032
3. Breaker RR (2012) Riboswitches and the RNA world. Cold Spring Harb Perspect Biol 4(2) pii: a003566
4. Suess B, Fink B, Berens C, Stentz R, Hillen W (2004) A theophylline responsive riboswitch based on helix slipping controls gene expression *in vivo*. Nucleic Acids Res 32:1610–1614

5. Nomura Y, Yokobayashi Y (2007) Dual selection of a genetic switch by a single selection marker. Biosystems 90:115–120
6. Lynch SA, Desai SK, Sajja HK, Gallivan JP (2007) A high-throughput screen for synthetic riboswitches reveals mechanistic insights into their function. Chem Biol 14:173–184
7. Fowler CC, Brown ED, Li Y (2008) A FACS-based approach to engineering artificial riboswitches. ChemBioChem 9:1906–1911
8. Topp S, Gallivan JP (2008) Random walks to synthetic riboswitches – a high-throughput selection based on cell motility. ChemBioChem 9:210–213
9. Fowler CC, Brown ED, Li Y (2010) Using a riboswitch sensor to examine coenzyme B_{12} metabolism and transport in *E. coli*. Chem Biol 17:756–765
10. Green MR, Sambrook J, MacCallum P (2012) Molecular cloning: a laboratory manual, 4th edn. Cold Spring Harbour Laboratory Press, Cold Spring Harbour NY
11. Winkler W, Nahvi A, Breaker RR (2002) Thiamine derivatives bind messenger RNAs directly to regulate bacterial gene expression. Nature 419:952–956
12. Nomura Y, Yokobayashi Y (2007) Reengineering a natural riboswitch by dual genetic selection. J Am Chem Soc 129:13814–13815
13. Shaner NC, Steinbach PA, Tsien RY (2005) A guide to choosing fluorescent proteins. Nat Methods 2:905–909
14. Chudakov DM, Matz MV, Lukyanov S, Lukyanov KA (2010) Fluorescent proteins and their applications in imaging living cells and tissues. Physiol Rev 90:1103–1163
15. Andersen JB, Sternberg C, Poulsen LK, Bjorn SP, Givskov M, Molin S (1998) New unstable variants of green fluorescent protein for studies of transient gene expression in bacteria. Appl Environ Microbiol 64:2240–2246
16. Close D, Xu T, Smartt A, Rogers A, Crossley R, Price S, Ripp S, Sayler G (2012) The evolution of the bacterial luciferase gene cassette (lux) as a real-time bioreporter. Sensors (Basel) 12:732–752
17. Fraga H (2008) Firefly luminescence: a historical perspective and recent developments. Photochem Photobiol Sci 7:146–158
18. Miller JH (1972) Experiments in molecular genetics. Cold Spring Harbor Laboratory Press, Cold Spring Harbor, NY
19. Dambach MD, Winkler WC (2009) Expanding roles for metabolite-sensing regulatory RNAs. Curr Opin Microbiol 12:161–169
20. Chatterji D, Ojha AK (2001) Revisiting the stringent response, ppGpp and starvation signalling. Curr Opin Microbiol 4:160–165
21. Nahvi A, Sudarsan N, Ebert MS, Zou X, Brown KL, Breaker RR (2002) Genetic control by a metabolite binding mRNA. Chem Biol 9:1043
22. Mandal M, Lee M, Barrick JE, Weinberg Z, Emilsson GM, Ruzzo WL, Breaker RR (2004) A glycine-dependent riboswitch that uses cooperative binding to control gene expression. Science 306:275–279
23. Winkler WC, Cohen-Chalamish S, Breaker RR (2002) An mRNA structure that controls gene expression by binding FMN. Proc Natl Acad Sci U S A 99:15908–15913
24. Sudarsan N, Wickiser JK, Nakamura S, Ebert MS, Breaker RR (2003) An mRNA structure in bacteria that controls gene expression by binding lysine. Genes Dev 17:2688–2697
25. Mandal M, Breaker RR (2004) Adenine riboswitches and gene activation by disruption of a transcription terminator. Nat Struct Mol Biol 11:29–35
26. Nahvi A, Barrick JE, Breaker RR (2004) Coenzyme B12 riboswitches are widespread genetic control elements in prokaryotes. Nucleic Acids Res 32:143–150
27. Sudarsan N, Lee ER, Weinberg Z, Moy RH, Kim JN, Link KH, Breaker RR (2008) Riboswitches in eubacteria sense the second messenger cyclic di-GMP. Science 321:411–413
28. Lee ER, Baker JL, Weinberg Z, Sudarsan N, Breaker RR (2010) An allosteric self-splicing ribozyme triggered by a bacterial second messenger. Science 329:845–848
29. Kim JN, Roth A, Breaker RR (2007) Guanine riboswitch variants from Mesoplasma florum selectively recognize 2′-deoxyguanosine. Proc Natl Acad Sci U S A 104:16092–16097
30. Baker JL, Sudarsan N, Weinberg Z, Roth A, Stockbridge RB, Breaker RR (2012) Widespread genetic switches and toxicity resistance proteins for fluoride. Science 335: 233–235
31. Winkler WC, Nahvi A, Roth A, Collins JA, Breaker RR (2004) Control of gene expression by a natural metabolite-responsive ribozyme. Nature 428:281–286
32. Ames TD, Breaker RR (2011) Bacterial aptamers that selectively bind glutamine. RNA Biol 8:82–89
33. Mandal M, Boese B, Barrick JE, Winkler WC, Breaker RR (2003) Riboswitches control fundamental biochemical pathways in Bacillus subtilis and other bacteria. Cell 113:577–586

34. Roth A, Winkler WC, Regulski EE, Lee BW, Lim J, Jona I, Barrick JE, Ritwik A, Kim JN, Welz R, Iwata-Reuyl D, Breaker RR (2007) A riboswitch selective for the queuosine precursor preQ1 contains an unusually small aptamer domain. Nat Struct Mol Biol 14:308–317
35. Meyer MM, Roth A, Chervin SM, Garcia GA, Breaker RR (2008) Confirmation of a second natural preQ1 aptamer class in Streptococcaceae bacteria. RNA 14:685–695
36. Wang JX, Lee ER, Morales DR, Lim J, Breaker RR (2008) Riboswitches that sense S-adenosylhomocysteine and activate genes involved in coenzyme recycling. Mol Cell 29: 691–702
37. Winkler WC, Nahvi A, Sudarsan N, Barrick JE, Breaker RR (2003) An mRNA structure that controls gene expression by binding S-adenosylmethionine. Nat Struct Biol 10: 701–707
38. Corbino KA, Barrick JE, Lim J, Welz R, Tucker BJ, Puskarz I, Mandal M, Rudnick ND, Breaker RR (2005) Evidence for a second class of S-adenosylmethionine riboswitches and other regulatory RNA motifs in alpha-proteobacteria. Genome Biol 6:R70
39. Poiata E, Meyer MM, Ames TD, Breaker RR (2009) A variant riboswitch aptamer class for S-adenosylmethionine common in marine bacteria. RNA 15:2046–2056
40. Ames TD, Rodionov DA, Weinberg Z, Breaker RR (2010) A eubacterial riboswitch class that senses the coenzyme tetrahydrofolate. Chem Biol 17:681–685

Chapter 15

Screening Assays to Identify Artificial glmS Ribozyme Activators

Christina E. Lünse and Günter Mayer

Abstract

Ribsowitches are putative drug targets as they often regulate the expression of essential bacterial genes. This finding necessitates the development of suitable assays, at best high-throughput (HT) compatible, which allow the screening of compound libraries for riboswitch activation. Here, we describe a HT-compatible fluorescence-based screening assay employing a minimal core motif of the *Bacillus subtilis* glmS riboswitch and the metabolite-induced self-cleavage assay using the full-length glmS ribozyme of *Staphylococcus aureus* for the identification of artificial molecules activating this regulatory RNA.

Key words glmS ribozyme, Screening, Ribozyme cleavage, Riboswitch, Fluorescence polarization

1 Introduction

The glmS ribowitch is found in the 5′ untranslated region (5′ UTR) of many Gram-positive bacteria and regulates expression of the enzyme D-fructose-6-phosphate amidotransferase, which generates glucosamine-6-phosphate (GlcN6P). When the glmS riboswitch binds GlcN6P a self-cleavage reaction is induced leading to enhanced degradation of the glmS mRNA [1, 2]. Consequently, the concentration of the cell-wall precursor glucosamine-6-phosphate is reduced and bacterial growth inhibited. This role predestines the glmS riboswitch as potential target RNA for developing novel antibacterial compounds. Here, we describe a HT-compatible fluorescence-based screening assay [3] and a metabolite-induced self-cleavage assay for the identification of novel glmS ribozyme activators [4].

The fluorescence-based screening assay makes use of a minimal ribozyme core motif, in which the cleavage reaction is followed by fluorescence polarization (Fig. 1). When the fluorescein-labeled glmS RNA is incubated with GlcN6P the ribozyme cleaves itself and fluorescence polarization decreases.

Daniel Lafontaine and Audrey Dubé (eds.), *Therapeutic Applications of Ribozymes and Riboswitches: Methods and Protocols*, vol. 1103, DOI 10.1007/978-1-62703-730-3_15,

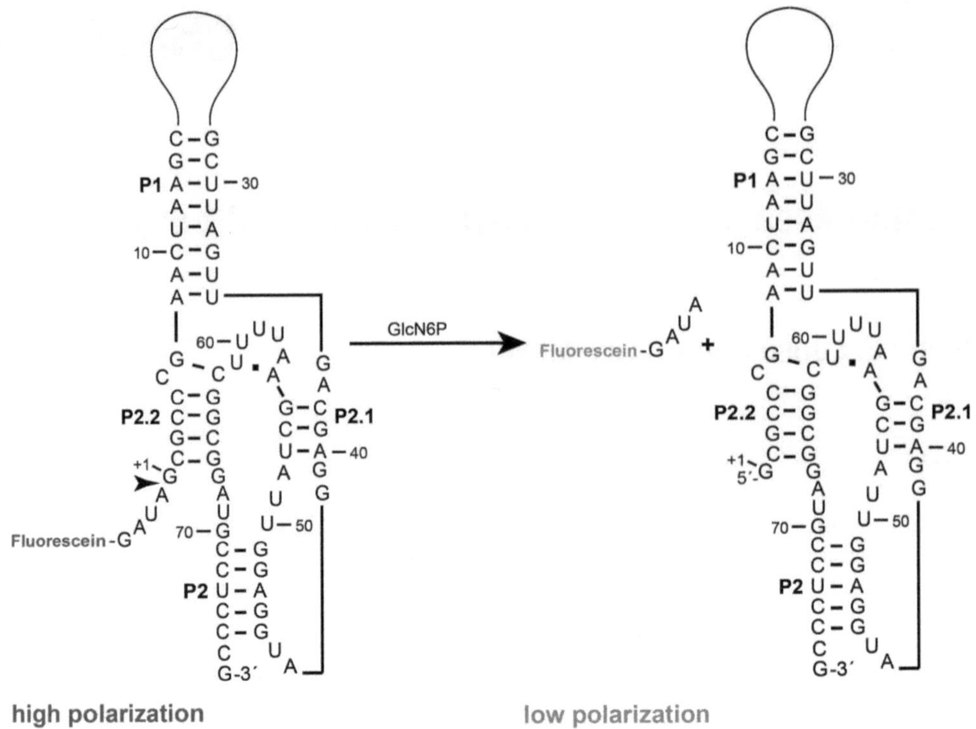

Fig. 1 Fluorescence polarization based screening assay. The fluorescein labeled glmS core motif has a higher polarization (*red*) compared to the glucosamine-6-phosphate (GlcN6P) induced cleavage product (*green*) (Color figure online)

Furthermore, a metabolite-induced self-cleavage assay can be used to screen for compounds activating the glmS ribozyme. For this assay, radioactively labeled glmS RNA is prepared by T7 RNA polymerase transcription from a PCR template. Subsequently, the RNA is PAGE purified, dephosphorylated at its 5′ end and labeled using ^{32}P-ATP. After purification the glmS RNA is used for the metabolite-induced self-cleavage assay in the presence of its natural metabolite GlcN6P or compounds to be investigated for ribozyme activation (Fig. 2).

2 Materials

Prepare all solutions using diethylpyrocarbonate treated ultrapure water and store at room temperature unless indicated otherwise. DNA templates and RNA is always stored at −20 °C. Make sure that all solutions and equipment are RNase free.

2.1 Transcription and RNA Workup Components

1. Transcription buffer: Hepes pH 7.9, 200 mM (*see* **Note 1**).
2. Magnesium chloride solution ($MgCl_2$): 100 mM.
3. Nucleoside triphosphate Mix: mix stock solutions of ATP, UTP, GTP, and CTP (100 μM) in a ratio of 1:1:1:1 to obtain a 25 μM NTP mix.

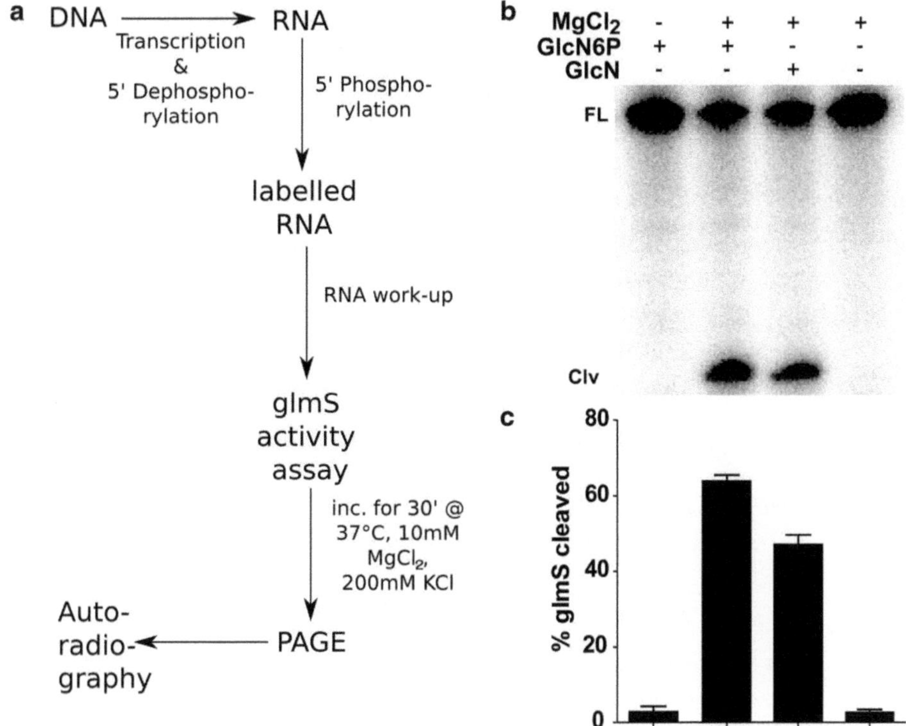

Fig. 2 (**a**) A T7 promoter containing glmS DNA template is used for in vitro transcription and subsequent RNA isolation. The glmS RNA is first dephosphorylated at its 5′ end and finally radioactively labeled. This RNA is used for the metabolite-induced self-cleavage assay to screen for glmS ribozyme activators. (**b**) Autoradiograph of a standard metabolite-induced cleavage reaction showing no RNA cleavage in the absence of $MgCl_2$, or metabolite and prominent RNA self-cleavage upon incubation with either its natural metabolite GlcN6P or the unphosphorylated aminosugar GlcN in the presence of $MgCl_2$. (**c**) PAGE quantification results obtained by evaluation of band intensities of full length RNA (FL) and cleavage product (Clv) using AIDA software

4. Dithiothreitol (DTT): DTT is dissolved in DEPC water to a final concentration of 100 mM, aliquots are stored at −20 °C (*see* **Note 2**).
5. RNasin (Promega, 40 U/μl), DNase I (Roche, 5 U/μl), T7 RNA polymerase (in house preparation, 50 U/μl) are stored at −20 °C and kept in a cooling block (−20 °C) when briefly removed from the freezer.
6. Inorganic pyrophosphatase (2 U/μl) is stored at 4 °C (*see* **Note 3**).
7. dsDNA template containing a T7 promoter for in vitro transcription (for *Staphylococcus aureus* glmS: 5′+T7 glmS Sau long: TCG TAA TAC GAC TCA CTA TAG GTA ATG ATT AAT GGA AAG GGG G; 3′ Sau glmS neu: ATC TTA TTA ACT TTG TCC ATT AAG TCA CCC).
8. 2× PAGE-loading buffer pH 8.0: Mix 9.5 ml formamide, 25 μl 10 % SDS, 10 μl 0.5 M EDTA (*see* **Note 4**) and add 0.5 ml

DEPC water. Add bromophenol blue and xylene cyanol to an aliquot of this buffer to obtain a size marker (*see* **Note 5**).

9. Silanized glass wool and syringe, sterile scalpel blade.
10. 6 % Preparative polyacrylamide gel.

2.2 Dephosphorylation Components

1. glmS RNA.
2. Calf intestine alkaline phosphatase (CIAP, Promega, 20 U/μl) and buffer containing 50 mM Hepes (pH 9.3 at 25 °C), 1 mM $MgCl_2$, 0.1 mM $ZnCl_2$, and 1 mM spermidine (*see* **Note 6**).
3. RNasin (Promega, 40 U/μl).
4. Bovine serum albumin (BSA, Sigma, 10 mg/ml).
5. Ethylenediaminetetraacetic acid solution: 0.5 M, pH 8.0.

2.3 Phosphorylation Components

1. T4 Polynucleotide kinase (PNK, 10 U/μl) and 1× buffer containing 70 mM Hepes, 10 mM $MgCl_2$, 5 mM DTT, pH 7.6 at 25 °C (*see* **Note 6**).
2. RNasin (Promega, 40 U/μl).
3. ^{32}P-ATP 10 mCi/ ml.
4. G25 columns (GE Healthcare).

2.4 Metabolite-Induced Self-Cleavage Components

1. Cleavage buffer 10×: Hepes (pH 7.5) 500 mM, KCl 2 M (*see* **Note 7**).
2. Magnesium chloride solution ($MgCl_2$): 100 mM.
3. Glucosamine-6-phosphate: 1 mM (Sigma), or glucosamine: 1 mM (Sigma).
4. Compounds to be tested in appropriate stock solution (5–10× of wanted final concentration).
5. Dimethylsulfoxide (DMSO) depending on the assay and compounds tested (*see* **Note 8**).

2.5 Fluorescence Polarization Assay: Preparation of Labeled RNA

1. Transcription buffer: Hepes pH 7.9, 200 mM.
2. Guanosine monophosphorothiate (GMPS, emp biotech, Germany).
3. dsDNA template containing a T7 promotor (glmS riboswitch *core*) (5′-TAATACGACTCACTATA*GATAGCGCCCGAACT AAGCGCCCGGAAAAAGGCTTAGTTGACGAGGA TGGAGGTTATCGAATTTTCGGCGGATGCCTCCCG*-3′).
4. 5-(Iodoacetamido)-fluorescein (Sigma).
5. G25 columns (GE Healthcare).

2.6 FP-Screening Setup

1. Fluorescein-labeled RNA.
2. Glucosamine-6-phosphate (Sigma).
3. Glucosamine (Sigma).

4. Fluorescence polarization reader (e.g., Tecan Ultra).
5. Half-well microtiterplates (black) (Corning).
6. FP-cleavage buffer (50 mM Tris pH 7.9, 200 mM KCl, 10 mM $MgCl_2$, 0.001 % Tween-20).

3 Methods

3.1 glmS Ribozyme Transcription and Workup

1. For in vitro transcription, components prepared as stated in Subheading 2.1 were mixed according to Table 1. The reaction was incubated at 37 °C overnight.
2. Treat samples with 0.5 μl of DNase I for 10–20 min at 37 °C.
3. Prepare a preparative, 6 % PA-gel and pre-run after polymerization for 15–30 min.
4. Mix samples with PAGE-loading buffer and load 50 μl per well (*see* **Note 5**).
5. Run gel at a constant power of 20 W and approximately 360 V.
6. Once the RNA is well separated, disassemble the apparatus and use UV shadowing at 254 nm to visualize appropriate bands and excise RNA with a sterile scalpel blade (*see* **Note 9**).
7. Crush gel slices with a 1 ml pipette tip, suspend it in 500 μl 0.3 M NaOAc (pH 5.4) and incubate at 65 °C for 90 min under vigorous shaking.
8. Pass gel suspension through a 5 ml syringe packed with silanized glass wool, followed by two to five washings with 100 μl 0.3 M NaOAc solution.

Table 1
Pipetting scheme for in vitro transcription

Reagent	Amount	Stock	Resultant
Hepes pH 7.9	20 μl	200 mM	40 mM
$MgCl_2$	25 μl	100 mM	25 mM
DTT	5 μl	100 mM	5 mM
NTP mix	10 μl	25 mM	2.5 mM
RNasin	1.24 μl	40 U/μl	50 U
Inorganic pyrophosphatase	0.2 μl	2 U/μl	0.4 U
dsDNA template	10 μl	150–300 pmol	1.5–3 μM
T7 RNA polymerase	5 μl	50 U/μl	250 U
DEPC water	ad 100 μl		

Table 2
Pipetting scheme for 5′ dephosphorylation

Reagent	Stock	Amount	Resultant
CIAP buffer	10×	5 μl	1×
BSA (10 mg/ml)	10×	5 μl	1×
RNA template			75 pmol
CIAP	20 U/μl	0.85 μl	17 U
RNasin	40 U/μl	0.5 μl	20 U
DEPC water	ad 50 μl		

9. Precipitate RNA with three volumes of EtOH abs. (*see* **Note 10**) and incubate for 20 min at −80 °C. Centrifuge samples at 4 °C for 30 min, take off supernatant and wash with 100 μl 70 % EtOH. Spin for 2 min at RT, remove supernatant and air-dry the pellets before dissolving in DEPC water (in 1/10 of original transcription volume, here 10 μl).
10. Determine RNA concentration by standard methods.

3.2 Dephosphorylation

1. Mix components stated in Table 2 for 5′ dephosphorylation reaction, however add enzymes last.
2. Incubate at 37 °C for 15 min.
3. Add 0.5 μl of CIAP and incubate at 55 °C.
4. Add 0.5 μl 0.5 M EDTA and inactivate the enzyme at 75 °C for 10 min.
5. Add 150 μl DEPC water and perform a phenol–chloroform extraction followed by a sodium acetate precipitation.
6. Resuspend pellet in 5 μl DEPC water, use 3 μl for subsequent phosphorylation reaction and run the remaining 2 μl on an agarose gel to check for RNA integrity.

3.3 Phosphorylation

1. Prepare reaction mixture as stated in Table 3 on ice.
2. Incubate at 37 °C for 30 min.
3. Desalt and remove excess labeled ATP by passing sample though a G25 column. For this, wash G25 column with 100 μl DEPC water, add 30 μl DEPC water to phosphorylation sample, mix by pipetting and pass the total amount of 50 μl sample through the column (follow manufacturer's instructions).
4. Purify radiolabeled RNA by PAGE as stated under Subheading 3.1, **steps 3–9**.

Table 3
Pipetting scheme for 5′ phosphorylation

Reagent	Stock	Amount	Resultant
T4 PNK buffer	10×	2 μl	1×
CIAP-RNA		3 μl	45 pmol
^{32}P-ATP	10 μCi/μl	2 μl	20 μCi
T4 PNK	10 U/μl	2 μl	20 U
RNasin	40 U/μl	0.3 μl	12 U
DEPC water		ad 20 μl	

Table 4
Pipetting scheme for metabolite-induced self-cleavage assay

Reagent	Stock	Amount	Resultant
Cleavage buffer	10×	1 μl	1×
$MgCl_2$	100 mM	1 μl	10 mM
RNA		0.5 μl	
GlcN6P	1 mM	2 μl	200 μM
DEPC water		ad 10 μl	

3.4 Metabolite-Induced Self-Cleavage Assay

1. Prepare an analytical 17 % PA-gel.
2. Every experiment should contain a positive control (addition of 200 μM glucosamine-6-phosphate), an RNase control (no addition of $MgCl_2$) and a negative control (no addition of metabolite or compound).
3. Prepare reactions to contain components according to Table 4 by first mixing the cleavage buffer, $MgCl_2$, DEPC water and metabolite.
4. Before addition, pre-fold labeled RNA by heating to 90 °C for 1 min (*see* **Note 11**).
5. Quick spin RNA and allow to slowly adjust to room temperature by incubation at 23 °C for 5 min so that RNA can fold into its active conformation (*see* **Note 12**).
6. Add folded, radioactively labeled glmS RNA to the reaction mix to initiate the cleavage reaction.
7. Incubate for up to 30 min at 37 °C for screening purposes.
8. During incubation period, pre-run analytical PA-gel at 400 V for about 30 min.

Table 5
Pipetting scheme for preparation of fluorescein-labeled RNA

Reagent	Amount	Stock	Resultant
Hepes pH 7.9	20 µl	200 mM	40 mM
$MgCl_2$	25 µl	100 mM	25 mM
DTT	5 µl	100 mM	5 mM
NTP mix	10 µl	25 mM	2.5 mM
GMPS	6.6 µl	150 mM	10 mM
RNasin	1.24 µl	40 U/µl	50 U
dsDNA template	10 µl	150–300 pmol	1.5–3 µM
T7 RNA polymerase	5 µl	50 U/µl	250 U
DEPC water	ad 100 µl		

9. Stop cleavage reaction by addition of 5 µl 2× sucrose loading buffer and short spin to collect the entire sample in vial bottom.
10. Load samples and run at 450–600 V for 2–3 h.
11. Disassemble gel chamber and expose wrapped gel to a phosphorimager screen overnight at −80 °C.
12. Read screen and quantify band intensities using AIDA (or other appropriate) software (*see* **Note 13**).

3.5 Fluorescence Polarization Assay: Preparation of Labeled RNA

1. For preparation of labeled RNA, first an in vitro transcription was performed using the components and materials as stated in Subheading 2.5, which were mixed according to Table 5. The reaction was incubated at 37 °C overnight (*see* **Note 14**).
2. Treat samples with 0.5 µl of DNase I for 15 min at 37 °C.
3. Sample was phenol–chloroform extracted and RNA precipitated with ethanol abs.
4. The RNA pellet was dissolved in 100 µl H_2O and passed twice through a G25 microspin column.
5. Labeling of GMPS-RNA was facilitated by incubation with a 200-fold molar excess of 5-(iodoacetamido)-fluorescein in coupling buffer (50 mM Tris pH 8.0, 50 mM EDTA, and 2 M Urea) for 2 h at 40 °C.
6. Prepare a preparative 6 % PA-gel and pre-run after polymerization for 15–30 min.
7. Mix samples with PAGE-loading buffer and load 50 µl per pocket (*see* **Note 5**).

Table 6 Pipetting scheme for FP-cleavage assay

Reagent	Amount	Stock	Resultant
FP-cleavage buffer pH 7.9	5 μl	5×	1×
GlcN (optional)	5 μl	1 mM	200 μM
GlcN6P (optional)	5 μl	1 mM	200 μM
Fluorescein-labeled RNA	2.5 μl	1 μM	400 nM
DEPC water	ad 25 μl		

8. Run gel at a constant power of 20 W and approximately 360 V (*see* **Note 15**).
9. Once the RNA is well-separated, disassemble the apparatus and use UV shadowing at 254 nm to visualize appropriate bands and excise RNA with a sterile scalpel blade.
10. Crush gel slices with a 1 ml pipet tip, suspend it in 500 μl 0.3 M NaOAc (pH 5.4) and incubate at 65 °C for 90 min under vigorous shaking.
11. Pass gel suspension through a 5 ml syringe packed with silanized glass wool followed by two to five washings with 100 μl 0.3 M NaOAc solution.
12. Precipitate RNA with three volumes of EtOH abs and incubate for 20 min at −80 °C. Centrifuge samples at 4 °C for 30 min, take off supernatant and wash with 100 μl 70 % EtOH. Spin for 2 min at RT, remove supernatant and air-dry the pellets before dissolving in DEPC water (in 1/10 of original transcription volume, here 10 μl).
13. Determine RNA concentration by standard methods.

3.6 FP Screening Setup

1. Components listed under Subheading 2.6 were mixed according to Table 6 (*see* **Note 16**).
2. Incubate the reaction for 30 min at 25 °C.
3. Read for fluorescence polarization with an appropriate reader (e.g., Tecan Ultra, Crailsheim, Germany).

4 Notes

1. Whenever working with glmS RNA, Tris buffer has to be avoided as this primary amine can induce glmS ribozyme cleavage. Therefore, buffers different from Tris have to be used for all reactions containing glmS ribozyme (transcription, dephosphorylation, and phosphorylation, cleavage assay).

2. DTT is very labile due to oxidation with atmospheric oxygen. To avoid inactivation, DTT aliquots thawed once are discarded after use.
3. To prevent inhibitory effects of accumulating pyrophosphate, which is released when a nucleoside triphosphate is incorporated into the growing chain, inorganic pyrophosphatase is added. It catalyzes the hydrolysis of inorganic pyrophosphate to form orthophosphate.
4. When preparing the EDTA solution the pH has to be adjusted to 8 while dissolving EDTA to ensure solubility.
5. Always load sample-free pockets with loading buffer to ensure optimal running conditions. As size marker RNA loading buffer containing bromophenol blue and xylene cyanol can be used.
6. Do not use buffers provided by the enzyme manufacturer as they contain Tris. Buffers were prepared with same concentrations of Hepes instead.
7. Cleavage buffer was designed according to protocols published by Winkler et al. [1].
8. Cleavage assay tolerates the addition of dimethylsulfoxide to up to 2 %.
9. Keep UV-exposure time to a minimum to avoid RNA damage [5].
10. Absolute ethanol for nucleic acid precipitation should be ice cold.
11. It has to be assured that the RNA solution does not contain divalent ions like Mg^{2+} as they would cause disintegration of RNA, when heating.
12. This step is crucial for efficient cleavage.
13. Before actual experiments are performed a cleavage assay test should be done, to control glmS ribozyme cleavage in the presence of 200 μM GlcN6P or GlcN and 10 mM $MgCl_2$ in 1× cleavage buffer. Moreover, these tests are to be used as an estimate on how much the stock RNA can be diluted for following experiments (usually 1:10–1:20).
14. Prevent excessive air exposure to avoid oxidation of GMPS and GMPS labeled RNA.
15. Prevent exposure to light to avoid bleaching of fluorescein.
16. Tris buffer was used for the FP-based cleavage assays as it performed best in these assays. Anyway, Hepes buffer and PBS will yield similar results.

References

1. Winkler WC, Nahvi A, Roth A, Collins JA, Breaker RR (2004) Control of gene expression by a natural metabolite-responsive ribozyme. Nature 428(6980):281–286
2. Collins JA, Irnov I, Baker S, Winkler WC (2007) Mechanism of mRNA destabilization by the glmS ribozyme. Genes Dev 21(24):3356–3368
3. Mayer G, Famulok M (2006) High-throughput-compatible assay for glmS riboswitch metabolite dependence. Chembiochem 7(4): 602–604
4. Lünse CE, Schmidt MS, Wittmann V, Mayer G (2011) Carba-sugars activate the glmS-riboswitch of *Staphylococcus aureus*. ACS Chem Biol 6(7):675–678
5. Kladwang W, Hum J, Das R (2012) Ultraviolet shadowing of RNA can cause significant chemical damage in seconds. Sci Rep 2:517

Chapter 16

Analysis of Riboswitch Structure and Ligand Binding Using Small-Angle X-ray Scattering (SAXS)

Nathan J. Baird and Adrian R. Ferré-D'Amaré

Abstract

Small-angle X-ray scattering (SAXS) is a powerful tool for examining the global conformation of riboswitches in solution, and how this is modulated by binding of divalent cations and small molecule ligands. SAXS experiments, which typically require only minutes per sample, directly yield two quantities describing the size and shape of the RNA: the radius of gyration (R_g) and the maximum linear dimension (D_{max}). Examination of these quantities can reveal if a riboswitch undergoes cation-induced compaction. Comparison of the R_g and D_{max} values between samples containing different concentrations of ligand reveals the overall structural response of the riboswitch to ligand. The Kratky plot (a graphical representation that emphasizes the higher-resolution SAXS data) and the $P(r)$ plot or pair-probability distribution (an indirect Fourier transform, or power spectrum of the data) can provide additional evidence of riboswitch conformational changes. Simulation methods have been developed for generating three-dimensional reconstructions consistent with the one-dimensional SAXS data. These low-resolution molecular envelopes can aid in deciphering the relative helical arrangement within the RNA.

Key words Riboswitch, Small-angle X-ray scattering, RNA folding

1 Introduction

SAXS can yield data describing the global conformation of biological macromolecules in dilute aqueous solutions [1, 2]. SAXS is a powerful tool for investigating RNAs under various conditions, for example, to examine the overall structural response of riboswitches to metal ions or cognate or non-cognate small molecules, changes in pH, temperature, etc. SAXS reports directly on the global structure of RNA and requires no labeling. This contrasts with biophysical or biochemical techniques that require labeling. For instance, fluorescence resonance energy transfer (FRET) employs fluorophores, and these report on their local environment, rather than on the global structure of the RNA. SAXS experiments do not suffer from the size and labeling limitations of NMR, or the requirement for well-ordered crystals of X-ray crystallography.

Daniel Lafontaine and Audrey Dubé (eds.), *Therapeutic Applications of Ribozymes and Riboswitches: Methods and Protocols*, vol. 1103, DOI 10.1007/978-1-62703-730-3_16, © Springer Science+Business Media New York 2014

The primary limitations of SAXS are that it provides only low-resolution structural information and that it does so on the ensemble of conformations present in solution.

SAXS is a rapid method for monitoring RNAs in solution. It has been employed to characterize the global collapse of catalytic RNAs as they fold in response to divalent cations, to characterize the folding intermediates of large RNAs (using time-resolved SAXS, for instance) and to characterize the ionic atmosphere surrounding RNAs (using anomalous SAXS) [3–6]. Several studies have employed SAXS to characterize the response of riboswitch aptamer domains to divalent cations and small molecule ligands. These studies revealed that the folding behavior of these RNA domains is highly idiosyncratic [7–10]. Specifically, while some aptamer domains such as those of the flavin mononucleotide (FMN) and class-I *S*-adenosylmethionine (SAM-I) riboswitches are largely preorganized in physiologic or higher concentrations of Mg^{2+} and display no additional overall compaction upon ligand binding, other aptamer domains, such as those of the class-I cyclic diguanylate (c-di-GMP-I) and thiamine pyrophosphate (TPP) riboswitches exhibit a large conformational change upon binding to their cognate ligands. SAXS studies on the glycine riboswitch have addressed the complex relationship between compaction of that RNA upon cation and ligand binding [11, 12]. In addition to global size parameters, for RNAs with well-defined conformations, it is possible to obtain low-resolution shape information from three-dimensional reconstructions based on the SAXS data [13]. These low-resolution reconstructions can in turn guide further studies that employ local probes [14].

SAXS can be carried out with X-rays generated in the home laboratory or using synchrotron X-radiation. Although home sources have the advantage of ready access to extensive biochemical laboratory facilities, the brightness and very high collimation of X-ray sources from synchrotron sources allow the use of considerably more dilute samples, both facilitating experiments with scarce material and reducing the likelihood of intermolecular interactions that complicate the analysis and interpretation of SAXS data. In addition, the use of synchrotron sources allows access to modalities such as time-resolved and anomalous SAXS, and makes it practical to collect radiation scattered at a higher Bragg angle (wide-angle X-ray scattering, or WAXS). In this chapter, we limit the discussion to techniques employed in basic (static, non-anomalous) SAXS characterization of riboswitch RNAs using synchrotron sources.

2 Materials

We assume that experiments will be carried out at a synchrotron beamline equipped for performing SAXS experiments with biological macromolecules. In addition to the materials listed below,

Table 1
SAXS R_g analysis of *T. tengcongensis* SAM-I riboswitch

	R_g (Å) - SAM	R_g (Å) + SAM
Non-SEC treated RNA	29.9	26.3
SEC-treated RNA	25.3	22.8
Crystal structure	n.a.	21.5

Experiments were performed in 20 mM Tris–HCl, pH 8, 40 mM NaCl, 1.5 mM $MgCl_2$

n.a.—not applicable. The crystal structure R_g was calculated from PDB ID 3GX5 using CRYSOL from the ATSAS suite

the beamline should provide all necessary hardware (X-ray optics, data collection and control electronics) and software, software for initial data reduction and analysis, and software and hardware for sample manipulation. SAXS data can be analyzed using the ATSAS suite of programs (http://www.embl-hamburg.de/biosaxs/software.html). In reducing and analyzing data obtained at beamlines ID-12 and ID-18 at the Advanced Photon Source (APS) at Argonne National Laboratory (Lemont, Illinois, USA), we have also used the Igor Pro (WaveMetrics) software package.

1. In vitro transcribed or synthetic riboswitch RNA. Methods for preparation of homogeneous RNA samples have been described elsewhere [15, 16].
2. Buffer, preferably Tris or HEPES which serve as free-radical scavengers. Data presented in this study were prepared in 20 mM Tris–HCl, pH 8 (SAM-I riboswitch, Table 1) or 20 mM HEPES-KOH, pH 7.5 (glycine riboswitch, Table 2).
3. $MgCl_2$, KCl, and/or NaCl solutions to fold the RNA. Final solution conditions contained 40 mM NaCl and 1.5 mM $MgCl_2$ (SAM-I riboswitch, Table 1) or 40 mM KCl and 10 mM $MgCl_2$ (glycine riboswitch, Table 2).
4. Riboswitch ligands. In this study, we collected data in the absence and presence of saturating ligand concentrations (2 mM S-adenosylmethionine (SAM) or 10 mM glycine).
5. Size-exclusion chromatography (SEC) column, liquid chromatography system (optional).
6. Syringe filters for dust and aggregate removal (e.g., 0.02 μm Anotop filters, GE Healthcare).
7. Centrifugal concentrators of suitable molecular weight cutoff (e.g., Amicon concentrators, EMD Millipore).

Table 2
SAXS analysis of *F. nucleatum* glycine riboswitch single aptamers

	Aptamer #1	Aptamer #2	Aptamer #1	Aptamer #2
	R_g (Å)	R_g (Å)	$I(0)$	$I(0)$
– Glycine	23.2	29	43.6	52.6
+ Glycine	24.1	33.6	52.2	71.2
+ Glycine, high Mg^{2+}	24.2	37.6	54.6	93.9

All experiments were performed in 20 mM Tris–HCl, pH 8, 40 mM KCl, 10 mM $MgCl_2$, excepting the high Mg^{2+} condition which contained 20 mM $MgCl_2$. All samples were prepared to ~0.5 mg/mL

3 Methods

3.1 General Considerations

The raw data obtained from a SAXS experiment is the scattered X-ray intensity as a function of momentum transfer, q (measured in reciprocal angstroms), where $q = 4\pi \sin \theta/\lambda$, 2θ is the scattering angle, and λ is the wavelength of X-rays used. The q-range, nominally the range of angles, is determined by the experimental geometry (distance from the sample to detector, beam-stop size and position, detector active area and angular resolution, etc.) and the energy of X-rays used. These parameters are generally set by the beamline scientist prior to data collection. The forward scattering intensity at zero angle, $I(0)$, is directly proportional to the molecular mass times the concentration (g/L) of the sample. Because of this relationship, scattering from dust particles and large aggregates, if present, will dominate the scattering signal from the macromolecules of interest. Thus, it is important to exclude aggregates from SAXS samples prior to data collection. Contributions from larger particles will dominate the observed scattering intensity, particularly at low q values where the radius of gyration (R_g) will be determined (see below), leading to errors. Finally, it is important to remember that riboswitch RNAs stability is a function of cation and ligand concentrations. We recommend monitoring the global conformational changes as a function of both of these.

3.2 Sample and Buffer Preparation

1. Each SAXS measurement will require ~100 μL of sample at concentrations between 0.3 and 1.0 g/L (*see* **Note 1**).
2. Fold riboswitch RNA by addition of $MgCl_2$ and incubation of the RNA at a suitable temperature (e.g., 37–50 °C). After folding, equilibrate the reaction by incubation at 37 °C for 15–30 min.
3. Samples must be homogenous. It is imperative to eliminate higher order oligomers which may result from misfolded or

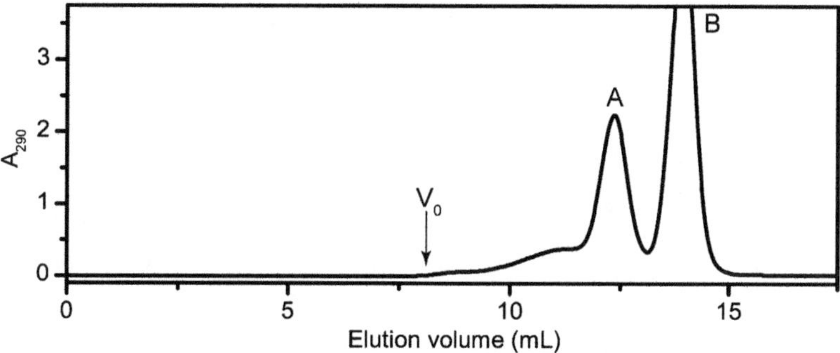

Fig. 1 Size-exclusion chromatogram of refolded SAM-I riboswitch. RNA was heated in 20.3 mM Tris–HCl, pH 8, 40.6 mM NaCl for 2 min at 85 °C followed by incubation at room temperature for 5 min. $MgCl_2$ was added to yield final concentrations of 20 mM Tris–HCl, 40 mM NaCl, and 1.5 mM $MgCl_2$. The sample was loaded onto a Superdex 200 HR10/30 column (GE), pre-equilibrated in a buffer matching the final sample conditions. Two distinct peaks are observed, labeled A and B, along with an early "shoulder" which may contain higher-order oligomers. The maximum absorbance of peak B is beyond the sensitivity range of the detector. Absorbance at 290 nm was monitored for the same reason; both peaks A and B were outside the range of the detector at 260 nm and 280 nm wavelengths

aggregated RNA. Demonstration of homogeneity can be accomplished at minimum by native gel electrophoresis or more explicitly by dynamic light scattering or SEC. The size-exclusion chromatogram of the SAM-I riboswitch (Fig. 1) demonstrates the presence of higher order complexes in a sample that was refolded following purification of the RNA by denaturing gel electrophoresis and desalting. The scattering data from the non-SEC-treated sample (Fig. 2b, top) do not demonstrate signs of aggregation and the Guinier plot (Fig. 2c) indicates only slight aggregation (see Initial SAXS data analysis below). More importantly, the chromatogram exhibits a significant peak (Fig. 1, peak A) that elutes earlier than the monomeric RNA. This peak comprises ~30 % of the total absorbance, indicating that the original sample contains a significant fraction of a higher order oligomer, possibly a dimer. Analysis of SAXS data for the non-SEC treated inhomogeneous sample, containing both monomer and dimer, results in larger radii of gyration compared to the R_g's obtained for the homogeneous monomeric samples, isolated from SEC peak B (Table 1). Furthermore, the smaller R_g of the monomeric sample bound to SAM is in good agreement with known crystal structures [17, 18] of the ligand-bound riboswitch. The discrepancy in R_g between the homogeneous and heterogeneous samples underscores the need for careful sample preparation (Table 1).

4. Samples must have matching blank buffer solutions. Close matching of blank and sample can be achieved either by SEC

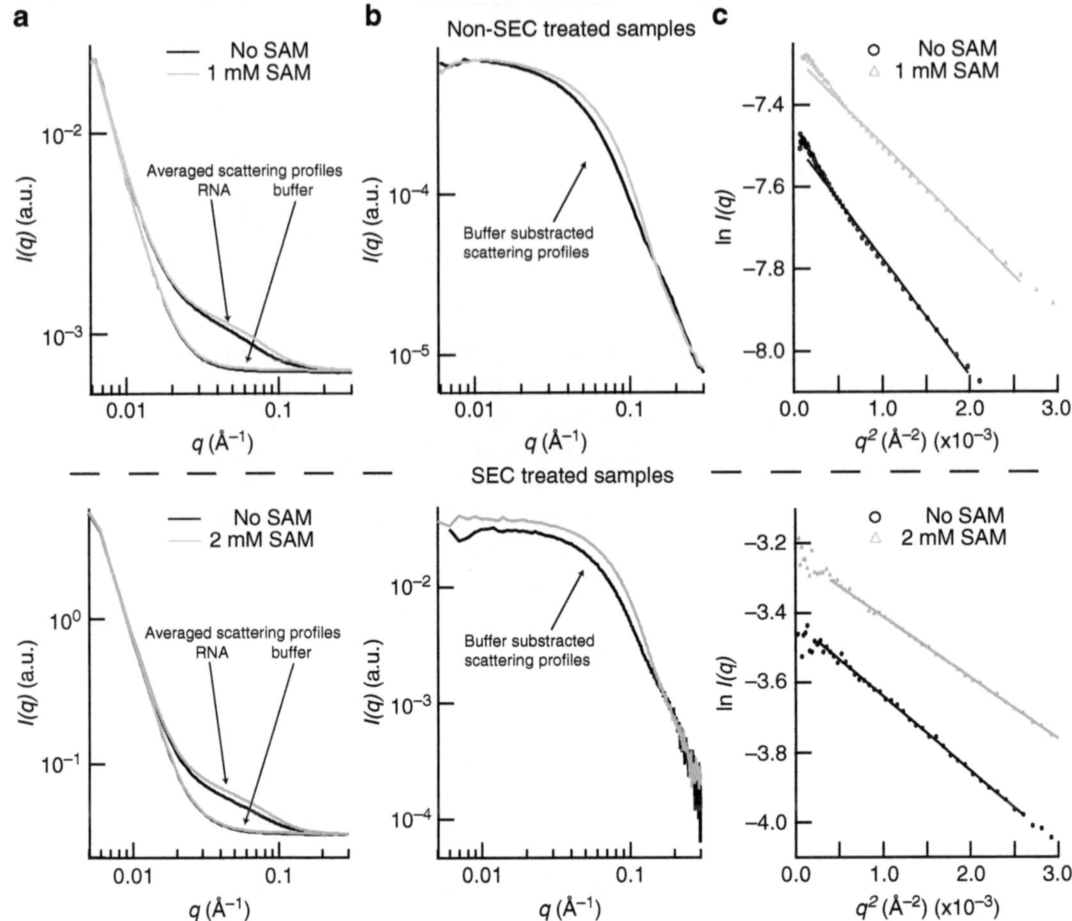

Fig. 2 SAXS data obtained for the SAM-I riboswitch aptamer. Data in top panels were collected on samples not purified by SEC. Data in bottom panels were collected from samples purified by SEC (RNA from peak B in Fig. 1 was used for SEC purified SAXS data collection). The ordinate in each plot is scaled to maximize differences between the apo (*black*) and bound (*grey*) samples. (**a**) Primary data in the form of integrated, averaged one-dimensional scattering profiles for sample and buffer are plotted as intensity vs. q. The buffer curves in the absence and presence of SAM are identical. (**b**) The background-subtracted primary data exhibit the typical plateau at low q. (**c**) Guinier plots of the SAM-I riboswitch from which R_g and $I(0)$ are obtained (see text). The curves and fits in the top panel have been manually offset for clearer inspection

of the sample RNA in a column equilibrated in the blank buffer, or by performing multiple washes (sequential concentrations and dilutions) with the blank buffer using centrifugal concentrators (e.g., Amicon concentrators). Extensive dialysis will assure precise thermodynamic equilibrium, but is time-consuming and may not be compatible with integrity of the RNA.

5. If samples have not been purified by SEC, they must be passed through 0.02 μm (i.e., 200 Å) Anotop filters to remove aggregates and dust (*see* Subheading 3.1, *above*).

6. Samples should be shipped to the beamline utilizing dry ice or cold packs or at room temperature, depending on sensitivity of samples to degradation or aggregation as a function of freezing or cooling.

3.3 SAXS Data Collection

The following instructions are based on our experience collecting SAXS data at two beamlines at the APS, ID-12 BESSRC and ID-18 BioCAT. Both beamlines utilize a similar experimental setup and data analysis software. Data are collected under continuous flow to limit the exposure of the sample to X-ray radiation. For simplicity, we refer to any data collection event as a "sample" regardless of whether the sample contains buffer or macromolecule. Although the specific unit operations will be somewhat different at different synchrotron beamlines, the overall procedures will still resemble those described below.

1. To remove any residual macromolecule from previous runs purge the sample flow cell using the Hamilton titrator syringe at the beamline with ≥ 2 mL of each of the following solutions in the given order: water, 20 % bleach, water, isopropanol, and water.
2. Wash the flow cell extensively with sample buffer, making sure to leave no air in the sample line.
3. Initiate the beamline-programmed titration program, which is used to fill the flow cell with sample and subsequently to control the sample flow during data collection. Continuous flow of sample reduces radiation damage.
4. First, aspirate 20 μL of air to generate a gap between the buffer-filled line and the sample being loaded. This air gap is more critical during the loading of the RNA sample in order to avoid mixing and dilution.
5. Aspirate 70–100 μL of sample for data collection.
6. Close and interlock the experimental hutch. Open the beamline shutter.
7. Using the beamline-specific control script, enter the desired exposure time, number of exposures, and time between exposures. Collect 20 individual 1-s exposures without any time delay between exposures. Provide succinct sample names. It might be helpful to indicate in the filename whether the sample is a blank or contains RNA.
8. Initiate the data collection script while simultaneously initiating the titrator program for continuous sample flow.
9. After data collection has been completed, enter the experimental hutch and recover the sample.
10. Repeat **steps 1–9** for each buffer/RNA combination.

11. If SAXS data will be measured on multiple RNA samples in the same buffer (e.g., a dilution series), collection of buffer data can be carried out less often but must be obtained at least prior to the first RNA sample and following the last RNA sample utilizing that buffer. Agreement between the initial and final buffer samples is an indication that there have been no aberrant changes in the delivered beam intensity or focus during data collection and no deposition of macromolecule on the capillary during exposure to X-rays.
12. Samples can be reused for data collection several times. However, after three or four collections, samples will begin to show signs of radiation damage (*see below*).

3.4 Initial SAXS Data Analysis

1. Radially integrate the data for each of the twenty 2-D scattering images, to produce twenty 1-D scattering profiles.
2. Import integrated data into data analysis software as scattering profiles of intensity vs. q (Å^{-1}). Confirm that the 20 replicate exposures from each sample overlay. A few aberrant curves are likely the result of an air bubble in the flow cell line. These may be safely deleted. If there exists a general spread of non-overlaying curves, it may indicate radiation damage to the sample and the sample must be repeated and/or discarded.
3. Separately average and save the well-overlaid curves for each buffer or RNA sample (Fig. 2a).
4. Perform background subtraction of the averaged buffer scattering curve from the averaged RNA scattering curve. The buffer corrected intensity should exhibit a plateau at low q for globular macromolecules (Fig. 2b). Aggregation is often observed as a monotonically increasing intensity while moving toward the low q region.
5. Evaluation of the Guinier plot, ln $[I(q)]$ vs. q^2, at low q is the next step in analyzing the data quality. This plot should be linear; upward curvature in the plot is indicative of aggregation. Note that Fig. 2c (top) demonstrates a slight curvature in the Guinier plot, indicating a small degree of aggregation, while the background subtracted data in Fig. 2b (top) did not immediately indicate aggregation, as discussed above. The data in the two panels of Fig. 2c are fit in a similar low q region. In the top panel (non-SEC purified sample), the data deviate from the linear fit at low values of q^2. This upturn is indicative of aggregation of the sample. Though only slight upturn is observed in this plot (top panel), samples suffering from severe aggregation will exhibit more considerable deviation from linearity. The data in the bottom panel are linear over the whole range of the fit. Aggregation may be removed by passing the sample through a 0.02 μm filter. Then the experiment can be repeated.

6. The Guinier analysis is valid in the low q region (as $q \rightarrow 0$), where the scattering data can be approximated as $I(q) \approx I(0)e(-q^2R_g^2/3)$. A fit to Guinier plot in the low q region ($qR_g < 1.3$) results in a slope equal to $-R_g^2/3$ and a y-intercept equal to ln $I(0)$. This extrapolated $I(0)$ scattering intensity can be compared with RNA standards which have been well-characterized, such as tRNA, to demonstrate that the sample is monomeric (or oligomeric, see next step below). Data at the lowest (and highest) q have increased noise, owing to the similarity in scattering intensity of the buffer and sample (Fig. 2a) and are much more sensitive to buffer mismatches. Hence, slightly mismatched buffer subtraction can also give rise to non-linearity at low q.
7. If a crystal structure of the riboswitch is available, the R_g can be calculated from the coordinates using CRYSOL, a program within the ATSAS suite [19]. In such a case, agreement of the experimental R_g with the crystal structure R_g will suffice to demonstrate that the solution sample is monomeric and homogeneous. If no structural information is available for the riboswitch, it is important to compare forward scattering intensities, $I(0)$, with a standard sample to demonstrate that the riboswitch is monomeric (or a specific oligomer) and homogeneous. Data should be collected for the standard sample during each trip to the synchrotron. Because the forward scattering is proportional to the product of the molecular weight (MW) and the concentration (C, in g/L), comparison with a standard allows for estimation of the RNA sample molecular weight, $MW_r \propto MW_sC_sI(0)_r/C_rI(0)_s$, where the standard and RNA are indicated by the subscripts s and r, respectively. Table 2 provides an example. The isolated second aptamer of the dual-aptamer glycine riboswitch from *Fusobacterium nucleatum* exhibits an increase in R_g upon binding glycine. This was initially unexpected as the generally expected outcome is global compaction (decrease in R_g) in the presence of ligand. A closer inspection revealed a concomitant increase in $I(0)$, suggesting the presence of an oligomer. Increasing the [Mg^{2+}] to 20 mM results in an additional increase in R_g and $I(0)$. The $I(0)$ nearly doubles under these conditions, indicating dimerization of this aptamer. In contrast, the first aptamer from this glycine riboswitch does not exhibit a significant increase either in R_g or $I(0)$ upon introduction of ligand, indicating that it does not dimerize. A closer look at the data from the first aptamer is useful as well. Not only does the first aptamer remain monomeric under all experimental conditions (no change in $I(0)$, but there is no change in R_g for this aptamer upon introduction of its cognate ligand, glycine. This could be the result of non-saturating ligand concentrations at the given [Mg^{2+}]. Alternatively, the riboswitch may not undergo a global conformational change upon ligand

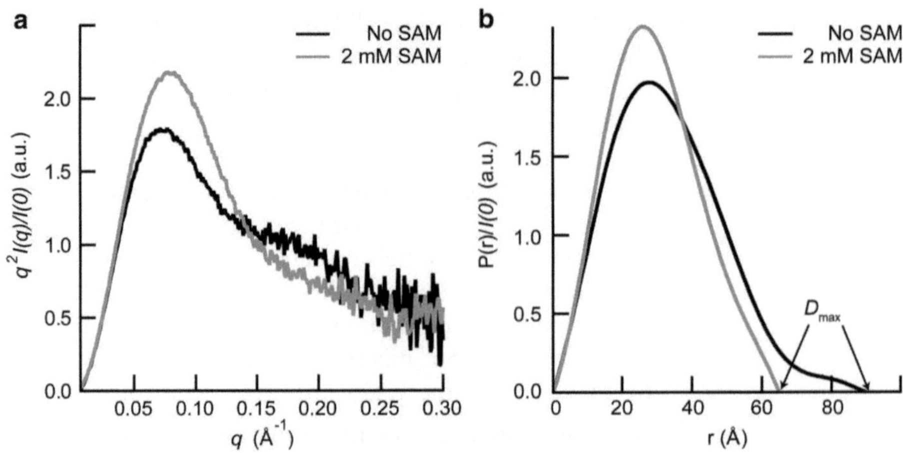

Fig. 3 Transformations of the primary SAXS curve (SEC-treated samples). (**a**) A Kratky plot is used to evaluate the compaction and relative degree of folding of the molecule. The SAM-I riboswitch in both the absence and presence of SAM is generally folded, exhibiting a characteristic peak and subsequent decline in the Kratky plot. The addition of SAM results in additional compaction and folding, as observed by the increased peak and steep decline (see text). (**b**) The $P(r)$ plot is a real-space transformation of the primary SAXS data and reports on the general shape changes of the riboswitch in the absence and presence of SAM. Additionally, the maximum distance within the molecule, D_{max}, is obtained from this plot

binding. A few additional examples have been reported in which ligand binding does not induce a global compaction. In the presence of saturating ligand concentrations neither the lysine riboswitch nor the FMN riboswitch undergo global conformation changes. The R_g's of these riboswitch aptamers are unchanged in the absence and presence of ligand [8, 20].

3.5 Further SAXS Data Analysis

Observing conformational changes by SAXS goes beyond measurements of R_g. The Kratky and pair-distance probability distribution function, or $P(r)$, plots are helpful in comparing data obtained under various conditions (e.g., riboswitch ± ligand, or low vs. high concentrations of Mg^{2+}).

1. The Kratky plot, $q^2 I(q)$ vs. q, is useful for comparing the compaction of the molecule under varying conditions. In such a plot, folded macromolecules exhibit a steep rise and peak followed by a pronounced decline [21, 22]. Unfolded molecules do not exhibit a steep decline, and may continue rising throughout the q range. Intermediate conformations may exhibit a decline in the Kratky plot after the peak but the shape of the curve will be distinct from the most folded conformation, exhibiting a broader peak. Fig. 3a depicts differences in a Kratky plot between apo- and ligand-bound SAM-I riboswitch samples. In the absence of SAM, the riboswitch is partially folded, as represented by a well-defined peak and decline.

The riboswitch undergoes further structural organization to a final compact conformation in the presence of SAM, where the shape of the curve becomes more pronounced both in its peak and its decline.

2. The *P*(*r*) plot (Fig. 3b) is a discrete Fourier transformation of the data which provides a measure of the likelihood of two electrons within the molecule being present at a distance *r* from each other. It is analogous to the power spectrum in physics or the Patterson function in crystallography. The curve provides information, in real-space, on the shape of the molecule and approaches zero at its maximum dimension, D_{max}. This plot is useful for comparison of different samples to demonstrate changes in shape and D_{max}.
3. Generation of the *P*(*r*) plot is commonly performed using the program GNOM from the ATSAS suite [23]. GNOM requires the background subtracted scattering profile as its input. An initial D_{max} estimate must be provided by the user.
4. The *P*(*r*) plot must be iteratively tested, with the user varying the initial D_{max}, to determine the most robust transformation of the raw data (*see* **Note 2**). A reasonable plot has three diagnostic features. First, the curve must approach zero at distances greater than or equal to D_{max}. This requirement is imposed by the software. Second, the curve must be positive at all points. Finally, the curve should be generally smooth, particularly near the D_{max}.
5. Because the scattering intensity is proportional to the concentration and molecular weight, Kratky and *P*(*r*) plots generated from samples containing the same RNA (e.g., apo-riboswitch RNA and ligand-bound riboswitch RNA) can be normalized to the forward scattering intensity of each sample. This normalization avoids the pitfall of over-interpreting differences in peak heights between the samples that arise from differences in concentration.

3.6 Three-Dimensional Shape Reconstructions Based on SAXS Data

Beyond one-dimensional scattering plots and numerical descriptions of RNA conformations (R_g, D_{max}), ab initio reconstructions based on the scattering data can provide low-resolution models of the riboswitch conformation in solution. Dummy atom models can be created using the program DAMMIF from the ATSAS suite [24]. This program uses simulated annealing procedures to generate models whose scattering profiles are consistent with the experimental data. Several previous SAXS studies of riboswitches have employed dummy atom models to interpret conformational changes induced by ligand binding. For instance, in the TPP and cyclic-diguanylate riboswitches large-scale reorientations of specific helical elements have been observed by examination of the reconstructed models in the absence and presence of ligand [8, 10, 25].

In these cases and others, general agreement has also been reported between solution SAXS reconstructions and available crystal structures.

1. Within the ATSAS suite of programs, the DAMMIN/F programs are used to reconstruct a dummy atom model containing an array of beads. The two programs differ in their algorithms (and required processing time) but yield comparable results.
2. The output from GNOM is the input required for DAMMIF. We recommend utilizing the output from three individual GNOM runs, which themselves differ by the user-input D_{max}, near the actual D_{max} and larger. Perform 10–15 DAMMIF reconstructions for each GNOM output file used. Use of the default parameters and "slow" mode is sufficient [13]. Running ten DAMMIF reconstructions simultaneously requires 50 min on an iMac equipped with a 2.66 GHz Intel Core 2 Duo processor and 4 GB of RAM.
3. Use DAMAVER [26] from the ATSAS suite to align, average, and filter the ten 3D reconstructions. The DAMAVER log file will list the normalized spatial discrepancy (NSD) for each model after comparison with all other models. NSD < 1 indicates similarity of each model to those against which it was compared. Visually inspect each of the three filtered dummy atom models (generated from the three GNOM output files used) to demonstrate convergence of the data to a consistent dummy atom model structure. This comparison also provides a visual evaluation of the general variance between the dummy atom models. Understanding the variance between the reconstructed models of the same data will help avoid over-interpretation of the structural changes observed between models from *different* solution conditions. For reference, Fig. 4 demonstrates the agreement between the filtered DAMMIF reconstruction of the SEC-treated SAM-I SAXS data in the presence of 2 mM SAM with an available crystal structure of the SAM-I riboswitch [17].

4 Notes

1. RNA purification can be performed under either denaturing or native conditions. Regardless, the sensitivity of SAXS data to higher order oligomeric contaminants still requires demonstration that the sample is homogeneous.
2. Jacques and Trewhella have recently published a helpful review regarding SAXS experimentation and data treatment. Their review provides additional descriptions that complement

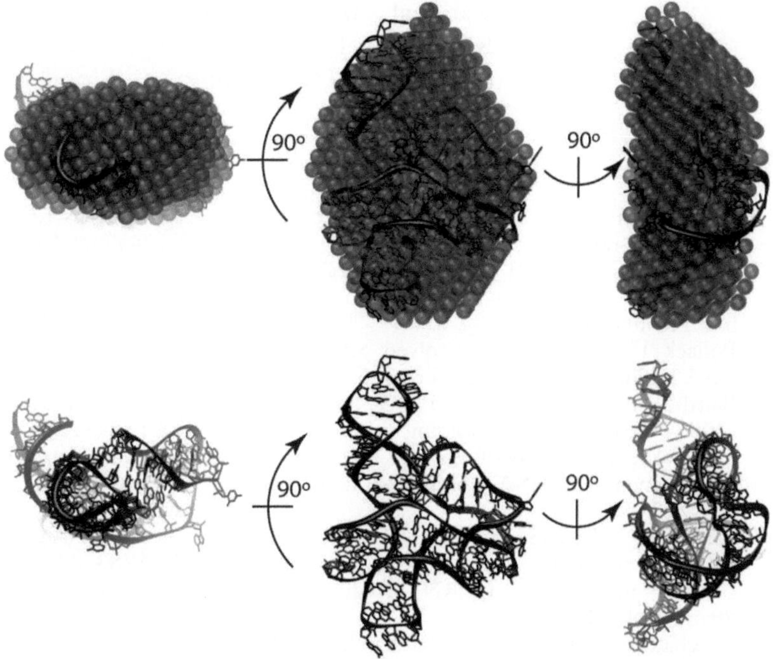

Fig. 4 Three-dimensional reconstructions of SAXS data. Ten independent DAMMIF reconstructions were run using GNOM output from the SEC-purified SAM-I riboswitch aptamer bound to SAM (+2 mM SAM). The dummy atom models were then aligned, averaged, and filtered using DAMAVER. The *top panel* illustrates the overlay of the dummy atom reconstruction and the SAM-I crystal structure (PDB ID 3GX5). The *bottom panel* shows the crystal structure in the same orientations as the *top panel*. The resulting filtered DAMMIF model matches the dimensions and general shape of the crystal structure well. Images were prepared using CHIMERA [28]

the protocol described here, particularly regarding data evaluation [27].

3. A useful forum for queries regarding general SAXS information, data analysis, and ATSAS software programs can be found at http://www.saxier.org/forum/.

Acknowledgments

The authors would like to thank K. Deigan and J. Zhang for helpful comments, L. Guo (sector 18-ID BioCAT) and S. Seifert and X. Zuo (sector 12-ID, BESSRC) for assistance with SAXS data collection. Use of the Advanced Photon Source was supported by the US Department of Energy, Basic Energy Sciences, Office of Science, under contract No. W-31-109-ENG-38. BioCAT is a

National Institutes of Health-supported Research Center RR-08630. This work was supported by the intramural program of the National Heart, Lung, and Blood Institute, NIH.

References

1. Guinier A, Fournet G (1955) Small-angle scattering of X-ray. Wiley, New York
2. Feigin LA, Svergun DI (1987) Structure analysis by small-angle X-ray and Neutron scattering. Plenum, New York
3. Pollack L (2011) Time resolved SAXS and RNA folding. Biopolymers 95:543–549
4. Baird N, Westhof E, Qin H, Pan T, Sosnick T (2005) Structure of a folding intermediate reveals the interplay between core and peripheral elements in RNA Folding. J Mol Biol 352:712–722
5. Russell R, Millett IS, Doniach S, Herschlag D (2000) Small angle X-ray scattering reveals a compact intermediate in RNA folding. Nat Struct Biol 7:367–370
6. Moghaddam S, Caliskan G, Chauhan S, Hyeon C, Briber RM, Thirumalai D, Woodson SA (2009) Metal ion dependence of cooperative collapse transitions in RNA. J Mol Biol 393:753–764
7. Baird NJ, Kulshina N, Ferré-D'Amaré AR (2010) Riboswitch function: flipping the switch or tuning the dimmer? RNA Biol 7:328–332
8. Baird NJ, Ferré-D'Amaré AR (2010) Idiosyncratically tuned switching behavior of riboswitch aptamer domains revealed by comparative small-angle X-ray scattering analysis. RNA 16:598–609
9. Stoddard CD, Montange RK, Hennelly SP, Rambo RP, Sanbonmatsu KY, Batey RT (2010) Free state conformational sampling of the SAM-I riboswitch aptamer domain. Structure 18:787–797
10. Ali M, Lipfert J, Seifert S, Herschlag D, Doniach S (2010) The ligand-free state of the TPP riboswitch: a partially folded RNA structure. J Mol Biol 396:153–165
11. Lipfert J, Das R, Chu VB, Kudaravalli M, Boyd N, Herschlag D, Doniach S (2007) Structural transitions and thermodynamics of a glycine-dependent riboswitch from Vibrio cholerae. J Mol Biol 365:1393–1406
12. Lipfert J, Sim AY, Herschlag D, Doniach S (2010) Dissecting electrostatic screening, specific ion binding, and ligand binding in an energetic model for glycine riboswitch folding. RNA 16:708–719
13. Lipfert J, Chu VB, Bai Y, Herschlag D, Doniach S (2007) Low-resolution models for nucleic acids from small-angle X-ray scattering with applications to electrostatic modeling. J Appl Crystallogr 40:229–234
14. Wood S, Ferre-D'Amare AR, Rueda D (2012) Allosteric tertiary interactions preorganize the c-di-GMP riboswitch and accelerate ligand binding. ACS Chem Biol 7:759–770
15. Batey RT, Kieft JS (2007) Improved native affinity purification of RNA. RNA 13:1384–1389
16. Milligan JF, Groebe DR, Witherell GW, Uhlenbeck OC (1987) Oligoribonucleotide synthesis using T7 RNA polymerase and synthetic DNA templates. Nucleic Acids Res 15:8783–8798
17. Montange RK, Mondragón E, Van Tyne D, Garst AD, Ceres P, Batey RT (2010) Discrimination between closely related cellular metabolites by the SAM-I riboswitch. J Mol Biol 396:761–772
18. Baird NJ, Zhang J, Hamma T, Ferré-D'Amaré AR (2012) YbxF and YlxQ are bacterial homologs of L7Ae and bind K-turns but not K-loops. RNA 18:759–770
19. Svergun DI, Bargerato C, Koch MHJ (1995) CRYSOL - a program to evaluate X-ray solution scattering of biological macromolecules from atomic coordinates. J Appl Crystallogr 28:768–773
20. Garst AD, Héroux A, Rambo RP, Batey RT (2008) Crystal structure of the lysine riboswitch regulatory mRNA element. J Biol Chem 283:22347–22351
21. Doniach S (2001) Changes in biomolecular conformation seen by small angle X-ray scattering. Chem Rev 101:1763–1778
22. Kratky O, Porod G (1949) Rontgenuntersuchung geloster fadenmolekule. Recueil des travaux chimiques des pays-bas Journal of the Royal Netherlands Chemical Society 68:1106–1122
23. Svergun DI (1992) Determination of the regularization parameter in indirect-transform methods using perceptual criteria. J Appl Crystallogr 25:495–503
24. Franke D, Svergun DI (2009) DAMMIF, a program for rapid ab-initio shape determination in small-angle scattering. J Appl Crystallogr 42:342–346
25. Kulshina N, Baird NJ, Ferré-D'amaré AR (2009) Recognition of the bacterial second messenger cyclic diguanylate by its cognate

riboswitch. Nat Struct Mol Biol 16: 1212–1217

26. Volkov VV, Svergun DI (2003) Uniqueness of ab initio shape determination in small-angle scattering. J Appl Crystallogr 36:860–864
27. Jacques DA, Trewhella J (2010) Small-angle scattering for structural biology-Expanding the frontier while avoiding the pitfalls. Protein Sci 19:642–657
28. Pettersen E, Goddard T, Huang C, Couch G, Greenblatt D, Meng E, Ferrin T (2004) UCSF Chimera–a visualization system for exploratory research and analysis. J Comput Chem 25:1605–1612

Chapter 17

Use of SHAPE to Select 2AP Substitution Sites for RNA–Ligand Interactions and Dynamics Studies

Marie F. Soulière and Ronald Micura

Abstract

Most regulatory RNA molecules must adopt a precise secondary fold and tertiary structure to allow their function in cells. A number of experimental approaches, such as the 2-Aminopurine-Based RNA Folding Analysis (2ApFold), have therefore been developed to offer insights into the folding and folding dynamics of RNA. A crucial requirement for this method is the selection of proper 2AP labeling positions. In that regard, we recently discovered that Selective 2′-Hydroxyl Acylation analyzed by Primer Extension (SHAPE) offers a reliable path to identify appropriate nucleotides for 2AP substitution on a target RNA. This chapter describes the straightforward procedure to select 2AP substitution sites in RNA molecules using SHAPE probing. The protocols detail the preparation of the target RNA by transcription, and the SHAPE steps including (1) probing of the RNA, (2) reverse transcription with a radiolabeled primer, (3) sequencing gel, and (4) analysis of the obtained band pattern.

Key words SHAPE, RNA probing, 2-Aminopurine, 2ApFold, *preQ₁cII* riboswitch, SAFA

1 Introduction

During the past decade, RNA molecules have been increasingly demonstrated to play a critical role in the control of gene expression and the catalysis of biological reactions [1–3]. Ribosomal RNAs (rRNAs), transfer RNAs (tRNAs), and riboswitches must adopt a precise secondary fold and tertiary structure to allow their function in cells [4–6]. Accordingly, many studies have recently been dedicated to the elucidation of RNA structures, dynamics, and folding pathways of RNAs. A variety of experimental approaches can offer insights into the folding dynamics of RNA: structural probing experiments, NMR spectroscopy, small-angle X-ray scattering (SAXS) analysis, and fluorescence spectroscopy (Reviewed in ref. 7).

This work was supported by a grant from the Austrian Science Foundation FWF (P21641, I317). Marie F. Soulière is the recipient of an EMBO long-term fellowship (ALTF 637–2010) from the European Molecular Biology Organization.

Daniel Lafontaine and Audrey Dubé (eds.), *Therapeutic Applications of Ribozymes and Riboswitches: Methods and Protocols*, vol. 1103, DOI 10.1007/978-1-62703-730-3_17, © Springer Science+Business Media New York 2014

One of our laboratory's research contributions is the 2-Aminopurine-Based RNA Folding Analysis (2ApFold) approach that makes use of the introduction of noninvasive 2-aminopurine modifications in RNA sequences [7]. We have successfully applied this method to the study of a number of riboswitches [8–11]. The 2-aminopurine (2AP) modified base is substituted in strategic positions of an RNA molecule to use its fluorescence response to address folding and kinetics during secondary and tertiary structure rearrangements taking place during, for example, ligand-induced RNA folding [7, 12, 13]. However, in the absence of high-resolution structural data, the selection of proper 2AP labeling positions can be problematic. In that regard, we recently discovered and reported a method to select appropriate nucleotide sites for 2AP substitutions on a target RNA using Selective 2′-Hydroxyl Acylation analyzed by Primer Extension (SHAPE) [14]. SHAPE reagents react with the 2′-hydroxyl group of bases, with little preference for base identity, with high sensitivity to conformational dynamics [15–17]. A correlation between SHAPE chemistry and the generalized NMR order parameter S2 was previously reported demonstrating that SHAPE monitors spatial disorder and structural dynamics of nucleotides [18]. We further proved, with 2AP and SHAPE experiments on three distinct riboswitch RNAs, that a direct correlation exists between single nucleotide flexibility detected by SHAPE and the fluorescence response of a 2AP base incorporated at the corresponding position in an RNA molecule [14].

This chapter describes a straightforward approach to select 2AP substitution sites in RNA molecules for 2AP binding and kinetics studies using SHAPE probing. The method requires the preparation of the target RNA by transcription, followed by SHAPE steps which include probing of the RNA, reverse transcription with a radiolabeled primer, resolving of the products on a sequencing gel, and analysis of the obtained band pattern. This protocol details, as a general example, the steps for the selection of 2AP substitution bases for the $\mathrm{preQ_1}$ class II riboswitch (*preQ*$_1$*cII*).

2 Materials

Prepare all solutions using nanopure water (prepared by purifying deionized water to attain a sensitivity of 18 MΩ cm at 25 °C) and analytical grade reagents.

The steps in Subheadings 3.3–3.5 include work with radioactive material. Proper training and licenses should be obtained before attempting these experiments.

1. Loading buffer: 95 % formamide, 10 mM EDTA, with 0.1 % bromophenol blue and/or 0.1 % xylene cyanol blue dyes.

2. 5× Tris-Borate-EDTA buffer (5× TBE): 108 g Tris, 55 g boric acid, and 40 mL of 0.5 M EDTA, pH 8.0. Complete to 2 L with nanopure water.
3. 1× TBE–7 M urea gel dilution buffer: 210 g Urea, 100 mL 5× TBE, 200 mL nanopure water. Dissolve by heating at 45 °C. Complete volume to 500 mL.
4. 20 % Polyacrylamide solution (19:1): 38 g Acrylamide, 2 g bisacrylamide, 84 g urea, 40 mL 5× TBE, 40 mL nanopure water. Dissolve by heating at 45 °C. Complete volume to 200 mL.
5. Elution buffer: 0.3 M NH_4OAc and 0.1 % SDS.
6. Precipitation solution: 5 μL 20 mg/mL Glycogen, 650 μL water, 75 μL 1 M NaOAc, pH 5.2.
7. BzCN solution (600 mM): 39 mg Benzoyl cyanide (BzCN) in 500 μL anhydrous DMSO.
8. 10× MOPS–KCl buffer (500 mM MOPS, 1 M KCl): 523.2 mg MOPS, 372.8 mg KCl, and 2 mL nanopure water. Adjust to pH 7.5 with 10 % KOH. Complete to 5 mL.

3 Methods

3.1 Preparation of RNA for SHAPE

The *preQ₁cII* RNA sequence for SHAPE was generated by purchasing two partially overlapping DNA oligonucleotides, one including a T7 promoter sequence, and subjecting them to a PCR filling reaction to obtain the double-stranded DNA which is then further transcribed. The final desired RNA sequence has to include spacer sequences in 5′ (5–10 nt) and 3′ (8–20 nt) of the analyzed RNA to allow quantification (*see* **Note 1**). Furthermore, it also requires a 3′ sequence (18 nt) to allow primer annealing for reverse transcription (Fig. 1).

1. Mix 10 μL 10× *Pfu* buffer with $MgSO_4$, 10 μL 2 mM dNTPs, 4 μL each of 100 mM DNA oligonucleotides, and 0.5 μL *Pfu* DNA polymerase in a Thermocycler tube. Adjust the volume to 100 μL with nanopure water.
2. Incubate the reaction mixture in a Thermocycler for the following cycles:

   Denaturing for 2 min at 94 °C; nine cycles of denaturing for 1 min at 94 °C, annealing for 45 s at 55 °C, and primer extension for 1 min at 72 °C; final extension phase of 2 min at 72 °C; cool to 16 °C.
3. Precipitate the reaction using 2 volumes (200 μL) 100 % ethanol and 0.1 volume (10 μL) of 5 M NaCl. Incubate at −20 °C for 30 min. Centrifuge at 4 °C with 11,300 × *g* for 30 min.

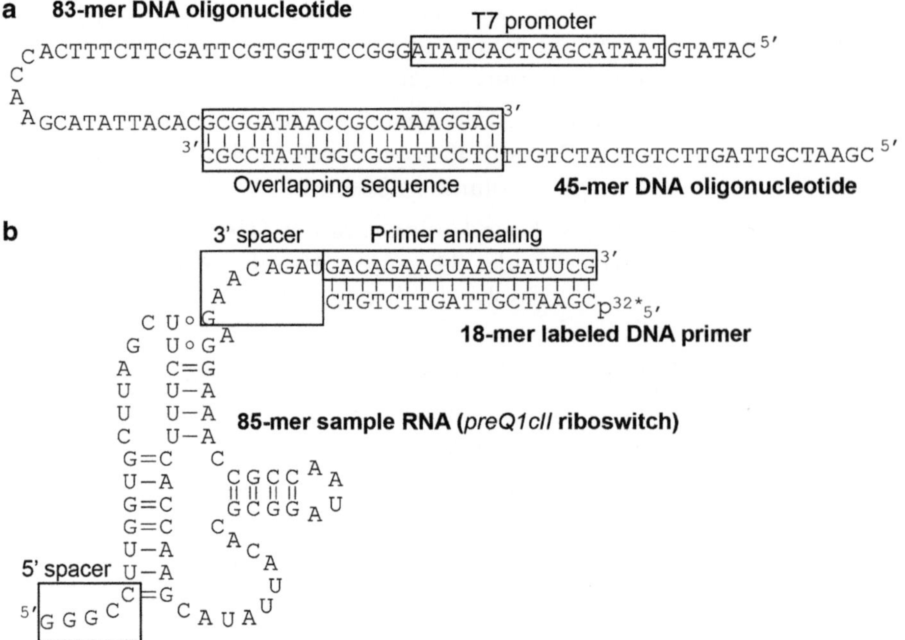

Fig. 1 Design of RNA sequence for SHAPE probing experiments. (**a**) DNA oligonucleotides for PCR filling reaction to obtain double-stranded DNA for transcription of the *preQ₁cII* riboswitch RNA. (**b**) Resulting *preQ₁cII* riboswitch RNA sequence for SHAPE, including 5′ and 3′ spacers, and reverse transcription primer annealing segment

Remove the supernatant and resuspend the precipitate in 100 μL nanopure water.

4. Wearing gloves, mix 50 μL of the DNA product, 20 μL 5× transcription buffer, 25 μL 25 mM rNTPs, 2 μL pyrophosphatase diluted 1:20, 1 μL RiboLock, and 2 μL T7 RNA polymerase, for a total volume of 100 μL. Incubate for 1–2 h at 37 °C. Add 1 μL more of T7 RNA polymerase and incubate for an added 1–2 h at 37 °C.
5. Add 1 μL of DNase (RNase-free). Incubate for 20 min at 37 °C.
6. Precipitate the reaction using 1 volume (100 μL) isopropanol and 0.2 volume (20 μL) of 1 M NaOAc, pH 5.2. Incubate at −20 °C for 30 min (*see* **Note 2**). Centrifuge at 4 °C with 11,300 × *g* for 30 min.
7. Remove the supernatant, let the pellet dry for 10 min, and then resuspend the pellet in 10 μL of water (*see* **Note 3**). Add 10 μL of loading buffer with bromophenol blue.
8. Purify the transcription product by loading the whole sample on an 8 % polyacrylamide gel (19:1 acrylamide:bisacrylamide, w/w), 7 M urea and 1× TBE (*see* **Note 4**). Add loading buffer

with xylene cyanol blue marker in a neighboring well. Migrate in 1× TBE at 25–30 W until bromophenol blue marker is 1 cm from the bottom of the gel.

9. Wrap the gel in transparent plastic foil and place on silica-coated TLC plateto visualize under 254 nm UV light. Mark the RNA bands and excise the corresponding gel pieces with a razor blade. Place the gel slices in a 1.5-mL Eppendorf tube.
10. Add elution buffer to cover the gel pieces and incubate at room temperature, with rotation, overnight.
11. Use a pipette to remove the liquid with flat gel-loading tips (*see* **Note 5**). Precipitate RNA in the liquid phase using 1 volume isopropanol and 0.2 volume 1 M NaOAc, pH 5.2. Incubate at −20 °C for 30 min (*see* **Note 2**). Centrifuge at 4 °C with 11,300 × *g* for 30 min.
12. Resuspend the RNA in 20 μL of nanopure water and dose its concentration using a spectrophotometer. Freeze at −20 °C for further use.

3.2 Probing of RNA with BzCN (2′-Hydroxyl Acylation)

A typical SHAPE experiment on an RNA that is responsive to magnesium and/or a ligand consists of six samples: two ladders (1, 2) and four probings (3, 4, 5, 6).

1. ddCTP ladder, **2**. ddGTP ladder.

3. DMSO control, **4**. BzCN RNA probing, **5**. BzCN RNA + $MgCl_2$ probing, **6**. BzCN RNA + $MgCl_2$ + ligand probing.

Ladder samples 1 and 2 will be prepared only during the reversion transcription step (Subheading 3.4). Samples 3, 4, 5, and 6 are subjected to the probing step described in this section prior to the reverse transcription steps described in Subheading 3.4.

1. Prepare samples 3, 4, 5, and 6 by mixing 1 μL of 10× MOPS–KCl buffer, 5 μL of 1 pmol/μL (1 μM) of sample RNA from transcription, and 2 μL water in 500 μL Thermocycler Eppendorf tubes. For samples 3 and 4, add two more microliters of water. For sample 5, add 1 μL of water and 1 μL 50 mM $MgCl_2$. For sample 6, add 1 μL 50 mM $MgCl_2$ and 1 μL of 500 μM $preQ_1$ ligand, for a final volume of 10 μL for each sample.
2. Incubate samples 3, 4, 5, and 6 for 2 min at 65 °C, followed by 5 min at 4 °C and 25 min at 37 °C in a Thermocycler.
3. During incubation, prepare the following solutions: 39 mg of BzCN in 500 μL anhydrous DMSO and the precipitation solution (5 μL glycogen 20 mg/mL, 650 μL water, 75 μL 1 M NaOAc, pH 5.2).
4. After incubation, add 1 μL of anhydrous DMSO to control sample 3 and mix thoroughly. Add 1 μL of BzCN to samples

4, 5, and 6, and mix thoroughly for each sample. The reaction with BzCN takes only a few seconds to occur.

5. Add 90 μL of the precipitation solution (prepared in **step 3**) to each sample tube. Then add 250 μL of 100 % EtOH to all tubes and mix by inverting several times. Incubate at −20 °C for 25 min.
6. Centrifuge at 11,300 × *g* for 25 min (*see* **Note 6**).
7. Remove the supernatant by pipetting out without disturbing the pellet (*see* **Note 2**). When most of the liquid is removed, invert the tubes and let stand upside down on an absorbent hand-paper for 10–15 min.
8. Resuspend the pellets in 8 μL of water. Store at −20 °C or proceed with the reverse transcription steps.

3.3 Preparation of Radiolabeled DNA Primer for Reverse Transcription

1. Mix 2 μL of 10 μM DNA primer (*see* **Note 7**), 2 μL of PNK buffer A, 2.5 μL of γ^{32}P-ATP (0.925 MBq), and 12.8 μL of nanopure water in a 500 μL Eppendorf tube. Add 0.8 μL T4 PNK kinase enzyme to the radioactive reaction mixture. The total of the reaction is 20 μL (*see* **Note 8**).
2. Incubate in a Thermocycler at 37 °C for 1 h, followed by enzyme inactivation 10 min at 70 °C.
3. After inactivation, add 20 μL of water to the mixture, followed by 40 μL of phenol/chloroform. Shake vigorously and centrifuge for 2 min at 11,300 × *g*.
4. Keep the top fraction for purification using a Mini Quick Oligo Column.
5. Get a Mini Quick Oligo Column out of the 4 °C refrigerator 5–10 min before use, and let warm at room temperature. Vigorously invert the columns five to six times. Remove the cap of the column (*see* **Note 9**) then snap off the bottom tip of the column.
6. Place the column in a 1.5 mL Eppendorf tube. Centrifuge for 1 min at 11,300 × *g* in a MiniQuick Eppendorf centrifuge or equivalent centrifuge (*see* **Note 10**). Discard the liquid in the Eppendorf tube and do a quick centrifugation to remove the remaining liquid in the tip of the column. Place your column in a clean 1.5 mL Eppendorf tube.
7. Apply your 40 μL sample from phenol/chloroform extraction slowly in the middle of the top of the column matrix (*see* **Note 11**).
8. Centrifuge at 11,300 × *g* for 4 min in a MiniQuick Eppendorf centrifuge.
9. Save the eluate, which contains your purified radiolabeled nucleic acid. Discard the used column as radioactive waste. Store at −20 °C or use directly for the reverse transcription steps.

3.4 Reverse Transcription on the Probed RNA

1. Prepare 2× 500 μL Eppendorf tubes with 5 μL of 1 pmol/μL (1 μM) sample RNA, and 3 μL water for the ladders (samples 1, 2 in Subheading 3.2).
2. Proceed with the following steps for all six samples: two ladders (1, 2) and four probings [3, 4, 5, 6 (from Subheading 3.2)].
3. Add 3 μL of the radiolabeled SHAPE DNA oligonucleotide (prepared in Subheading 3.3) to all RNA samples.
4. Incubate the six samples in a Thermocycler for 5 min at 65 °C, followed by 5 min at 35 °C, and 1 min at 4 °C.
5. During incubation prepare the reverse transcription pool (28 μL 5× first strand buffer, 7 μL DTT 0.1 M, 7 μL 10 mM dNTPs, 14 μL DMSO) and the precipitation solution (5 μL glycogen 20 mg/mL, 650 μL water, 150 μL 1 M NaOAc pH 5.2).
6. When incubations from **step 4** are completed, add 8 μL of the reverse transcription pool to each sample. Mix well by pipetting up and down at least five times as well as rotations with the pipette. For samples 1 (ddCTP ladder) and 2 (ddGTP ladder), add 1 μL of the ddCTP or ddGTP nucleotide, respectively, to the reaction tubes. Incubate samples 1 min at 61 °C.
7. Add 0.4 μL of the SuperScript III enzyme to each sample. Mix by pipetting up and down at least five times as well as rotations with the pipette (*see* **Note 12**). Incubate 10 min at 61 °C.
8. Add 1 μL of 4 N NaOH (*see* **Note 12**). Incubate 5 min at 95 °C, followed by incubation at 4 °C for 10 min.
9. Add 90 μL of the precipitation pool and 250 μL of 100 % EtOH to each sample. Mix by inverting several times. Incubate at −20 °C for 25 min.
10. Centrifuge at 11,300 ×*g* for 25 min.
11. Remove the supernatant by pipetting out the liquid and transferring it to an empty tube (the liquid may be radioactive). When most of the liquid is removed, let the Eppendorf tubes stand upside down on an absorbent hand-paper for 10–15 min.
12. Resuspend the radioactive pellets in 8 μL of loading buffer with xylene cyanol. Store at −20 °C or run the samples on a sequencing gel directly.

3.5 Sequencing Gel

1. Prepare a sequencing 10 % polyacrylamide gel (19:1 acrylamide:bisacrylamide, w/w), 7 M urea and 1× TBE, in 40× 20 cm glass plates with 0.4 mm spacers (*see* **Note 13**). Prerun the gel in 1× TBE buffer at 45 W for 1 h.
2. Prepare samples of your probed RNA and controls with equal radioactive counts (typically from 300 to 600 cpm) (*see* **Note 14**). Complete the samples to 4 μL with formamide-EDTA Loading buffer without blue markers.

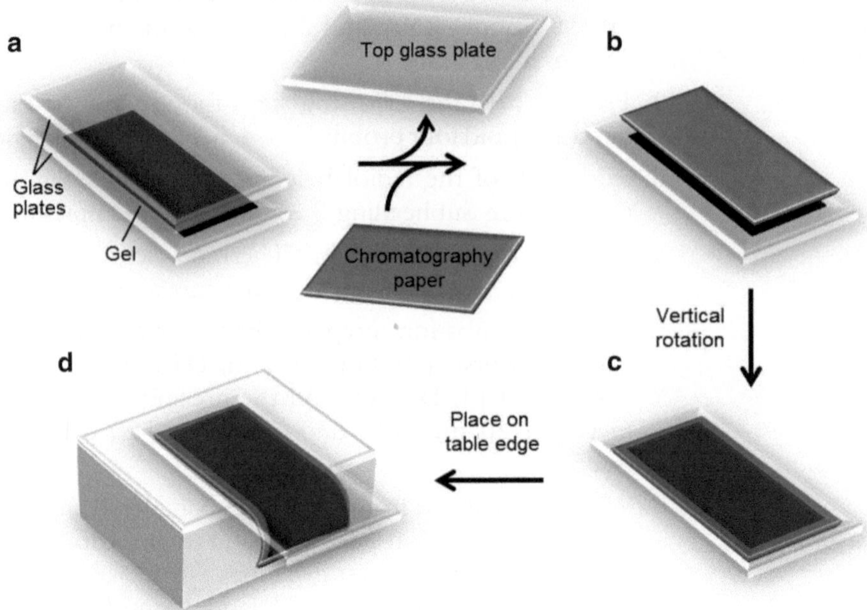

Fig. 2 Extraction of a sequencing gel from its glass plate casing

3. Clean the gel wells (*see* **Note 4**). Carefully load the samples on the gel with flat gel-loading tips (*see* **Note 3**). Add a bromophenol blue marker in a neighboring well. Migrate in 1× TBE at 45 W until bromophenol blue marker is 1 cm from the bottom of the gel.
4. Remove one glass plate and cover the gel with a Chromatography paper larger than the gel (Fig. 2). Flip the remaining glass plate with the gel and Chromatography paper (Fig. 2). Place on a table edge and pull down the Chromatography paper, making sure that the gel is sticking to it, until the whole gel is separated from the glass plate (*see* **Note 15**) (Fig. 2).
5. Put the gel on the Chromatography paper in a gel dryer. Cover gel with plastic wrapping, and dry gel for 1 h at 75 °C.
6. Wrap the dried gel on Chromatography paper in plastic wrapping. Expose on a ^{32}P-sensitive phosphorscreen overnight. Scan the screen on a Phosphorimager or equivalent instrument.

3.6 Analysis of the SHAPE Results

Band intensities visualized by gel electrophoresis are quantified using SAFA v.1.1 (Semi-Automated Footprinting Analysis Software) (https://simtk.org/home/safa) [19]. The software allows to quantify every band of each lane of the scanned gel (*see* **Note 16**). The output data will be saved as a .txt file. The values can then be imported in Microsoft Office Excel for further processing. Data sets are normalized by dividing all intensities by the intensity of the

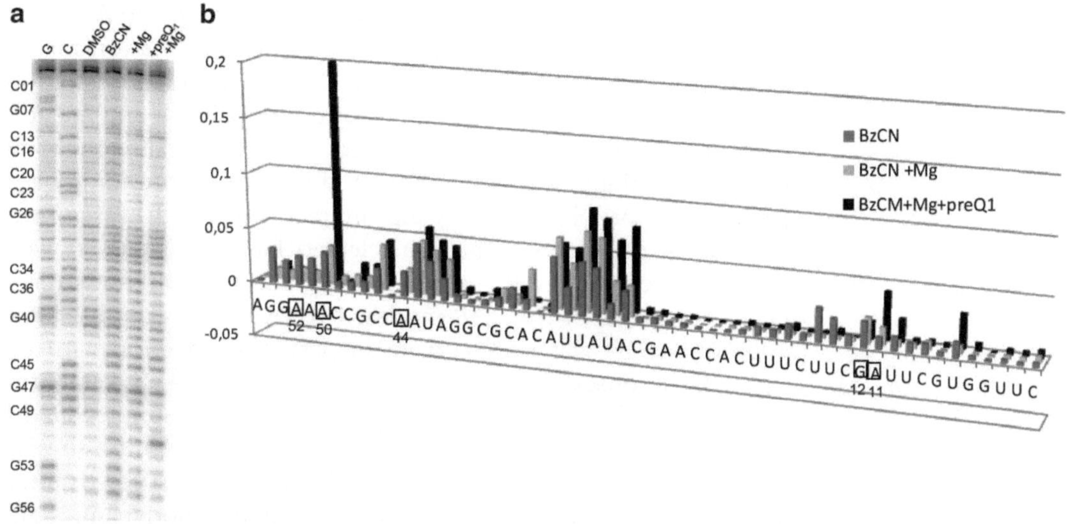

Fig. 3 Analysis of the SHAPE results. (**a**) The assignment of the bands on the gel is done using the ddC and ddG ladders (*lanes 1* and *2*). (**b**) Graphical representation of the 2′-OH acylation reactivity of all bases of the *preQ$_1$cII* RNA alone (*dark grey*), with magnesium (*light grey*) and in the presence of magnesium and preQ$_1$ ligands (*black*), after quantification using the SAFA software, as well as normalization and background subtraction in Excel

full-length transcript for each lane. The resulting data is then further processed by subtracting background cleavage for each nucleotide from the DMSO control lane (sample 3). You can assign your bands using the ddC and ddG ladders (Fig. 3a). Note that since the 2′-hydroxyl acylation from BzCN probing prevents the addition of nucleotides at the modification sites, reverse transcription products are one nucleotide shorter than actual reaction sites. Therefore, the intensity of a band on the sequencing gel represents the degree of modification of the preceding base in the primary sequence of the probed RNA.

After proper normalization, background subtraction, and assignment of the bands, you will obtain a graphical representation of the 2′-OH acylation reactivity of all bases of your RNA in the absence and presence of ligands (Fig. 3b). From the bases that are reported as flexible by SHAPE, we have demonstrated that you can make your selection of 2AP substitution sites [14]. The best substitution sites are typically the bases presenting the widest flexibility variations in SHAPE. You do not want to select a base that is very flexible in all tested conditions, but rather a base that presents a change upon ligand(s) addition. We also suggest to select positions that are compatible with a 2AP substitution. 2AP is a purine analog and can base pair with U. Therefore, while it can be substituted for any base in regions that are demonstrated to be single-stranded, we would favor A or G substitution sites. Finally, substitution sites should be selected in regions of the RNA that are of interest. For the *preQ₁cII* riboswitch, bases A11, G12, A44, A50, and A52 were selected (Fig. 3b). A50 has the widest flexibility variation for the whole riboswitch, while A11 and A52 have the widest flexibilities variation in their respective regions of the riboswitch and are adenine bases. Bases G12 and A44 were selected because they had opposite flexibility responses compared to their surrounding bases (decrease in probing upon ligand interaction), which was considered to be of interest. The five selected bases for *preQ₁cII* were demonstrated to be valid targets for 2AP substitution: the fluorescence responses of the 2AP-modified RNA variants upon ligands addition (Mg^{2+} and $preQ_1$) were in perfect correlation with the SHAPE responses, thus validating this approach [14]. Overall, this protocol provides a straightforward method using a simple SHAPE probing experiment to select appropriate 2AP substitution sites for fluorescence assays of your target RNA, in order to study its interaction with other molecules, their binding thermodynamics and kinetics, as well as the particular RNA folding pathway and its corresponding kinetics.

4 Notes

1. Generally, the 10–20 nucleotides adjacent to the reverse transcription primer binding site cannot be quantified due to the presence of fragments that reflect pausing by the Reverse Transcriptase enzyme during the initiation phase of reverse transcription [15]. A short sequence (5–10 nt) at the 5′ end of the RNA also usually cannot be quantified because of the intensity of the band corresponding to the full-length extension product after reverse transcription.
2. Precipitate with 2 volumes of 100 % EtOH if your RNA is shorter than 50 nucleotides, and 1 volume of 100 % isopropanol if your RNA is over 50 nucleotides. The precipitation can be kept at −20 °C for up to 2–3 days before continuing with the next steps.
3. Be careful not to lose the pellet, especially when precipitation is done with isopropanol. If you wait more than 5–10 min after centrifugation is completed, the pellet might be loose in the supernatant, not attached to the bottom of the tube. If you notice that this is the case, the best course of action is to centrifuge again for at least five more minutes to get the pellet to the bottom of the tube again.
4. Make sure that the wells of the gels are clean by injecting buffer with a syringe to wash the wells. Unwashed wells create bad migration bands and should be avoided.
5. A number of ways can be used to recover the RNA eluted in the buffer without the undesired acrylamide residues. The suggested

way is quick, inexpensive, and the resulting samples have been shown to be of sufficient purity and quality to be used for SHAPE probing and for mass spectrometry.

6. Some centrifuges require special adaptors for 500 μL tubes. If you do not possess a centrifuge that accepts 500 μL tubes, nor the adaptors, you can place the 500 μL tubes in 1.5 mL Eppendorf tubes, from which you have previously cut-away the caps, for the centrifugation.
7. Use the DNA primer designed for RT on the RNA that you are performing SHAPE on. Typically, an 18-mer DNA oligonucleotide is used (*see* Fig. 1).
8. This reaction can be scaled down or up depending on the number of reverse transcription (RT) experiments that you plan on. You can manage an approximation of the required amount by calculating that at the end of a 20 μL labeling reaction you will obtain a volume of approximately 45 μL, and that a typical RT reaction utilizes 3 μL of the labeled primer solution. Therefore, with this labeling you can perform 15 RT reactions. If more than one half-life of the radioactive isotope has passed, you can use 5 μL of the labeled primer solution.
9. When you remove the cap, usually some matrix is left in the cap. Pipette it out and onto the column before centrifugation.
10. Centrifugation is performed at 1,000 × g. It is crucial to verify the rpm value before centrifugation. DO NOT centrifuge at 11,300 × *g*. To ensure no problem with the centrifugation, direct the open cap of the Eppendorf tube towards the middle of the centrifuge before starting the centrifugation.
11. The column is effective for volumes between 20 and 50 μL. Be aware of this restriction if you scale the reaction up (*see* **Note 8**). If you prepare 40–50 μL of reaction, do not add water to the sample for the phenol/chloroform extraction step. Additionally, DO NOT apply sample on the sides of the column, it can lead to bypass of the column and recovery of nucleotides in the eluate.
12. The timing for reagents addition during **steps** 7 and **8** must correspond. The pipetting and mixing for the addition of SSIII in **step** 7 takes longer (because of the small volume and the viscosity) than the addition of NaOH in **step 8**. If you terminate the reactions normally with NaOH in **step 8**, the incubation time between the six samples will vary. It is therefore recommended to time the addition of the enzyme in **step** 7 to occur at an interval of 30 s between the samples and to use the same timing for **step 8**, so that all samples are equally incubated for 10 min.
13. It can be tricky to prepare thin 0.4 mm gels. The glass plates need to be very clean and we suggest to load the gel solution

from one corner of the gel (not in the middle) with a syringe or a pipette to avoid bubble formation.

14. The counts can be evaluated with a radioisotope counter. If you do not have access to such an apparatus, you can directly measure the counts with a radiation monitor (Geiger). Make sure to position the samples in the same place and orientation in front of the Geiger to manually evaluate the respective counts in each tube.
15. This step is best done immediately after the migration is finished, while the gel is still warm. When removing the top glass plate, if the gel is adhering to the plate, you can put the plate back in place, flip the glasses and gel, and try removing the other plate. Additionally, when pulling down the gel on the chromatography paper, if the gel is adhering to the glass plate, you can help yourself with a spatula and/or by injecting some water between the gel and the plate.
16. Instead of a fasta file, you can upload your sequence from a .txt file with a ">Name" heading. When you assign bands, change the starting band number to 1.

References

1. Strobel SA, Cochrane JC (2007) RNA catalysis: ribozymes, ribosomes, and riboswitches. Curr Opin Chem Biol 11:636–643
2. Serganov A, Patel DJ (2007) Ribozymes, riboswitches and beyond: regulation of gene expression without proteins. Nat Rev Genet 8:776–790
3. Inui M, Martello G, Piccolo S (2010) MicroRNA control of signal transduction. Nat Rev Mol Cell Biol 11:252–263
4. Batey RT, Rambo RP, Doudna JA (1999) Tertiary motifs in RNA structure and folding. Angew Chem Int Ed 38:2326–2343
5. Roth A, Breaker RR (2009) The structural and functional diversity of metabolite-binding riboswitches. Annu Rev Biochem 78:305–334
6. Schwalbe H, Buck J, Furtig B, Noeske J, Wohnert J (2007) Structures of RNA switches: insight into molecular recognition and tertiary structure. Angew Chem Int Ed 46:1212–1219
7. Haller A, Souliere MF, Micura R (2011) The dynamic nature of RNA as key to understanding riboswitch mechanisms. Acc Chem Res 44:1339–1348
8. Haller A, Rieder U, Aigner M, Blanchard SC, Micura R (2011) Conformational capture of the SAM-II riboswitch. Nat Chem Biol 7:393–400
9. Rieder U, Kreutz C, Micura R (2010) Folding of a transcriptionally acting preQ$_1$ riboswitch. Proc Natl Acad Sci U S A 2107:10804–10809
10. Rieder R, Lang K, Graber D, Micura R (2007) Ligand-induced folding of the adenosine deaminase A-riboswitch and implications on riboswitch translational control. Chembiochem 8:896–902
11. Lang K, Rieder R, Micura R (2007) Ligand-induced folding of the thiM TPP riboswitch investigated by a structure-based fluorescence spectroscopic approach. Nucleic Acids Res 35:5370–5378
12. Jean JM, Hall KB (2001) 2-Aminopurine fluorescence quenching and lifetimes: role of base stacking. Proc Natl Acad Sci U S A 98:37–41
13. Sinkeldam RW, Greco N, Tor Y (2010) Fluorescent analogs of biomolecular building blocks: design properties and applications. Chem Rev 110:2579–2619
14. Souliere MF, Haller A, Rieder R, Micura R (2011) A powerful approach for the selection of 2-aminopurine substitution sites to investigate RNA folding. J Am Chem Soc 133:16161–16167
15. Merino EJ, Wilkinson KA, Coughlan JL, Weeks KM (2005) RNA structure analysis at single nucleotide resolution by selective 2′-hydroxyl acylation and primer extension (SHAPE). J Am Chem Soc 127:4223–4231
16. Wilkinson KA, Merino EJ, Weeks KM (2006) Selective 2′-hydroxyl acylation analyzed by primer extension (SHAPE): quantitative RNA

structure analysis at single nucleotide resolution. Nat Protoc 1:1610–1616

17. Wilkinson KA, Vasa SM, Deigan KE, Mortimer SA, Giddings MC, Weeks KM (2009) Influence of nucleotide identity on ribose 29-hydroxylreactivity in RNA. RNA 15: 1314–1321
18. Gherghe CM, Shajani Z, Wilkinson KA, Varani G, Weeks KM (2008) Strong correlation between SHAPE chemistry and the generalized NMR order parameter (S2) in RNA. J Am Chem Soc 130:12244–12245
19. Laederach A, Das R, Vicens Q, Pearlman SM, Brenowitz M, Herschlag D, Altman RB (2008) Semi-automated and rapid quantification of nucleic acid footprinting and structure mapping experiment. Nat Protoc 3(9): 1395–1401

Chapter 18

Cell Internalization SELEX: In Vitro Selection for Molecules That Internalize into Cells

Amy Yan and Matthew Levy

Abstract

Aptamer technology allows for the selection of nucleic acids that can bind to and enter cells. By establishing conditions during the selection that eliminate cell-surface binders as well as non-internalizing RNAs, only extremely tightly bound aptamers or aptamers that have internalized are recovered. We describe a general scheme for selecting RNA molecules that are capable of internalizing into cells and discuss the factors that can affect a successful selection. Much like standard cell-surface selections, these types of selections should be possible independent of detailed knowledge of the cell surface.

Key words Aptamer, SELEX, In vitro selection, Cell internalization, Targeted delivery

1 Introduction

Aptamers are nucleic acid binding species generated by iterative rounds of in vitro selection, or SELEX, to specifically recognize a given target (reviewed in refs. 1–4). Briefly, randomized pools of RNA or ssDNA are incubated with the target under carefully chosen selection conditions. Binding species are partitioned away from nonbinders, amplified to generate a new pool, and the process is repeated until a desired "phenotype" is achieved or until sequence diversity is significantly diminished.

Initially, selections were performed against protein or small molecule targets. However, especially with the development of nuclease-stabilized nucleic acids (i.e., 2′*O*-methyl-, 2′-fluoro-, and 2′-amino-modified RNA) and the enzymes to generate them in reasonable quantities [5, 6], selection techniques have expanded dramatically to include cell-surface targets, whole cells and even in vivo targets. Some aptamers selected against specific cell-surface receptors have been successfully used for the "hitchhiking" of cargoes into cells and are being adapted for in vivo targeting and delivery [7, 8].

The same principles that guide the success of cell-surface selections can also be used to select for molecules that not only bind to

Daniel Lafontaine and Audrey Dubé (eds.), *Therapeutic Applications of Ribozymes and Riboswitches: Methods and Protocols*, vol. 1103, DOI 10.1007/978-1-62703-730-3_18, © Springer Science+Business Media New York 2014

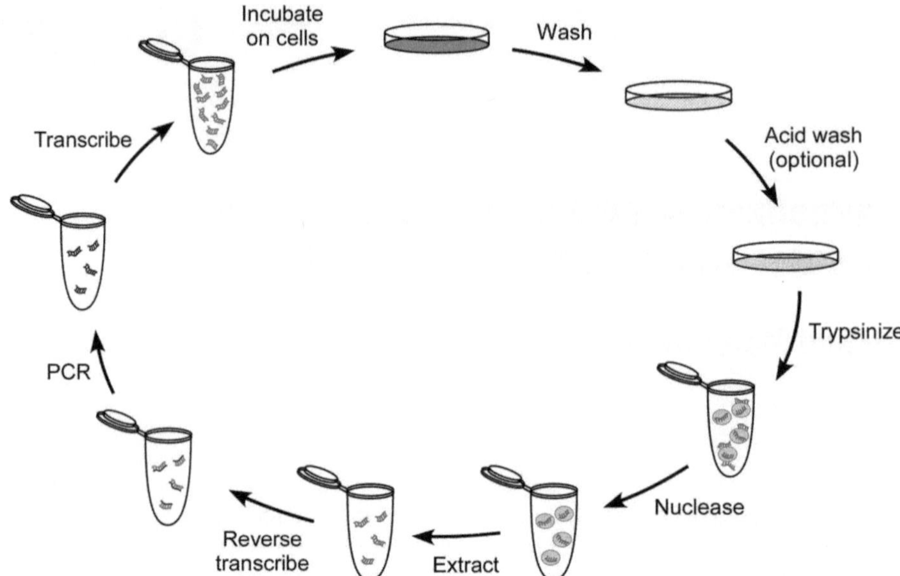

Fig. 1 The internalization selection. To select for molecules that will bind to and internalize into cells, an RNA pool is first incubated on cells. After several washes, an optional acid wash step and trypsinization, the cells are incubated with Riboshredder, a potent nuclease cocktail capable of cleaving 2′-fluoro-modified RNA thereby removing externally bound RNA molecules. Internalized RNA molecules are extracted along with total cellular RNA and re-amplified to generate an enriched pool for the next round. After many rounds of selection, aptamers that internalize into cells can be identified. Negative selection steps can be added to confer specificity of the aptamers for given target cell types

cells but also directly internalize into them. Such a procedure would bypass the need to identify a suitable cell-surface target and could find great utility in targeting cells without defined cellular markers. This chapter outlines a generic protocol for preparation of the starting pool, the internalization selection itself (Fig. 1), and, finally, one way to assess the success of the selection. As these methods build upon general selection techniques, readers may also wish to consult similar protocols [9–14]. Also, because the methods associated with basic tissue culture techniques and pool design and synthesis [15, 16] have been well described elsewhere, this protocol assumes that a pool has already been made and characterized and that cells and cell culture techniques are already in place.

A great number of factors and options are available for a selection. For simplicity, a 2′-fluoro-modified RNA pool selection against HeLa cells will be used as an example. However, the protocol should apply to any number of other pool types, cell lines, buffers, and variations. For example, we have successfully used this technique to identify an aptamer which is capable of being internalized by numerous cell lines [13], while Xiao et al. report using a similar approach to identify 2′*O*-methyl-modified aptamers which can distinguish prostate cancer cells from normal prostate cells [17]. Additionally, this approach can be used to aid in the identification of aptamers to specific cell targets. For example, by combining four rounds of a traditional in vitro selection against recombinant human transferrin receptor with a single round of internalization selection, we were able to identify aptamers which robustly bind this cell-surface receptor and are endocytosed by cells [14]. An internalization selection approach has also recently been used to identify aptamers which bind to rat HER2. For this, the authors combined negative selections against a cell line which did not express this receptor with positive, internalization selections against an engineered cell line expressing high levels of this protein [18].

Finally, as a point of precaution, we note that the ultimate fate and localization of the internalized RNA selected through this approach cannot be predicted and must be determined emperically, as different binding sites could lead to significantly different pathways. For example, this approach does not guarantee RNA access to the cytosol, as molecules will likely be retained in endosomes.

1.1 Planning Your Selection/Factors to Consider

Because of the inherent uncertainty involved in cell internalization selections, great care must be taken in defining the parameters and conditions used in the selection. Variations in some key factors could have a significant effect on the types of molecules that are eventually selected. We recommend reading through the protocol and the accompanying notes before starting. Even so, it is still possible that some cell types or conditions may not yield internalizing aptamers or may require many rounds of selection before a desirable product emerges.

1.2 Choosing Which Buffers to Use

Generally, the pool is heated and allowed to cool before using in a selection round to allow molecules to refold. Most sterile RNase/DNase-free buffers should be suitable for this step, but downstream applications may dictate which to use. For example, for molecules that may eventually be used in vivo, a buffer that reflects the delivery solvent might be chosen (e.g., PBS or HBSS). For the purpose of this protocol, we will use PBS (PBS without Mg^{2+} or Ca^{2+}) as the refolding buffer.

1.3 Amount of Pool to Use

The amount of pool to use in the initial round of selection should be based on the complexity of the pool. Ideally, the first round should make use of the best representation of as many unique sequences as possible. The number of sequences that represents the most amplifiable unique species is sometimes referred to as the pool genome. Thus, the first round should make use of at least one genome's worth of pool RNA; often, a few copies are recommended to prevent loss of unique sequences during handling. For a thorough discussion of pool design, synthesis, and complexity, see Piasecki et al. [16] and Hall et al. [15].

Subsequent rounds can use substantially less RNA, as pool complexity will be significantly reduced already, and relatively modest amounts of the amplified pool could represent all the species of the pool. A good starting point might be to drop the input sequence number by tenfold, or down to approximately 1–5 μg RNA. However, the amount to use each round can be partially based on practical parameters, such as the yield of transcriptions.

1.4 Blocking Agents

Cells can bind nucleic acids nonspecifically. We find that adding blocking agents in the form of ssDNA and tRNA to cells before and/or concurrently with the addition of pool can significantly reduce background binding and uptake. Levels of nonspecific binding and uptake vary between different cells, so it may be beneficial to assess background levels before performing a selection. This is most easily accomplished by flow cytometry using a fluorescently labeled naive library as described in Subheading 3.5 below.

1.5 Negative Selections

The protocol below describes the selection of molecules that internalize into any cell type. If cell specificity by the selected aptamer is desired, negative selection steps should be incorporated. That is, cell selections can be tailored such that the selected aptamer can recognize one cell type but not another. In this approach, the pool is first exposed to the undesired cell type. After incubation, the media from this negative cell type is transferred to the positive cell type. In this way, molecules that have bound to and internalized into the undesired cell type are removed before exposure to the desired cell type. Traditionally, the first round does not include any negative steps to allow for exposure to the maximal complexity of the pool, and negative selection steps begin at the second round.

Negative selections have been used to produce aptamers that could distinguish rat glioblastoma cells from normal brain cells [19], differentiated PC12 cells from parental cells [20], B-lymphocytes from T-lymphocytes [21], and more. Details for performing negative selections are below (*see* Subheading 3.4).

1.6 Adding Stringency

The core premise of the iterative rounds of this selection is to winnow away nonbinders and less robust internalizers at each round. To abet this process, selection rounds can include added levels of stringency at each round. Stringency can be introduced by reducing the number of target cells, increasing the number or volume of washes, increasing the amount of blocking agents, decreasing the incubation time of pool on cells, increasing the number of negative selection steps, or decreasing the concentration of library added to the cells in a given round. As noted in Subheading 1.5, the first round typically does not include any added stringency to maximize exposure to all possible species in the pool. Stringency can be added gradually as the selection progresses and can be partially based on the number of rounds the selection has undergone and how quickly the pool re-amplifies after a round or based on functional assessment for binding and uptake as determined by flow cytometry (*see* Subheading 3.5 below).

1.7 Contamination

For selections making use of 2′-fluoro- or 2′-*O*-methyl-modified RNA or ssDNA, nucleases are not as much of a concern as with canonical RNA. However, care should be taken to not introduce any possible nucleases either directly by the researcher or through poor solution maintenance.

A bigger concern for selection labs is the emergence of contaminations by pool molecules themselves. The high number of amplification events over the course of a selection sometimes allows for takeover by errant amplicons that settle in solutions. This may make itself apparent as a product in the (−) RT PCR (*see* Subheading 3.3). A "solutions control" PCR can also be run to diagnose this, wherein all the components of the PCR except for the template are cycled.

To reduce the chances of amplicon contamination, sterile and clean techniques should be adopted at all times, and tubes should remain closed as often as possible. All solutions should be briefly spun down in a benchtop nanofuge before opening to reduce aerosolization or inadvertent transfer by gloves of the molecules. It is also strongly recommended that all solutions for the selection should be divided into aliquots before starting the selection, with one aliquot as the working stock. In this way, if the current stock becomes contaminated, it can be discarded without loss of the entire stock.

1.8 Archiving

It is also a good idea to store parts of each round throughout the progress of the selection. This serves the dual purposes of having material at different rounds of the selection for assaying later on and of being a backup point should something go wrong with the selection at some point. Typically, we archive 1/3 to 1/2 of the RNA recovered from cells following each round of selection. However, note that in order to maximize the number of sequences carried through the first round of selection, the recovered RNA from cells in this initial round is used in its entirety as any mistake requires restarting the whole selection (*see* Subheading 3.3).

2 Materials

2.1 Pool Preparation

1. Y639F T7 RNA polymerase (*see* **Note 1**).
2. 2′-Fluoro-NTP mix: 25 mM each of 2′-fluoro-CTP (Trilink, San Diego, CA, USA), 2′-fluoro-UTP (Trilink, San Diego, CA, USA), GTP, and ATP.
3. 10× Transcription buffer [22]: 400 mM Tris–HCl, pH 8.3, 10 mM spermidine, 0.1 % Triton X-100, 500 mg/mL PEG-8000, 250 mM $MgCl_2$.
4. Model V16 upright gel electrophoresis system with 0.75 mm spacer and 10-well comb set (Labrepco, Horsham, PA, USA).

5. 10× TBE (Tris-borate-EDTA) buffer: 890 mM Trizma base, 890 mM boric acid, 20 mM EDTA, pH = 8.0.
6. 8 % Denaturing polyacrylamide solution, filtered: 200 mL 40 % 19:1::acrylamide:bis-acrylamide (*see* **Note 2**), 7 M urea, 100 mL 10× TBE, dH_2O up to 1 L.
7. 2× Denaturing dye: 95 % formamide, 20 mM EDTA, pH 8.0, 0.1 % bromophenol blue.
8. 20 × 20 cm K6F 60 Å pore size normal phaseTLC plate with F254nm fluorescence indicator (Whatman-GE Healthcare, Piscataway, NJ, USA).
9. Handheld compact UV lamp (UVP, Upland, CA, USA).
10. 14 mL Screw top test tubes 16 × 100 mm (USA Scientific, Ocala, FL, USA).
11. 0.45 μm Ultrafree-MC spin filter (Millipore, Billerica, MA, USA).

2.2 Cell Culture

1. HeLa cells (ATCC, Manassas, VA, USA).
2. D10: DMEM media supplemented with 10 % fetal bovine serum (FBS).
3. D10 blocked media: DMEM media supplemented with 10 % FBS, 1 mg/mL tRNA, and 1 mg/mL ssDNA.
4. 0.05 % Trypsin in 0.53 mM EDTA.
5. DPBS without Mg^{2+} or Ca^{2+} (Invitrogen, Carlsbad, CA, USA).
6. 5 % CO_2 incubator at 37 °C and 99 % humidity.
7. 24-Well tissue culture treated plates.
8. 25 cm^2 Tissue culture treated flasks
9. (Optional) penicillin/streptomycin solution (*see* **Note 3**).

2.3 Internalization Selection

1. Acid Wash solution: 200 mM cold glycine, 150 mM NaCl, pH 4, chilled to 4 °C.
2. PBS + (DPBS with calcium and magnesium; Invitrogen, Carlsbad, CA).
3. Hank's Buffered Saline solution (HBSS) with 0.1 % sodium azide (NaN_3).
4. Riboshredder RNase cocktail (Epicentre, Madison, WI, USA).
5. Trizol (Invitrogen, Carlsbad, CA).
6. 2 mL Phase-lock tubes (5 Prime, Gaithersburg, MD, USA).
7. Nanodrop (Nanodrop, Wilmington, DE, USA).

2.4 Pool Amplification and Regeneration

1. 20 μM Forward primer.
2. 20 μM Reverse primer.
3. Taq polymerase (Thermo Scientific, Waltham, MA, USA).

4. dNTP mix: equimolar mix of dATP, dTTP, dCTP, and dGTP to a final concentration of 4 mM each (Thermo Scientific, Waltham, MA, USA).
5. Moloney Murine Leukemia Virus reverse transcriptase (MMLV-RT; Invitrogen, Carlsbad, CA, USA).
6. Nusieve 3:1 agarose (Lonza, Walkersville, MD, USA).
7. SB buffer: 5 mM sodium borate.
8. 20 bp O'RangeRuler DNA ladder (Thermo Scientific, Waltham, MA, USA).
9. 0.5 μg/mL ethidium bromide.
10. 6× Orange G gel loading dye (*see* **Note 4**): 0.15 % Orange G (Sigma Aldrich, St. Louis, MO, USA), 10 mM Tris–HCl, pH = 7.6, 60 mM EDTA, 60 % glycerol.

2.5 Assaying the Progress of the Selection

1. Fluorescently labeled 20 μM Reverse primer (*see* **Note 5**).
2. FACS buffer: 1 % bovine serum albumin (BSA) and 0.1 % sodium azide in HBSS.
3. FACS buffer with 5 μg/mL bisbenzimide.

2.6 Identifying Aptamers from the Selection

1. TOPO TA Cloning Kit for Sequencing (Invitrogen, Carlsbad, CA).
2. Luria broth (LB) plates with 50 μg/mL kanamycin.

3 Methods

3.1 Pool Preparation

1. Transcribe at least one genome's worth of double-stranded pool DNA (where one genome corresponds to the number of molecules that represents the complete complexity of the pool; *see* **Note 6**). We perform modified transcriptions using a T7 RNA polymerase prepared in our lab bearing the Y639F mutation (*see* **Note 1**). Alternately, modified RNA can be generated using the commercially available Durascribe T7 Transcription Kit (Epicentre, Madison, WI) following the manufacturer's protocol. We typically use 1 μg dsDNA template per 20 μL transcription reaction and scale up the number of reactions to encompass the total amount of DNA to be used. Per reaction, mix:

   2 μL 10× Transcription buffer.

   2 μL 2′-Fluoro-NTP mix.

   2 μL 0.1 M DTT.

   3 μL Y639F T7 polymerase.

   1 μg dsDNA pool.

   dH_2O up to 20 μL.

2. Let the reaction proceed at 37 °C from 4 to 24 h.
3. Near the endpoint of the transcription, assemble the glass plates as per manufacturer's instructions for polyacrylamide gel electrophoresis (*see* **Note 7**).
4. Set up polyacrylamide solution by mixing:

   25 mL 8 % Denaturing polyacrylamide solution.

   100 μL 10 % APS.

   25 μL TEMED.
5. Mix well and pour or pipet slowly into the assembled glass plates (*see* **Note 2**).
6. Insert a 10-well comb and allow the gel to polymerize completely (from 30 to 45 min).
7. Add 2 μL DNase I per 20 μL transcription to the transcription reaction and incubate for >15 min at 37 °C.
8. Add an equal volume of 2× denaturing dye to the transcription reaction.
9. Denature the RNA:dye mix at 70 °C for 3 min.
10. Remove the comb from the polymerized gel and rinse briefly with water to remove unpolymerized acrylamide and acrylamide that may have deposited near wells. Assemble the polymerized gel into the Model V16 electrophoresis system as per manufacturer's instructions and fill lower and upper chambers with 1× TBE.
11. Purge leaked urea from each well with a large volume pipet and load upwards of 75 μL RNA/dye mix per well (*see* **Note 8**).
12. Electrophorese until the bromophenol blue nears the bottom of the gel.
13. Cover the TLC plate with several layers of plastic wrap, making sure the wrap lays flat.
14. Disassemble the gel apparatus and separate the two glass plates. If the glass has been properly treated (*see* **Note 7**), the gel should stick to the untreated plate.
15. Turn the plate over close to the covered TLC plate with the gel facing down. Begin at one corner with a clean spatula or finger and separate the gel from the glass. Gravity should peel the gel off the plate and onto the TLC plate.
16. UV shadow the gel by briefly shining UV from the handheld lamp over the gel on the TLC plate (Note: wear UV eye protection). RNA will be seen as a dark spots on the otherwise bright surface (*see* **Note 9**).
17. Use a clean razor blade to excise the RNA from the gel. Cut close to the dominant species while avoiding lower-sized species that may be seen.

18. Transfer the gel pieces to clean plastic wrap or parafilm. Cut the gel into smaller pieces if desired, and transfer the pieces to a 14 mL screw top tubes. Depending on the size of the transcription, multiple tubes may be needed. Do not overfill the tubes with gel chunks in order to allow for movement of the gel pieces.
19. Elute RNA out of the gel pieces by adding 6 mL 0.3 M NaCl to each tube. Incubate overnight with agitation or rotation (*see* **Note 10**). In later rounds, when lower volume transcriptions are prepared, smaller tubes and volumes of precipitation reagents may be used.
20. Transfer eluate from each tube to fresh tubes at 3 mL per tube.
21. Add 1–2 μL glycogen and 7.5 mL 100 % ethanol to each tube (*see* **Note 11**).
22. Mix well, and incubate at −80 °C for >15 min or −20 °C for several hours.
23. Precipitate the nucleic acid by centrifuging at >17,000 × *g* for 30 min.
24. Remove supernatant, taking care not to disturb the pellet.
25. Rinse the pellet by adding 1 mL 70 % ethanol and centrifuging for an additional 15 min.
26. Remove supernatant and air dry pellets or dry in vacuum chamber (*see* **Note 12**).
27. Resuspend the dried pellets in an appropriate volume of sterile RNase/DNase-free water (*see* **Note 13**). Resuspending pellets in the same volume as the transcription reaction often provides a good working concentration of RNA for the next round.
28. During this process, there may be small amounts of acrylamide that have co-precipitated with the RNA. Remove insoluble debris by transferring the RNA solution to a 0.45 μm Ultrafree-MC tube and spinning at 10,000 × *g* for 2 min. Transfer the RNA solution to a new tube.
29. Quantitate the RNA by Nanodrop (or other UV spectrometer) and calculate the number of genomes that has been transcribed. Determine the number of genomes to use for the selection (*see* **Note 14** and Subheading 1.3).

3.2 Cell Internalization Selection, First Round

1. Seed 50,000 cells in a well of a 24-well plate in 1 mL D10. Allow the cells to grow overnight at 37 °C with 5 % CO_2 and 99 % humidity.
2. Check the cells to make sure they are healthy. Remove the media and add 270 μL blocked D10 media to the well and return the plate to the incubator.
3. Mix the starting library in refolding buffer (PBS or other chosen buffer; *see* Subheading 1.2) with 1.5 M excess of reverse

primer, if desired (*see* **Note 15**), and denature at 70 °C for 3 min. Allow the pool to slow cool to room temperature (>15 min.). The volume of this solution should be held to 1/10 the volume of media/buffer in cells (i.e., 30 μL total for this example).

4. Add the library to the cells and mix by rocking the plate back and forth in multiple directions or by gentle pipetting and incubate for 1 h.
5. Remove the media from the well and wash three times with 1 mL HBSS + 0.1 % NaN_3.
6. Add 1 mL cold Acid Wash solution to cells and incubate no more than 10 s (*see* **Note 16**). Remove immediately and wash cells three times with 1 mL HBSS + 0.1 % NaN_3.
7. Add 250 μL trypsin solution to the cells and return to the incubator.
8. When cells have completely lifted off the substrate, inactivate the trypsin by adding 1 mL HBSS + 0.1 % NaN_3. Transfer contents to a 1.5 mL microcentrifuge tube.
9. Spin the cells at 300 × *g* for 5 min.
10. Gently resuspend the cells in 1 mL HBSS + 0.1 % NaN_3 and spin again at 300 × *g* for 5 min.
11. Gently resuspend the cell pellet in 100 μL HBSS and add 5 μL Riboshredder nuclease cocktail.
12. Mix gently by pipetting and incubate for 15 min at room temperature.
13. Wash the cells three times by gentle resuspension in 1 mL HBSS + 0.1 % NaN_3 and spinning at 300 × *g* for 5 min each time.
14. To lyse cells, resuspend the cell pellet in 500 μL Trizol. Mix well by pipetting or vortexing, and allow the cells to sit at room temperature for 5 min.
15. Centrifuge a 2 mL phase-lock tube at 7,000 × *g* for 2 min to spin down the wax.
16. Transfer the Trizol solution to the phase-lock tube.
17. Add 100 μL chloroform to the tube and mix by vigorous shaking for 15 s. DO NOT vortex the tube.
18. Centrifuge at 12,000 × *g* for 10 min at 4 °C. The solution will distribute into two phases: a lower, organic, pink phase separated by the waxy interphase from an upper, clear aqueous phase.
19. If the aqueous solution appears cloudy or if the phases are not distinctly separated, transfer the aqueous phase to a clean, pre-spun phase-lock tube and re-extract with another 100 μL chloroform. Transfer the aqueous upper phase to a clean microcentrifuge tube.

20. Add 1–2 μL glycogen and 250 μL isopropanol and incubate for 10 min at room temperature.
21. Centrifuge at 12,000 × *g* for 10 min at 4 °C.
22. Remove the isopropanol and wash the pellet by adding 1 mL 75 % ethanol and centrifuging again at 12,000 × *g* for 5 min at 4 °C.
23. Remove the ethanol and either air dry the pellet or dry under vacuum.
24. Resuspend the pellet in 50 μL dH_2O (*see* **Note 13**). This is the enriched product from the first round of selection, or the Round 1 or R1 selected RNA.
25. Quantitate the recovered RNA by UV spectroscopy (e.g., Nanodrop). The amount of total RNA recovered will vary with cell type and is no measure of the actual amount of pool recovered, but the value can be useful to ensure consistent recovery of material from round to round and to check for the absence of phenol contamination, as indicated by an absorbance peak centered at 270 nm (*see* **Note 17**).

3.3 Pool Amplification and Regeneration

In a typical selection in our lab, half of the product from each selection is used to make the pool for the next round, and the other half is archived. This ensures that should anything go amiss in later rounds, we can return to an earlier round of selection and try again. As the only backup point for the first round is to restart the entire selection anyway, we use the entirety of the Round 1 selected RNA to generate the pool for Round 2. As such, the following steps are for the first round regeneration only. Subsequent rounds should use half the volume of reagents for the reverse transcription. For example, for the R2 selected RNA, only use 22.5 μL instead of 45 μL of the RNA in the reverse transcription reaction and scale the entire reaction to half. For the PCR reactions, continue to run 100 μL reactions, but use 5 μL of the reverse transcription reaction instead of 10 μL for the PCR steps. No more than 5 μg of total cell RNA should be added per 20 μL reverse transcription reaction, and reactions should be scaled accordingly.

Once the enriched RNA has been converted to cDNA through reverse transcription (RT), the cDNA is amplified by PCR to produce enough template for transcription. Given the possibly low number of molecules that survive each round of selection, quantitation of the enriched RNA or cDNA is often difficult or requires so much of the material that diversity is depleted. How large or how many cycles to run a PCR, therefore, presents a problem. Too few cycles or too small a PCR results in low copies of enriched sequences or not enough template for transcription. Too many cycles or overloading of template can yield amplification artifacts or over-PCR products that have the potential to confound a selection.

To optimize PCR parameters, a cycle course using a small portion of the RT reaction is performed to determine how many cycles of PCR are necessary to amplify the enriched pool to maximum yield before over-PCR products appear. Once the cycling conditions are determined, the remaining cDNA can be amplified in a large-scale reaction. A reverse transcription and cycle course can be performed as follows:

1. For the reverse transcription, mix together 45 μL of the enriched RNA with 9 μL dNTP mix and 4.5 μL 20 μM reverse primer and heat to 65 °C for 5 min. Then cool on ice for 5 min.
2. Add 18 μL 5× First Strand Buffer and 9 μL 0.1 M DTT.
3. Remove 9 μL of the reaction to a separate tube. This serves as the (−) RT control reaction (*see* **Note 18**).
4. Add 4 μL MMLV-RT to the remaining reaction to yield an 80 μL RT reaction. This will be the (+) RT reaction.
5. Incubate both (−) RT and (+) RT at 37 °C for 50 min, followed by incubation at 70 °C for 15 min to inactivate the RT.
6. Set up a two separate PCR reactions for the cycle course, each with the following components:

   70 μL dH_2O (use 75 μL for subsequent rounds).

   10 μL 10× PCR buffer.

   5 μL dNTP mix.

   2 μL 20 μM Forward primer.

   2 μL 20 μM Reverse primer.

   1 μL Taq polymerase.
7. Add 10 μL (+) RT reaction to one reaction and 10 μL (−) RT reaction to the other tube (use 5 μL of each should be used in subsequent rounds).
8. Divide the (+) RT PCR into five 0.2 mL tubes.
9. Aliquot 20 μL (−) RT reaction to a separate 0.2 mL tube. Store the remainder on ice in case the cycle course needs to be repeated.
10. Run all five of the (+) RT PCR tubes and the one aliquot of the (−) RT PCR under PCR conditions appropriate for the pool and primer in a thermocycler. Choose the number of cycles to run the PCR and which cycles to analyze. It is wiser to run fewer cycles than more, as any reaction without product can be further cycled, but once a reaction has over-PCR'd, it cannot be resurrected. Therefore, for the initial round when a lot of pool may have nonspecifically carried through, perhaps start by choosing lower cycles to analyze such as 6, 8, 10, 12,

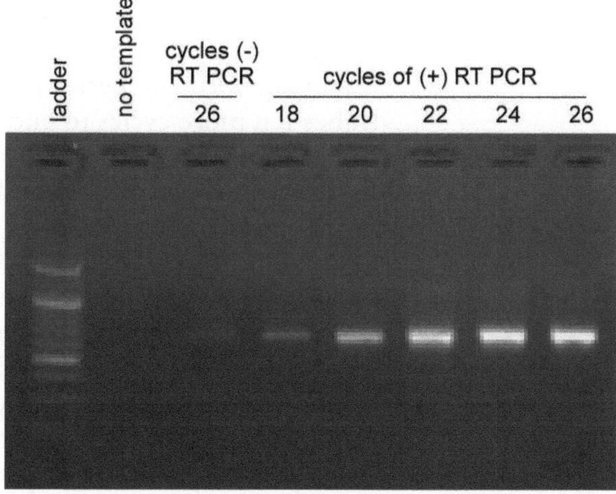

Fig. 2 Cycle course. An example cycle course is shown demonstrating the increase in PCR product with an increase in thermocycling. A (−) RT product appears late, as expected, >8 cycles behind the (+) RT reactions. This indicates that the product in the (+) RT reaction is due to amplification of the RNA and not due to carryover DNA. No product in the no template solutions control verifies that none of the solutions used in the PCR have been contaminated. Data provided by Keith Maier

and 14. Typically, these PCRs involves several cycles of some variation of:

94 °C for 30 s.

Annealing temperature from 45 to 65 °C (depending on the primers) for 30 s.

72 °C for 1 min.

11. At the chosen analysis cycles during thermocycling, remove one aliquot of the (+) RT reaction from the thermocycler once the cycle nears completion of the 72 °C step and store the tube on ice. Caution: only remove the tubes towards the end of the 72 °C step (*see* **Note 19**). Let the (−) RT aliquot run to the maximum number of cycles (*see* **Note 20**).
12. Pour an agarose gel appropriate for resolving the pool size from potential artifacts. For most pools currently in use (ranging from 60 to 150 bp), a 4 % 3:1 agarose gel with a nucleic acid stain (e.g., ethidium bromide) running in SB buffer is recommended.
13. Mix 5 μL of each PCR aliquot with 1 μL 6× Orange G dye and load onto the gel with a ladder. It is helpful for analysis to load the samples in order by cycle to monitor the progress of the PCR.
14. Analyze the gel to see which cycle gives the maximum yield of the correct size pool product without over-PCR (Fig. 2). If no products are seen, and the reactions have been set up correctly,

then perhaps not enough cycles have been run. All the aliquots should be run the number of cycles to continue the course. For example, with the cycles proposed in **step 10** above, run all tubes ten more cycles to encompass cycles 16, 18, 20, 22, and 24. If some products are visible, but not bright, do not run the last one in the cycle course and re-run all the other aliquots. With the example above, the cycle 14 tube would not be run, and the next gel would load the same 14 cycle tube as well as newly recycled 16, 18, 20, and 22 samples. At some point, the (−) RT reaction will also start showing products. Often, this can be an artifact of the PCR and may be the wrong size. Other times, it may be due to carryover DNA (*see* **Note 18**). Make sure the (−) RT reaction products comes up >8 cycles past the (+) RT reaction cycle equivalent before proceeding. For example, if the (−) RT product at 14 cycles looks about the same brightness at the (+) RT product at 10 cycles (a difference of only 4 cycles), the enriched RNA contained DNA carryover or reagents have become contaminated by pool molecules, and steps should be taken to correct this before going ahead.

15. Once the cycle course is satisfactory, set up a large-scale version of the cycle course using the remaining RT reaction. In subsequent rounds, only half of the RT reaction should be used for the large scale, and the other half should be archived. Divide the reaction into 100 μL aliquots for thermocycling.
16. Thermocycle the large-scale PCR the chosen number of cycles to maximize yield.
17. Mix 5 μL from one tube of the large-scale reaction with 1 μL 6× loading dye and run the sample into an agarose gel to check the size and integrity.
18. Once the product has been confirmed, pool the aliquots of large-scale PCR into a tube and add 1–2 μL glycogen, sodium acetate to 0.3 M and 2.5× volumes 100 % ethanol and incubate at −80 °C for >15 min or −20 °C for several hours.
19. Precipitate the PCR product by centrifuging at >17,000 ×*g* for 30 min.
20. Remove the supernatant, taking care not to disturb the pellet.
21. Rinse the nucleic acid pellet by adding 1 mL 70 % ethanol and centrifuging for an additional 5 min.
22. Remove supernatant and air dry the pellet or dry in a vacuum chamber (*see* **Note 12**).
23. Resuspend the dried pellet in 1/20 the volume of the large-scale PCR in H_2O or 20 mM Tris buffer, pH = 7.5 (*see* **Note 13**). Because the precipitated PCR contains nucleotides and primers, the actual amount of PCR product present cannot be determined

by Nanodrop. Quantitate the dsDNA by comparison with known mass standards on an agarose gel (*see* **Note 21**) or after purifying by PCR clean-up kits such as the QIAquick PCR Purification Kit (QIAGEN, Valencia, CA, USA) followed by Nanodrop.

24. Set up a transcription as described above (*see* Subheading 3.1, **step 1**). Because the pool has now been amplified by PCR, it now contains multiple copies of each sequence in the library. We typically use 1/3 to 1/2 of the total recovered dsDNA as our input for transcription, even if this will yield more RNA than needed for the next round to prevent bottlenecking of the population. Be sure to use at least 1 μg template per 20 μL reaction, as lower input of template results in significantly reduced yield. Scale the reactions accordingly.
25. Allow the transcription reaction to proceed at 37 °C for >4 h to overnight. We have found that yield improves with longer incubations.
26. Purify the pool RNA as described in Subheading 3.1, **steps 2–27**.
27. Repeat all steps from Subheading 3.2 for Round 2 and further rounds. Remember to scale down the reactions and input (*see* Subheading 1.3) and to store portions of each round as described in Subheading 1.8 and at the beginning of Subheading 3.3. If desired, add stringency as described in Subheadings 1.6 and 3.5 below or include negative selection steps as described in Subheading 3.4.

3.4 Negative Selections

If a baseline cell is available, and a negative selection step is desired, replace **steps 1–4** of Subheading 3.2 with the following.

1. Seed 50,000 negative and positive cells each in separate wells in a 24-well plate in 1 mL D10. Allow the cells to grow overnight at 37 °C with 5 % CO_2 and 99 % humidity.
2. Check the cells to make sure they are healthy. Remove the media from the negative cells and add 250 μL blocked D10. Return to the incubator.
3. Mix the starting library in the chosen selection buffer and denature at 70 °C for 3 min (*see* **Note 15**). Allow the pool to slow cool to room temperature (>15 min.)
4. Add the library to the negative cells and mix by rocking the plate back and forth in multiple directions.
5. Remove the media from the positive cells and add 250 μL blocked D10.
6. Incubate the plate for 1 h.

7. Remove the media from the positive cells and immediately transfer all the media + pool from the negative cells to the positive cells.
8. Incubate 1 h.
9. Proceed from Subheading 3.2, **step 5** with washing of the cells.

3.5 Assessing the Progress and Success of the Selection

A selection can take many rounds before successful aptamers are readily detectable. How many rounds are required varies, as each selection is different.

If the same amount of pool is used for each round (other than the initial round), a rough approximation of how many molecules are surviving the round can be made by comparing the number of PCR cycles needed for product to reach a certain mass. For example, say that the same mass of pool is used to initiate Rounds 3 and 4 of a selection and that no conditions are changed between the two rounds. If the resulting pool from Round 3 that initiates Round 4 is more enriched for internalizers than the pool that initiated Round 3, it would follow that more pool would be recovered from cells during Round 4, and the cycle courses between the rounds should reflect this. That is, the PCR product from Round 4 should appear some number of cycles earlier than that for Round 3.

This cycle course comparison can be used to approximate enrichment and, in some cases, judge whether greater stringency needs to be applied. However, this should not be used as a gauge for selection progress or success, as slight changes in conditions could also affect the cycle courses. Instead, the selection should be assayed at various points during the selection. A good range to perform assays might be every three or four cycles.

One method of assaying the selection rounds is detailed in the flow cytometry protocol below. Other methods may also be used (*see* **Notes 22** and **23**); however, in our experience, the flow cytometry analysis is relatively straightforward and reliable. Note that the protocol below will assess cellular uptake and is performed at 37 °C. Variations of the experiment can be performed in which individual steps are omitted (e.g., acid wash, Riboshredder) or altered (substituting treatment with EDTA to lift the cells instead of trypsin) to assess the effects of each step. Additionally, experiments can be performed at 4 °C which inhibits internalization via most cellular pathways. Assays may also be performed on trypsinized or cells lifted with EDTA to assess levels of cell binding.

1. The day before the assay, determine the number of experimental samples to be tested. Seed 50,000 cells per well using a 24-well plate in 500 μL media.
2. The following day, mix 1 μM round RNA for each round to be assayed (include Round 0) with 1.5 μM fluorescently labeled

reverse primer (*see* **Note 5**) in a total volume of 20 μL PBS or other selection buffer.

3. Heat RNA solutions to 70 °C for 3 min. Then allow the tubes to slow cool to room temperature.
4. Remove the media on the cells to be assayed and replace it with 180 μL of D10 blocked media.
5. Add 20 μL cooled RNA solution to the cells and mix by rocking the plate back and forth.
6. Incubate 1 h at 37 °C with 5 % CO_2 and 99 % humidity.
7. Wash 3× with 1 mL FACS buffer, 1× with 1 mL PBS and then trypsinize to lift the cells by adding 150 μL of trypsin followed by incubation at 37 °C for ~5 min (cells undergoing acid wash should be acid washed prior to the addition of trypsin).
8. Once the cells have released from the plate, quench the trypsin reaction by the addition of 1 mL FACS buffer (if desired, treat samples with Riboshredder after transferring cells to a microcentrifuge tube).
9. Centrifuge at 300 × *g* for 5 min.
10. Resuspend cells in 500 μL FACS buffer with bisbenzimide and transfer contents to FACS tubes.
11. Run samples on flow cytometer to assess degree of uptake (Fig. 3).

3.6 Identifying Aptamers from the Selection

Once an acceptable degree of internalization has been achieved or further rounds of selection fail to increase the signal as observed by flow cytometry or other analysis method, the sequences of individual aptamers/internalizing RNA can be determined by cloning and sequencing. The TOPO TA Cloning Kit is particularly useful for this application, as the PCR product from a given round can be cloned directly into the TOPO plasmid following manufacturer's instructions and then submitted for sequencing after purification or colony PCR.

Typically, anywhere from 20 to 100 individual sequences are sequenced and compared by alignment programs. Sequencing data can be sorted by families (consisting of sequences very similar to each other) and by frequency (how many times the sequence appears). Often, "winners"—those sequences that have dominated the selection—appear multiple times in the sequencing data, although multiple occurrences in the population do not in and of itself guarantee a functional sequence. At a minimum, sequences representing the dominant clones or containing the dominant motifs in the selection should be synthesized and tested individually for binding and internalization and compared to the round RNA from which it came and the Round 0 pool. However, we and others have found that often, the "odd-ball" sequence can prove

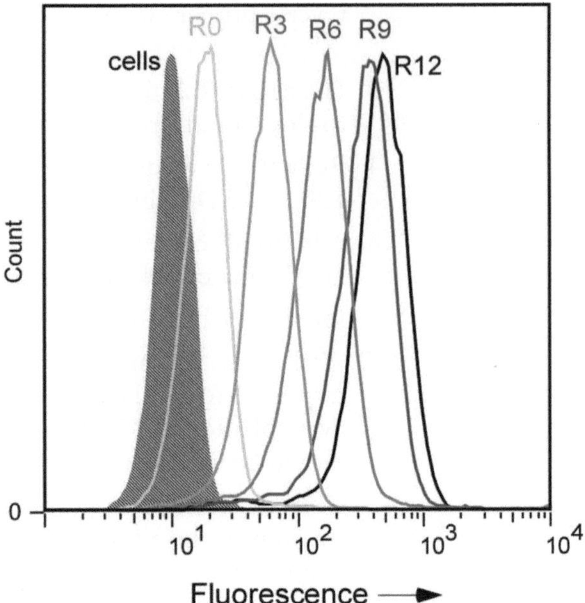

Fig. 3 Flow cytometry assay. An example of what a flow cytometry assay of a cell selection might look like is shown. Because of background binding and uptake, Round 0 (R0) will likely give some degree of signal and should always be included as a control in these assays. As the selection progresses, and, presumably, the pool becomes more enriched for internalized molecules, a concomitant increase in fluorescence signal can be detected with each round's pool RNA. In this example, the cells incubated with Round 12 (R12) RNA does not appear significantly more labeled than that of Round 9 (R9), suggesting no further improvement of the pool. At this point, stringency can be added to tease out further activity, or individual aptamers can be isolated and tested for activity. Data provided by Keith Maier

functional, sometimes more so than the dominant clones, and so it is advisable to test as many clones as possible.

Individual clones can be amplified from the plasmids or PCR products which were prepped for cloning and sequencing using the pool primers. For this, we typically recommend performing at least a 400 μL PCR reaction containing ~50 ng of plasmid input or 10 ng of PCR input and cycled ~14–18 cycles (~10 cycles if using clean PCR input) using conditions appropriate for the pool. The PCR reactions should be checked on an agarose gel to ensure efficient amplification. Precipitate the PCR reaction, resuspend in 20 μL dH_2O, and proceed with dsDNA quantitation and transcription as described above in Subheading 3.3, **step 23**. Following gel purification, the purified RNA templates can be assayed as described in Subheading 3.5.

4 Notes

1. *Y639F T7 polymerase.* The Y639F mutant of T7 RNA polymerase [6] allows for incorporation of dNTPs into RNA transcripts and is, therefore, useful for incorporation of modified nucleotides such as the 2′-fluoro-pyrimidines used in this example protocol. The Y639F used by our lab is made in the laboratory, but a commercial variant, R&DNA, is available (Epicentre, Madison, WI, USA). Alternately, the Durascribe T7 Transcription Kit (Epicentre, Madison, WI, USA) can be used.
2. *Percentage of acrylamide to use.* The concentration of acrylamide should be chosen to best resolve the pool from aborted transcripts or alternate products. For most pools in the literature (which have random regions ranging from 30 to 100 bases), an 8 % acrylamide gel is sufficient. If pools are significantly shorter or if heterogeneous transcriptions are a problem, a higher percentage gel might be chosen to better isolate the correct product.
3. *Antibiotics in cell culture.* Most cell culture protocols incorporate the use of penicillin/streptomycin solutions to keep contamination at bay. These should not interfere with the selection and may not be a bad idea, as pool and selection components may not necessarily be sterile during preparation. However, we do not include antibiotics in our cell culture protocol, as some of the cells are used for injection in vivo, and a subdued contamination in tissue culture could confound in vivo studies.
4. *Dyes for agarose electrophoresis.* Many pools used for selections (roughly 60–150 bp) co-migrate near bromophenol blue, one of the more standard dyes for electrophoresis, which runs at approximately 200 bp on 2 % agarose gels or 100 bp on 3 % agarose gels. Orange G runs smaller than bromophenol blue and well below most pools, thus not interfering with visualization.
5. *Choice of fluorescent label for flow cytometry.* The choice of fluorescent label used for flow cytometry will depend on the lasers available on your flow cytometer. In general, most cytometers have a 488 nm laser, a 633/642 nm laser, or both. For excitation with a 488 nm laser, we prefer to perform assays using an Alexa Fluor 488 labeled primer. While assays can also be conducted using the less expensive FAM- or FITC-labeled primer, the fluorescence of fluorescein is significantly diminished at $pH < 7$, which can negatively affect results, especially in acidified endocytic cell compartments. For excitation with the 633/642 nm laser, we recommend Alexa Fluor 647 or Cy5. We have found some dyes to show much stronger levels of nonspecific background binding than others. This can vary with cell type. Therefore, in addition to always running a

Round 0 RNA or nonbinding aptamer control, we also recommend running a fluorescent primer only control.

Perhaps the best and strongest signal for flow cytometry can be observed using the fluorescent protein phycoerytherin (PE), which is excited off of the 488 nm laser. This can be readily appended to aptamer libraries or individual clones using a biotinylated primer and a streptavidin–PE (SA–PE) conjugate [13, 23]. Additional care should be taken in setting up experiments with SA, as streptavidin itself is a tetrameric protein and capable of binding up to four biotinylated aptamers, and SA–PE conjugates are often multimeric, containing multiple SA (tetramers) and PE proteins. A monomorphic SA–PE conjugate is available from Prozyme (Hayward, CA, USA) which contains ~1 SA (tetramer) and ~1 PE molecule. When using SA–PE conjugates, we use a 1:1.5:1.5 M ration of aptamer:primer:SA–PE.

As with the fluorescently labeled primers described in Subheading 3.5, mix the aptamer or library sample with primer and heat to 70 °C for 3 min. Allow this to cool on the benchtop for 15 min and then add SA–PE followed by an additional 15 min incubation. Care should be taken with regard to interpreting the binding results, especially in making assessments of apparent binding affinity as SA–PE conjugates bearing multiple aptamers may show enhance binding due to avidity effects.

6. *Pool complexity and design.* The design and synthesis of pools appropriate for selections involves a discussion that is outside the scope of this chapter. For explanations and protocols related to understanding pool complexity, see Hall et al. [15] and Piasecki et al. [16].

7. *Preparing electrophoresis plates for pool purification.* To help the transfer of gels to the TLC plate for shadowing, we treat one side of the smaller plate used for electrophoresis with strong base and then a slicking agent. To do this, mark one side of the plate with permanent tape or other marking to indicate the non-treated side of a plate. Saturate approximately 30 mL isopropanol with potassium hydroxide pellets and shake well to mix. Pour a small amount on the unmarked side of the plate and rub with pressure in circular motions with a Kimwipe or paper towel, as if waxing or buffing. Keep pouring more solution and rubbing until the entire surface has been treated. Allow the isopropanol on the plate to dry, then wash the plate with soap and water and allow it to dry again. Treat the surface of the unmarked side with Gel-Slick (Lonza, Walkersville, MD, USA) by rubbing Gel-Slick on the entire surface until a faint haze is visible. Treat a second time. Then, rub away the haze with clean towels. Rinse briefly and dry before using. Alternately, we have found that Rain-X also works well. If the

smaller plate is treated, once the gel has transferred to the TLC plate, the orientation of the gel will be the same as when the gel was loaded, if the Model V16 system is used. Note: do not treat both plates.

8. *Urea seepage in polyacrylamide gels.* 7 M urea is used in these gels to help denature nucleic acids so that they run into the gel as true to their sizes as possible. Once immersed in buffer, the urea will start seeping out of the gel into the wells. This can be problematic, as the urea in the wells will occlude loading of sample. Once the gel is set up for running and before loading samples, the wells should be purged of urea by squirting out the wells thoroughly with buffer by pipet or syringe. This should be done immediately before loading the gel.
9. *Overexposure of nucleic acids to UV.* UV light can damage nucleic acids in a number of ways, and it is advisable to minimize exposure as much as possible. As cutting out a band from a gel should require no more than 4 cuts with a razor, we only flash the UV briefly on the gel to get quick glimpses of where to align the razor at each cut. The few glimpses taken together should be enough to gauge the quality of the transcription.
10. *Temperature of elution from gel chunks.* Protocols for elution of RNA from polyacrylamide gels suggest temperatures ranging from room temperature to 80 °C, with many recommending a 37 °C incubation overnight. In our experience, higher temperatures do seem to help elute nucleic acids if incubation times are short. However, for routine elution, an incubation overnight at room temperature on a rotator or other means of agitation is sufficient.
11. *Glycogen as a carrier for ethanol precipitation.* We regularly add glycogen to ethanol precipitations to act as a carrier and to help visualize the pellet after precipitation. If glycogen could interfere with your experiments, precipitation can be performed without it or an alternate carrier could be used such as tRNA.
12. *RNA pellets may dry clear.* Relatively pure pellets of RNA and DNA often dry clear and become hard to see, especially if carriers are not used in a precipitation. A pellet that isn't visible or seems to have disappeared may be seen as a slight outline once water is added (*see* **Note 13**). Regardless, even if the pellet isn't visible, proceed as if it were there. Centrifuging tubes in a certain orientation or marking the side of the tube where the pellet is expected may also help to locate the pellet.
13. *Resuspending nucleic acid pellets.* Nucleic acids are very soluble in water; however, when pellets are extremely dry, they may not immediately hydrate. Some protocols recommend resuspension of pellets by pipetting up and down. If a pellet is not fully hydrated and gets stuck to the pipet tip, a substantial

amount of the pellet could be unknowingly lost. To prevent this, we recommend adding water or the resuspension buffer to the tube at the site of the pellet and vortexing and spinning down with a nanofuge repeatedly to dislodge and hydrate the pellet.

14. *Quantitating how many genomes of RNA are produced.* It is important to calculate how many copies of a pool were made during the initial Round 0 transcription in order to know how much RNA should be used in the first round to represent as much diversity as possible. If one genome's worth of DNA went into the transcription, the amount of RNA produced can be calculated as follows:

    ($A_{260\ pool}$)/(extinction coefficient of pool) = concentration of RNA pool in mol/L

    (concentration of pool in mol/L) × (volume of RNA in L) × (6.022×10^{23} molecules/mol) = # molecules of RNA in transcription

    (# of RNA molecules in transcription)/(# of DNA sequences put in transcription) = # copies made = # genomes

    (volume RNA in μL)/(# genomes) = volume in μL to use for one genome's worth of RNA into the selection round.

15. *Blocking the constant region.* The random region of a pool has the potential to interact with the constant regions at either end. Also, some selections have yielded aptamers that end up requiring parts of the constant region to be functional. By pre-hybridizing the reverse primer to the pool before each round, the constant region is effectively blocked and is less likely to participate in potential binding activity. Especially if the flow cytometry assay described in the protocol will be employed, this blocking should be considered. That said, selections have been successfully performed without this addition, and assays can be performed by appending on a hybridization region post-selection.

16. *Acid wash of cells.* The 10 s acid wash step of this protocol was determined for HeLa cells, which are robust and tolerant of this treatment. Any other cell type should be first tested with the Acid Wash for different lengths of time to check for cell damage. This can be achieved by subsequently staining with trypan blue and inspecting the cells on a microscope or by treatment with a live/dead stain such as bisbenzimide followed by flow cytometric analysis.

17. *Phenol contamination of recovered RNA.* Phenol contamination following the Trizol steps can be spotted by spectroscopy as a peak centering around 270 nm. Excessive phenol may appear as the dominant peak in a UV spectrum, whereas lower

amounts of contamination may show up merely as a shoulder of the 260 nm peak. As phenol can have dire effects on reverse transcription, if phenol is suspected, it should be removed by performing an additional ethanol precipitation of the total recovered RNA.

18. *(−) RT reactions and the appearance of product in the (−) RT reaction.* (−) RT reactions are performed because of a small but inevitable amount of DNA that carries over in spite of the DNase step. If not outcompeted by nonspecific products, a product of the right size will likely eventually appear in a (−) RT reaction. More often, though, products appearing in the (−) RT reaction, especially after many PCR cycles, are of the wrong size and are likely attributed to nonspecific amplification artifacts. Even if the product is the right size, this may not be an issue as long as the product appears several cycles behind when the product appears in the (+) RT reaction, suggesting that most of the product in the cycle course is from the reverse-transcribed RNA and not carryover DNA. We recommend >8 cycles separating the appearance of product in the (+) RT reaction from the (−) RT reaction. If a right-sized product appears in the (−) RT PCR reaction within 8 cycles of the (+) RT PCR reaction, conditions should be reviewed to identify the source of template. Solutions used in the reactions should be assayed independently of the template to see if contamination has occurred or replaced by untouched aliquots. If the solutions are not a problem, pool RNA may be DNase treated again to ensure removal of all carryover DNA.

19. *Removing tubes at the 72 °C extension step.* During the cycle course, it is imperative that tubes are only removed from the thermocycler at the 72 °C extension step and towards the latter part of this step (≥45 s). This ensures that species in the tube are double stranded and full length. Removing the tubes at the annealing or denaturing step can lead to mixed populations that run at different apparent sizes on a gel.

20. *Reverse cycle course.* An alternate way to perform cycle courses involves putting tubes into the thermocycler at certain time points rather than taking tubes out. In this approach, the (−) RT and longest cycling (+) RT tubes are put into the thermocycler first for the cycling program, while the other tubes remain chilled on ice. At the chosen cycles to analyze, tubes are put into the thermocycler in the reverse order that they would have in the traditional cycle course. By the end of the cycling program, the last tube put into the machine has run the fewest cycles. In this way, if a 4 °C step is included at the end of the thermocycling program, the hands-on portion of the cycle course is completed early, and the completed reactions can be left at 4 °C until ready for analysis.

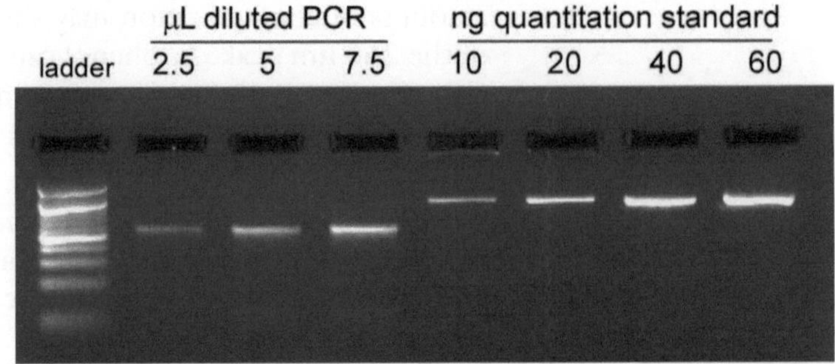

Fig. 4 Quantitation by gel electrophoresis. By running a dilution of the PCR into an agarose gel compared to known mass standards, an estimate of the concentration of the PCR can be made. In this example, 5 μL of the PCR sample appears to be of an intensity between 10 and 20 ng and can be approximated to be about 10 ng for the purpose of transcription. It is helpful to visualize the samples within a couple centimeters after running into the gel to minimize diffusion of the bands

21. *Gel quantitation*. Nucleic acid concentrations can be estimated by visualization in gel by comparison to mass standards. To do this, resuspend the large-scale PCR pellet into 1/20 the volume of the large PCR. Dilute 1 μL of this resuspended solution into 49 μL of resuspension buffer or water. Run 5 μL of this latter dilution on an agarose gel alongside mass standards. We typically load 10, 20, 40, and 60 ng of a 200 bp quantitation standard (Thermo Scientific). Run the samples approximately 1–2 cm into the gel before visualizing (once bands have migrated too far, they diffuse and make quantitation harder). An approximation of mass can be made by comparing the sample band to the quantitation standards (Fig. 4).
22. *Radiactive assay of rounds*. Radioactive assays are another way to assay uptake of pool RNA into cells. In this approach, pool can be radiolabeled during the transcription or end labeled, then incubated with cells as described in the protocol. After significant washing and nucleasing, cells can be trypsinized or scraped for scintillation counting. Round 0 controls should be included to assess improvement of pool or individual sequences over the course of the selection.
23. *Microscopy assays*. As with the flow cytometry assay described in the protocol, pool molecules can be fluorescently tagged by hybridization of a fluorescently labeled reverse primer sequence, incubated with cells as described above, then analyzed by microscopy. We do not recommend this for most selections, as microscopy data, unless performed correctly with the strictest of controls, can lead to false impressions of uptake.

Acknowledgement

This work was supported by a grant from Stand Up 2 Cancer (SU2C-AACR-IRG0809).

References

1. Ellington AD, Conrad R (1995) Aptamers as potential nucleic acid pharmaceuticals. Biotechnol Annu Rev 1:185–214
2. Famulok M, Mayer G (1999) Aptamers as tools in molecular biology and immunology. Curr Top Microbiol Immunol 243:123–136
3. Marshall KA, Ellington AD (2000) In vitro selection of RNA aptamers. Methods Enzymol 318:193–214
4. Yan AC, Bell KM, Breeden MM et al (2005) Aptamers: prospects in therapeutics and biomedicine. Front Biosci 10:1802–1827
5. Padilla R, Sousa R (2002) A Y639F/H784A T7 RNA polymerase double mutant displays superior properties for synthesizing RNAs with non-canonical NTPs. Nucleic Acids Res 30:e138
6. Sousa R, Padilla R (1995) A mutant T7 RNA polymerase as a DNA polymerase. EMBO J 14:4609–4621
7. Zhou J, Rossi JJ (2011) Cell-specific aptamer-mediated targeted drug delivery. Oligonucleotides 21:1–10
8. Ni X, Castanares M, Mukherjee A et al (2011) Nucleic acid aptamers: clinical applications and promising new horizons. Curr Med Chem 18:4206–4214
9. Cerchia L, Giangrande PH, McNamara JO et al (2009) Cell-specific aptamers for targeted therapies. Methods Mol Biol 535:59–78
10. Codrea V, Hayner M, Hall B et al (2010) In vitro selection of RNA aptamers to a small molecule target. Curr Protoc Nucleic Acid Chem Chapter 9: Unit 9.5 1–23
11. Hall B, Arshad S, Seo K et al (2009) In vitro selection of RNA aptamers to a protein target by filter immobilization. Curr Protoc Mol Biol Chapter 24: Unit 24.23
12. Knudsen SM, Robertson MP, Ellington AD (2002) In vitro selection using modified or unnatural nucleotides. Curr Protoc Nucleic Acid Chem Chapter 9: Unit 9.6
13. Magalhaes ML, Byrom M, Yan A et al (2012) A general RNA motif for cellular transfection. Mol Ther 20:616–624
14. Wilner SE, Wengerter B, Maier K et al (2012) An RNA alternative to human transferrin: a new tool for targeting human cells. Mol Ther 1:1–14
15. Hall B, Micheletti JM, Satya P et al (2009) Design, synthesis, and amplification of DNA pools for in vitro selection. Curr Protoc Mol Biol Chapter 24: Unit 24.22
16. Piasecki SK, Hall B, Ellington AD (2009) Nucleic acid pool preparation and characterization. Methods Mol Biol 535:3–18
17. Xiao Z, Levy-Nissenbaum E, Alexis F et al (2012) Engineering of targeted nanoparticles for cancer therapy using internalizing aptamers isolated by cell-uptake selection. ACS Nano 6:696–704
18. Dastjerdi K, Tabar GH, Dehghani H et al (2011) Generation of an enriched pool of DNA aptamers for an HER2-overexpressing cell line selected by cell SELEX. Biotechnol Appl Biochem 58:226–230
19. Blank M, Weinschenk T, Priemer M et al (2001) Systematic evolution of a DNA aptamer binding to rat brain tumor microvessels. Selective targeting of endothelial regulatory protein pigpen. J Biol Chem 276: 16464–16468
20. Wang C, Zhang M, Yang G et al (2003) Single-stranded DNA aptamers that bind differentiated but not parental cells: subtractive systematic evolution of ligands by exponential enrichment. J Biotechnol 102:15–22
21. Shangguan D, Li Y, Tang Z et al (2006) Aptamers evolved from live cells as effective molecular probes for cancer study. Proc Natl Acad Sci U S A 103:11838–11843
22. Milligan JF, Groebe DR, Witherell GW et al (1987) Oligoribonucleotide synthesis using T7 RNA polymerase and synthetic DNA templates. Nucleic Acids Res 15:8783–8798
23. Li N, Ebright JN, Stovall GM et al (2009) Technical and biological issues relevant to cell typing with aptamers. J Proteome Res 8: 2438–2448

Chapter 19

DNA Electronic Switches Based on Analyte-Responsive Aptamers

Jason M. Thomas, Hua-Zhong Yu, and Dipankar Sen

Abstract

Aptamers have proven to be very useful as high-affinity and -specificity molecular recognition elements in analytical sensors of various forms. Herein, we describe a general process for creating an aptamer-based sensor that functions as an analyte-responsive, nano-sized, electronic switch. These sensors can provide an electrochemical readout, by switching through-DNA charge transfer across a DNA three-way junction from "off" to "on" in response to the binding of a target ligand to the sensor's aptamer domain. We detail the general design principles for such sensors, as well as the biochemical charge transfer assays used to identify functional sensors. In these gel electrophoresis-based assays, analyte-responsive conductivity switching is detected conveniently through biochemical experiments that characterize oxidative DNA damage patterns in sequencing PAGE gels.

Key words Aptamers, DNA switches, DNA charge transfer, Biomarker detection, Analyte sensors

1 Introduction

DNA aptamers are robust and inexpensive receptors that can be easily selected in vitro (using SELEX methodologies) from random sequence libraries for binding to almost any type of target molecule (ligand, analyte), ranging from small molecules to biomacromolecules [1]. Selected aptamers usually bind their targets with both high affinity and specificity [1]; they have therefore attracted a great deal of attention in recent years for their utility as molecular recognition elements in new biosensor technologies [2–4]. Many aptamer-based sensor designs have been reported, that provide a variety of analytical readouts, most often based on changes in optical [2], fluorescent [2, 5], or electrochemical properties [6–8]. The operation of such sensors often depends on large-scale conformational changes of the DNA, such as global folding or secondary structure rearrangements [6–10]. For example, analyte binding-induced structural changes can be transduced into an optical readout by altering the distance between two fluorophores

Daniel Lafontaine and Audrey Dubé (eds.), *Therapeutic Applications of Ribozymes and Riboswitches: Methods and Protocols*, vol. 1103, DOI 10.1007/978-1-62703-730-3_19, © Springer Science+Business Media New York 2014

(in FRET-based experiments) or between a fluorophore and a quencher (in fluorescence quenching experiments). A conceptually similar approach has been successful for electrochemical aptamer-based sensors. Therein, analyte binding-induced aptamer structural changes cause a redox label, appended to the aptamer, to be drawn closer to an electrode surface upon which the aptamer is immobilized [6, 7]. The decrease in electrode-to-redox label distance leads to an increase in the rate of direct electron transfer between them, leading to increased electrochemical signal in the presence of the analyte.

Over the past few years, we have been developing a unique class of electrochemical aptamer-based sensors, whose function relies on the switching of through-DNA charge flow, rather than direct charge transfer, between a redox label and an electrode [11–15]. The "coupled ligand deoxyribosensor" (CLD) design, which we describe here, is based on a DNA three-way helical junction formed by attaching an aptamer domain to the junction between two conductive double-helical stems [called the reporter and detector stems, respectively (Fig. 1)]. Conformational changes within the aptamer domain, triggered by analyte binding, induce inter-helical π-stacking between the reporter and detector stems. Such inter-helical stacking enhances electrical contact and thus facilitates electron–hole transfer between the helices, allowing charge to flow through the three-way junction between the reporter and detector stems (*see* Fig. 1). We have found that even subtle conformational changes induced by analyte binding in the aptamer domain can trigger conductivity switching at the three-way junction [14]. In contrast to this, conformational changes and/or secondary structure rearrangements on a more global scale are required in systems designed to function on the basis of modulation of direct electron transfer between the electrode and the redox label tethered to the DNA [6]. In the case of CLDs, successful analyte-responsive conductivity switching of a three-way junction is first identified in solution, through convenient biochemical charge transfer assays (vide infra); functional sensors can then easily be adapted to an electrode-bound form in which conductivity switching is monitored as an electrochemical readout (a description of the electrochemical CLD format is beyond the scope of this article and is described elsewhere [13, 14]). To date, we have been successful in applying the CLD paradigm to create sensors for a wide range of analytes, including proteins [13, 14], small molecules [11, 12], and metal cations (namely Hg^{2+}) [15].

Herein, we describe the general design principles as well as the biochemical charge transfer experiments used to identify functional CLDs. With respect to the choice of aptamer to be incorporated into a CLD, three properties are of particular importance: (1) The minimized functional aptamer sequence should not exceed ~40 nt,

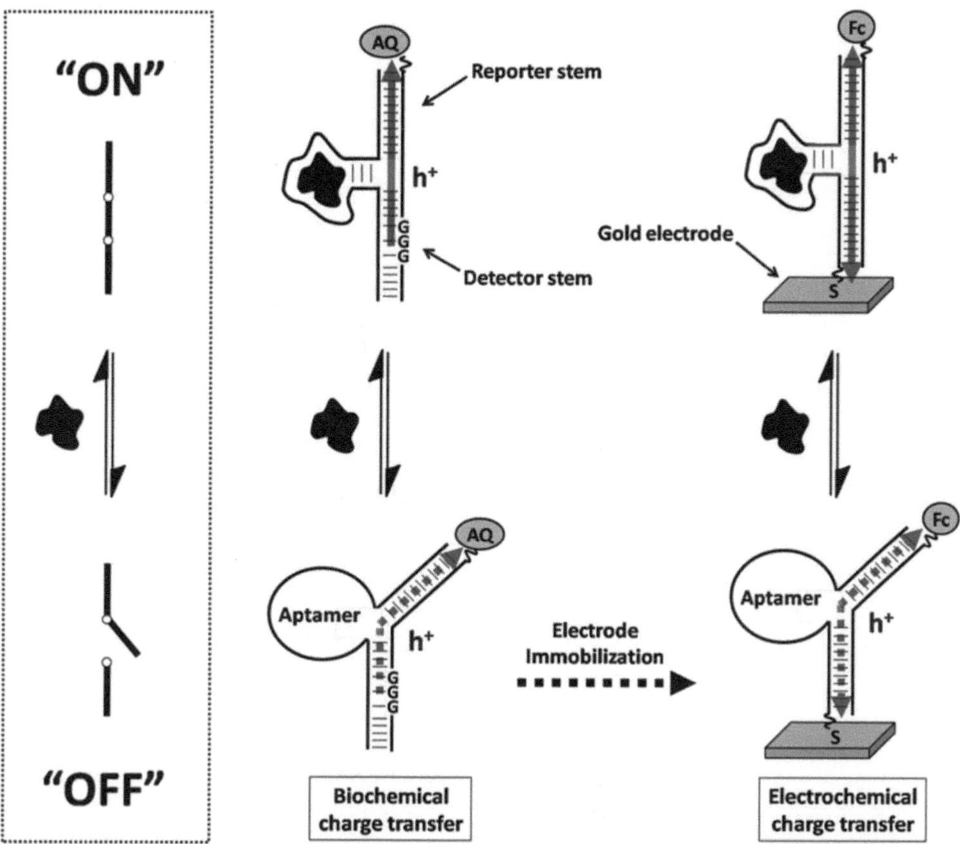

Fig. 1 Illustration of the function of coupled ligand deoxyribosensors (CLDs), in both biochemical and electrochemical charge transfer configurations. The reporter and detector stems comprise the through-DNA charge conduction pathway. Conformational changes that occur in the aptamer domain upon analyte binding induce π-stacking between the reporter and detector stem, so that the CLD functions as analyte-responsive conductivity switches. AQ = anthraquinone, Fc = ferrocene. Reprinted with permission from ref. 13. Copyright (2008): American Chemical Society

so that the CLD oligonucleotide that incorporates the aptamer domain is of a manageable length. (2) The aptamer domain must be joined to the CLD construct so as to form a three-way helical junction; therefore, the termini of the minimized aptamer should ideally form a short Watson–Crick paired helix [11, 12, 14]. If such a secondary structure element is not present in the aptamer, at the very least, the termini should be oriented in proximity to one another, in which case such termini can be extended, such that they form a short base-paired helix capable of functioning as the third arm of the three-way junction. This approach was successful in the creation of a thrombin-detecting CLD based on the anti-parallel G-quadruplex aptamer for thrombin (*see* Fig. 2) [13]. (3) The aptamer should undergo a detectable conformational change upon analyte binding, although even subtle conformational

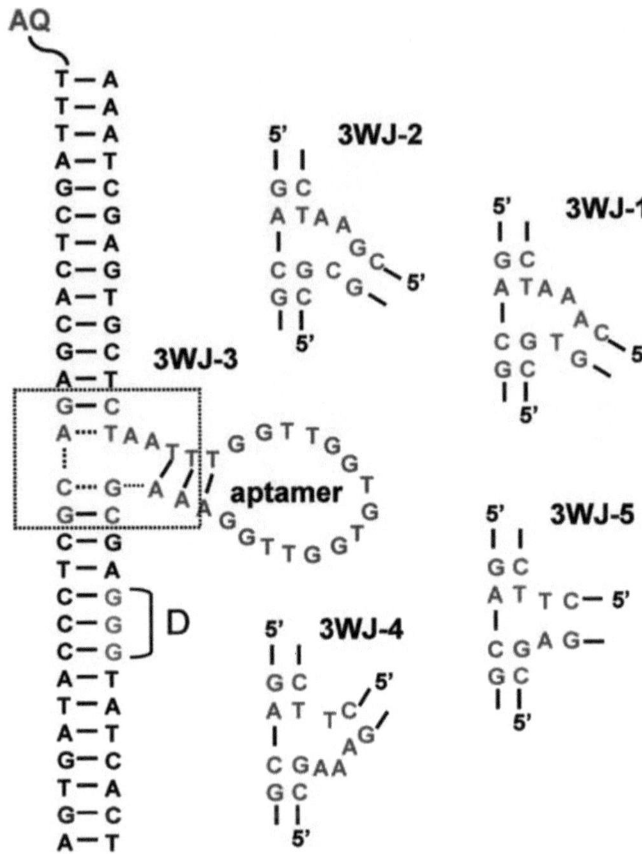

Fig. 2 The thrombin CLD secondary structure and the junction sequences (within the *dotted box*) tested in its development [13]. Two bulged adenosines are included at the junction at the 5′-end of the aptamer domain in constructs 3WJ-1, 3WJ-2, and 3WJ-3. These constructs differed in the identity of the base pairs that form the short helical stem through which the aptamer is attached to the three-way junction. Two bulged adenosines are included at the junction at the 3′-end of the aptamer domain in construct 3WJ-4, and bulged nucleotides are omitted in construct 3WJ-5. Increased oxidative damage to the detector guanines (denoted by "D") upon anthraquinone irradiation is indicative of enhanced charge transfer and junction conductivity. Reprinted with permission from ref. 13. Copyright (2008): American Chemical Society

changes have proven to be sufficient to enable CLD function. We recommend the use of standard reactivity protection assays as a convenient method for identifying conformational changes induced by analyte binding (for instance, ligand-induced conformational changes in the aptamer can be monitored by reactivity of individual DNA bases and nucleotides to damaging agents, such as dimethylsulfate and $KMnO_4$) [16]. More sophisticated methods, such as NMR or X-ray crystallography, can also be used to identify suitable analyte-induced aptamer conformational changes.

Once one or more potentially suitable aptamer sequences have been identified, they are incorporated into CLD constructs to generate several junction variants, whose conductivity switching can be assayed biochemically. Figure 1 illustrates the principle of monitoring conductivity switching in the biochemical CLD format. Fundamentally, an electron–hole is injected into the reporter stem of the CLD by UV photo-excitation of an anthraquinone moiety appended to the terminus of this DNA stem (here, the photo-excited triplet state of the anthraquinone serves as a photo-oxidant). If sufficient inter-helical stacking between the reporter and detector stems exists at the three-way junction, the electron–hole will migrate through the double-helical conduction path to the "detector" guanine triplet in the detector stem. Because these three tandem guanines are the most easily oxidized residues in the conduction pathway, the electron–hole will persist long enough on these bases such that they will be oxidatively damaged (by reaction with water and/or oxygen) [17]. Such damage can be revealed by piperidine-induced strand cleavage at these sites, which can be visualized in sequencing PAGE gels and quantified by densitometry. Successful CLD constructs should show increased detector guanine damage as the concentration of analyte is increased (i.e., increased junction conductivity in the presence of analyte) (*see* Fig. 3).

Much work has been devoted to understanding the structure of junctions formed at the intersection of three fully Watson–Crick paired arms. These studies have established general guidelines that predict the coaxial stacking of two of the three arms in certain circumstances [18–22]. However, because only a short helix is incorporated in the aptamer arm of a CLD three-way junction, these guidelines are not expected to apply strictly, as deviations from ideal duplex structure are likely at the point of attachment of the aptamer domain. Therefore, five or more CLD junction sequence variants are typically tested, chosen to represent as comprehensive a range of structural space accessible to a three-way junction as possible. As illustrated by the example of the thrombin CLD (Fig. 2), junction variants incorporate two-nucleotide bulges at either the 5′-terminus (variants 3WJ-1, 3WJ-2, and 3WJ-3), the 3′-terminus (variant 3WJ-4) of the aptamer domain, or a completely base-paired junction (3WJ-5) [13]. Other sequence variations are also possible (for example different sequences for the bulged nucleotides or base-paired aptamer stem), but from our experience in developing five different CLDs [11–15] we have found that a functional junction sequence can be identified from the variants shown in Fig. 2. Other experimental parameters that may be varied to optimize junction conductivity switching include buffer composition (particularly divalent metal presence/concentration)

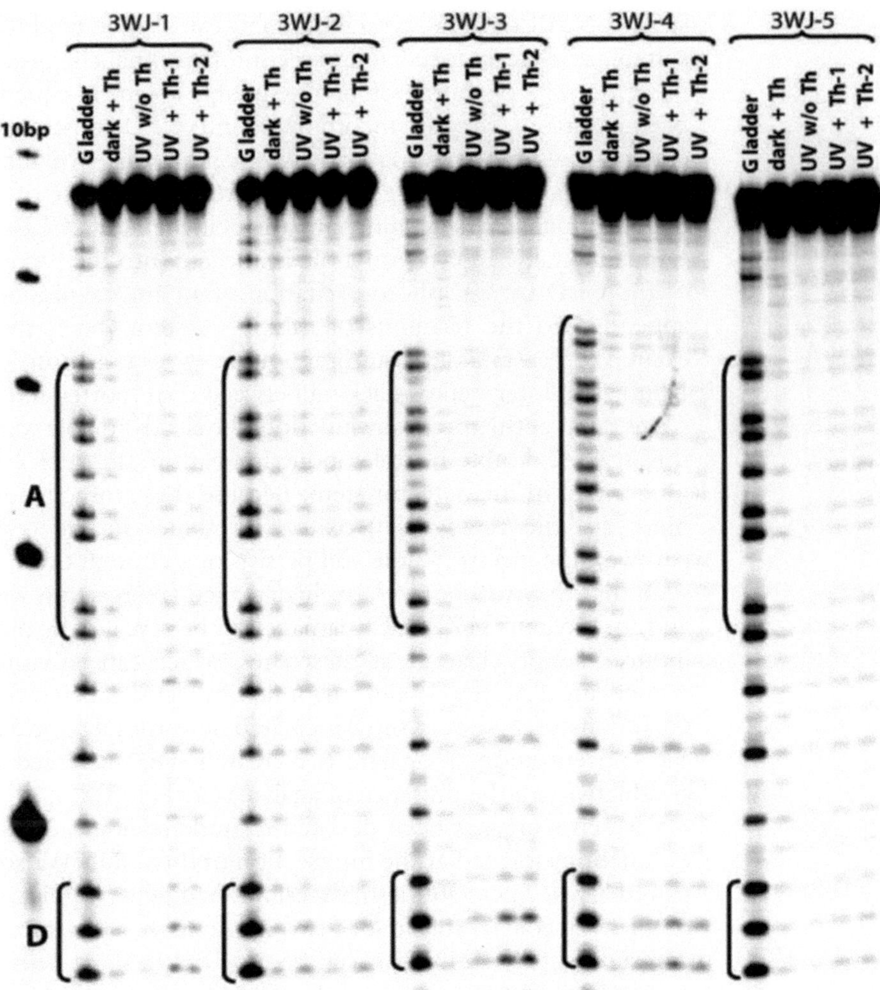

Fig. 3 Biochemical charge transfer assays for candidate CLDs for thrombin detection. For each construct, the *lanes* correspond to samples treated to: no irradiation but added thrombin ("dark+Th"), irradiation but no added thrombin ("UV w/o Th"), and irradiation plus added thrombin ("UV+Th-1" and "UV+Th-2"). The concentration of thrombin present in the replicate samples loaded into the lanes marked "Th-1" and "Th-2" was 160 nM. G-specific sequencing ladders are shown for each construct in the lanes marked "G ladder". "A" and "D" indicate, respectively, the aptamer domain sequence and the guanine triplet in the detector stem, as shown in the sequences in Fig. 2. Note the increase in detector guanine band intensity for 3WJ-3 upon addition of thrombin. This indicates increased charge transfer to anthraquinone, and thus increased three-way junction conductivity, in the presence of thrombin

and the length of the helix connecting the aptamer domain to the junction. We recommend the conduction path sequences described below, as these have proven to be successful in several different CLDs. A full discussion of the sequence dependence of DNA charge transfer is beyond the scope of this article.

2 Materials

2.1 Oligonucleotides

Complementary oligonucleotide: $H_2N(CH_2)_6O$-TTTAG CTCACGAGACGCTCCCATAGTGA.

Control oligonucleotide: TCACTATGGGAGCGTCTCGTGAG CTAAA.

CLD with 5′-bulge at junction: TCACTATGGGAGCG **AA** (APTAMER) TCTCGTGAGCTAAA.

CLD with 3′-bulge at junction: TCACTATGGGAGCG (APTAMER) **AA** TCTCGTGAGCTAAA.

CLD with no bulge at junction: TCACTATGGGAGCG (APTAMER) TCTCGTGAGCTAAA.

Adenosine residues highlighted in bold text are unpaired, bulged nucleotides located where the aptamer domain is joined to the three-way junction.

2.2 Oligonucleotide Preparation and Purification

Except for the complementary oligonucleotide, which is HPLC purified as described in Subheading 3.1 below, all sensor and control oligonucleotides must be pretreated with hot piperidine to remove (by breakage of the DNA strand) of any preexisting lesions that may interfere with DNA damage analysis experiments that follow.

1. Prepare the unlabelled oligonucleotides in 10 % piperidine (v/v, aqueous) and heat at 90 ° C for 30 min. Evaporate the samples to dryness by lypophilization (or in a speed-vac), resuspend in 100 μL water, and evaporate to dryness again.
2. Purify the oligonucleotides using appropriate standard denaturing (TBE/urea) PAGE gels. Sensor and control oligonucleotides must be 5′-^{32}P-labelled (using γ-[^{32}P]-ATP and polynucleotide kinase; standard labelling kits/supplies are widely available). Repurify these oligonucleotides by standard denaturing PAGE following ^{32}P-labelling.

3 Methods

3.1 Anthraquinone Oligonucleotide Bioconjugate

Anthraquinone must be protected from light as much as possible. Carry out all manipulations under low light conditions.

1. The *N*-hydroxysuccinimidyl (NHS)-ester of anthraquinone-2-carboxylic acid (AQ-NHS) can be purchased commercially (EMP Biotech) or prepared according to literature precedent [23]. Prepare 20 nmol of the C6-amino conjugated complementary oligonucleotide in 50 μL of 200 mM Na-borate buffer (pH 8.5) and add 50 μL of a 50 mM solution of AQ-NHS

in DMF. Allow this mixture to react at room temperature overnight on a bench-top shaker or tube roller.

2. Add 100 μL water to the reaction mixture, and extract three times with chloroform, retaining the aqueous phase each time. Precipitate the DNA from the aqueous phase by adding 20 μL of 3 M NaOAc (pH 5.2), and 600 μL of absolute ethanol. Mix thoroughly and chill on dry ice until the solution becomes highly viscous. Centrifuge at 13,000 × *g* (in a microcentrifuge) for 30 min at 4 °C. Decant and discard the liquid, taking care not to disturb the DNA pellet. Wash the pellet briefly with ice-cold 70 % ethanol. Briefly dry the pellet in a speed-vac and resuspend in 0.1 M triethylammonium acetate (pH 7.0) for HPLC purification.
3. The anthraquinone-conjugated oligonucleotide can be readily separated from unconjugated oligonucleotide by reverse-phase HPLC [11]. The conditions are as follows: solvent A: 0.1 M triethylammonium acetate (pH 7.0)/CH_3CN (20:1) and solvent B: CH_3CN; 20 min linear gradient from 5 % solvent B to 30 % solvent B; flow rate of 1.0 mL/min; 250 × 4.6 mm, 5 μm-C18 column. Monitor the eluate composition at 260 and 340 nm. The anthraquinone-conjugated DNA elutes at ~18 min and shows characteristic absorbance at both 260 and 340 nm. The unreacted AQ-NHS also absorbs at 260 and 340 nm, but elutes as a broad peak ~5 min after the DNA conjugate.
4. The HPLC-purified anthraquinone–DNA conjugate is then dried in a speed-vac, resuspended in 10 mM Tris–HCl (pH 8)/0.1 mM EDTA, and stored at −20 °C. Oligonucleotide concentration is determined from the absorbance at 260 nm. The extinction coefficient of the anthraquinone is 29,000 M^{-1} cm^{-1}.

3.2 Assembly and Testing of Candidate Sensor Constructs

1. Anneal the sensor or control constructs by heating to 100 °C for 2 min the anthraquinone-conjugated complementary oligonucleotide (300 nM) and sensor oligonucleotide (<10 nM 5′-^{32}P-labelled + nM unlabelled carrier) in 10 mM Tris (pH 8)/0.1 mM EDTA. Enough labelled oligonucleotide should be added such that each experimental aliquot (**step 2** below) taken from the annealing mixture should contain ~1,000 cpm. Cool the samples over the course of ~30 min to room temperature in a PCR machine or appropriate water bath. Then, add an equal volume of 2× concentrated aptamer-specific binding buffer (*see* **Note 1**).
2. Transfer three 49 μL aliquots for each sensor or control construct to the wells of a polystyrene 96-well ELISA plate; one aliquot is held as a dark control in a foil covered tube. Add 1 μL of 50× concentrate of the aptamer target (at two different

concentrations) to two of the samples. Equilibrate the plate in contact with a water bath at 4 °C inside a cold room.

3. Irradiate the samples inside a cold room (4 °C) by placing the ELISA plate 4 cm below a standard hand-held laboratory UV lamp (for example UVP Black-Ray UVL56 lamp) at 365 nm for 30 min. Ensure uniform irradiation as much as possible by loading samples into only two adjacent rows of an ELISA plate wells that are positioned equidistant from the UV lamp.
4. Following irradiation, transfer the samples to Eppendorf tubes and ethanol precipitate by adding 1/10th volume 3 M NaOAc (pH 5.2) and 3 volumes ethanol. Chill, centrifuge, wash, and dry as described above in Subheading 3.1. Resuspend the dried DNA pellets in 10 % aqueous piperidine, heat to 90 °C for 30 min, then speed-vac to dryness. Resuspend the samples in 100 μL water and speed-vac to dryness. Resuspend the samples in standard denaturing PAGE loading solution (*see* **Note 2**), vortex well, and heat at 95 °C for 3 min. Load the samples onto standard denaturing (TBE/urea) PAGE gels (10–15 % 19:1 acrylamide:bisacrylamide, depending on the length of the labelled aptamer strand used).
5. Analyze the autoradiography images by quantifying the band intensity of the detector guanines as a fraction of the total intensity in the lane. Successful sensor variants should show significant, reproducible increase in detector guanine band intensity with increasing analyte concentration (such as for construct 3WJ-3 in Fig. 3). Once CLD construct(s) have been identified that increase their junction conductivity in the presence of analyte, a titration experiment (to measure conductivity at many analyte concentrations) should be undertaken to assess the sensitivity of each construct. Experimental conditions, such as buffer composition, may also be varied to optimize the sensor response (*see* **Note 1**).

4 Notes

1. The composition of the buffer may affect junction folding and structure switching. For fully helical junctions, it has been observed that at least 1 mM Mg^{2+} is required for efficient coaxial stacking. Despite this finding, Mg^{2+} was not required for optimal function of the thrombin CLD. Thus it is advisable to assay the CLD variants under various buffer conditions, for example in the presence and absence of divalent metal cations.
2. Standard denaturing gel loading solution consists of: 5 mM EDTA (pH 8.0)/0.01 % bromophenol blue/0.01 % xylene cyanole/95 % formamide.

Acknowledgments

This work has been supported by the Natural Sciences and Engineering Research Council of Canada (NSERC), the Canadian Institutes of Health Research (CIHR), and the Canadian Institute for Advanced Research (CIFAR). D.S. is a Fellow of CIFAR.

References

1. Famulok M, Mayer G, Blind M (2000) Nucleic acid aptamers – from selection in vitro to applications in vivo. Acc Chem Res 33(591):599
2. Liu J, Cao Z, Lu Y (2009) Functional nucleic acid sensors. Chem Rev 109(1948):1998
3. Song S, Wang L, Li J, Zhao J, Fan C (2008) Aptamer-based biosensors. Trends Anal Chem 27(108):117
4. O'Sullivan CK (2002) Aptamers – the future of biosensing? Anal Bioanal Chem 372(44):48
5. Wang RE, Zhang Y, Cai J, Cai W, Gao T (2011) Aptamer-based fluorescent biosensors. Curr Med Chem 18(4175):4184
6. Lubin AA, Plaxco KW (2010) Folding-based electrochemical biosensors: the case for responsive nucleic acid architectures. Acc Chem Res 43(496):505
7. Cheng AKH, Sen D, Yu H-Z (2009) Design and testing of aptamer-based electrochemical biosensors for proteins and small molecules. Bioelectrochemistry 77(1):12
8. Willner I, Zayats M (2007) Electronic aptamer-based sensors. Angew Chem Int Ed Engl 46(6408):6418
9. Li D, Song S, Fan C (2010) Target-responsive structural switching for nucleic acid-based sensors. Acc Chem Res 43(631):641
10. Nutiu R, Li Y (2003) Structure-switching signaling aptamers. J Am Chem Soc 125(4771):4778
11. Fahlman RP, Sen D (2002) DNA conformational switches as sensitive electronic sensors of analytes. J Am Chem Soc 124(4610):4616
12. Sankar CG, Sen D (2004) DNA helix-stack switching as the basis for the design of versatile deoxyribosensors. J Mol Biol 340(459):467
13. Huang YC, Ge B, Sen D, Yu H-Z (2008) Immobilized DNA switches as electronic sensors for picomolar detection of plasma proteins. J Am Chem Soc 130(8023):8029
14. Thomas JM, Chakraborty B, Sen D, Yu H-Z (2012) Analyte-driven switching of DNA charge transport: de novo creation of electronic sensors for an early lung cancer biomarker. J Am Chem Soc 134(13823):13833
15. Thomas JM, Yu H-Z, Sen D (2012) A mechano-electronic DNA switch. J Am Chem Soc 134(13738):13748
16. Maxam AM, Gilbert W (1980) Sequencing end-labeled DNA with base-specific chemical cleavages. Methods Enzymol 65:499–560
17. Gasper SM, Schuster GB (1997) Intramolecular photoinduced electron transfer to anthraquinones linked to duplex DNA: the effect of gaps and traps on long-range radical cation migration. J Am Chem Soc 119(12762):12771
18. Assenberg R, Weston A, Cardy DLN, Fox KR (2002) Sequence-dependent folding of DNA three-way junctions. Nucleic Acids Res 30(5142):5150
19. Lilley DMJ (2000) Structures of helical junctions in nucleic acids. Q Rev Biophys 33(109):159
20. van Buuren BN, Overmars FJ, Ippel JH, Altona C, Wijmenga SS (2000) Solution structure of a DNA three-way junction containing two unpaired thymidine bases: identification of sequence features that decide conformer selection. J Mol Biol 304(371):383
21. Leontis NB, Hills MT, Piotto M, Ouporov IV, Malhotra A, Gorenstein DG (1995) Helical stacking in DNA three-way junctions containing two unpaired pyrimidines: proton NMR studies. Biophys J 68(251):265
22. Welch JB, Walter F, Lilley DMJ (1995) Two inequivalent folding isomers of the 3-way DNA junction with unpaired bases: sequence dependence of the folded conformation. J Mol Biol 251(507):519
23. Telser J, Cruickshank KA, Morrison LE, Netzel TL, Chan K (1989) DNA duplexes covalently labelled at two sites: synthesis and characterization by steady-state and time-resolved optical spectroscopies. J Am Chem Soc 111(7226):7232

Index

Daniel Lafontaine and Audrey Dubé (eds.), *Therapeutic Applications of Ribozymes and Riboswitches: Methods and Protocols*, vol. 1103, DOI 10.1007/978-1-62703-730-3, © Springer Science+Business Media New York 2014

MIX
Papier aus verantwortungsvollen Quellen
Paper from responsible sources
FSC® C105338

If you have any concerns about our products,
you can contact us on
ProductSafety@springernature.com

In case Publisher is established outside the EU,
the EU authorized representative is:
Springer Nature Customer Service Center GmbH
Europaplatz 3, 69115 Heidelberg, Germany

Printed by Libri Plureos GmbH
in Hamburg, Germany